Air Pollution Impacts on Plants in East Asia

Takeshi Izuta
Editor

Air Pollution Impacts on Plants in East Asia

Springer

Editor
Takeshi Izuta
Professor
Institute of Agriculture
Tokyo University of Agriculture and Technology
Fuchu, Tokyo, Japan

ISBN 978-4-431-56436-2 ISBN 978-4-431-56438-6 (eBook)
DOI 10.1007/978-4-431-56438-6

Library of Congress Control Number: 2016962203

Printed on acid-free paper

This Springer imprint is published by Springer Nature
The registered company is Springer Japan KK
The registered company address is Chiyoda First Bldg. East, 3-8-1 Nishi-Kanda,
Chiyoda-ku, Tokyo 101-0065, Japan

Preface

Plants can be regarded as life-support equipment, because they can provide the oxygen necessary for respiration to living organisms, alleviate global warming by the absorption and fixation of atmospheric CO_2, and purify a polluted atmosphere by the absorption and/or adsorption of air pollutants. However, anthropogenic emission of pollutants into the atmosphere causes air pollution and its adverse effects on many plant species all over the world. This means that we are turning off this life-support equipment by ourselves. For sustainable development, therefore, we must reduce the emissions of air pollutants into the atmosphere and protect plant ecosystems against air pollutants.

In Asia, air pollution is one of the most serious environmental problems in the twenty-first century. Recently, transboundary air pollution has become an environmental problem, especially in East Asia. In this region, there is a possibility that gaseous air pollutants such as photochemical oxidants, acid deposition, and aerosols cause detrimental effects on crops, trees, and natural vegetation. In Japan, experimental studies on the effects of air pollutants on crops and trees, and field surveys of air pollution and its effects on trees in forest ecosystems have been conducted since the 1970s and 1990s, respectively. Recently, Chinese researchers have actively studied the effects of ozone, a main component of photochemical oxidants, on crops and trees. However, limited information is available on the effects of air pollutants on plants in other Asian countries.

The aim of this book is to provide information on air pollution in East Asia; information on the effects of gaseous air pollutants, acid deposition, and aerosols on Asian crops and trees; and case studies of air pollution in Japanese forests. This information should be of interest to environmental scientists, plant scientists, government officials, industrialists, environmentalists, undergraduate and graduate students, and others concerned with the air pollution problem in Asia and other parts of the world.

I hope that this book will contribute both to an increased awareness of the air pollution problem and to the protection of plants against air pollutants in Asia.

Last, but not least, I would like to sincerely thank the Springer Editorial Team, Dr. Mei Hann Lee and Ms. Madonna Samuel, for their editorial assistance, advice, and support during work on this book project.

Fuchu, Tokyo, Japan Takeshi Izuta

Contents

Part VI Effects of Aerosol on Plants

Part I
Air Pollution in East Asia

Chapter 1
Gaseous Species

Shiro Hatakeyama

Abstract Gaseous pollutants in East Asia are explained in this chapter. Sulfur dioxide (SO_2) emissions in this region of the world are now decreasing, although these emissions in China are still the highest in the world. SO_2 emission in South Korea decreased very quickly; in Japan, it took about 20 years to reduce the atmospheric SO_2 concentration from 50 ppb to less than 5 ppb, whereas it took only 10 years in Korea to reduce the SO_2 concentration by the same amount, owing to a switch in fuel sources from coal to natural gas.

The emission of nitrogen oxides ($NOx = NO + NO_2$) is still increasing in China, and this emission causes an increase in tropospheric ozone. The transboundary transport of ozone and its precursors is affecting not only the countries surrounding China, but also North American countries.

Trends in ozone concentrations over the East China Sea were analyzed, based on aerial observations for up to 20 years. A clear increase in ozone was found, particularly in the boundary layer, lower than 1,500 m above sea level.

Keywords SO_2 • NO_2 • Ozone • Air quality standards • Yearly trends • Long-range transport

1.1 Introduction

The economy in Asia is among the most rapidly developing in the world, and economic growth in East Asia is particularly rapid. For example, China's gross domestic product (GDP) was the second highest in the world in 2010. Such economic growth has brought about an increase in fossil fuel consumption and inevitably caused increasing emission of air pollutants. Thus, at the end of the twentieth century, East Asia was responsible for the largest emissions of sulfur dioxide (SO_2) and nitrogen oxides ($NOx = NO + NO_2$) in the world, exceeding those of the United States and also those of Europe Akimoto 2003; Ohara et al. 2007).

S. Hatakeyama
Center for Environmental Science in Saitama, Kazo, Saitama 347-0115, Japan
e-mail: hatashir@cc.tuat.ac.jp

T. Izuta (ed.), *Air Pollution Impacts on Plants in East Asia*,
DOI 10.1007/978-4-431-56438-6_1

During the twentieth century, acid rain was often implicated as a cause of forest decline and was considered a global environmental problem. However, the true problem associated with acid rain is not the acidification of rainwater, but rather, the anthropogenic emission of acid raw materials such as SO_2 and NOx as exhaust gases from factories and automobiles. The dissolution of these acidic substances in rainwater forms acid rain, and thus acid rain is a result of air pollution. From this viewpoint, acid rain can be considered a form of long-range, transboundary air pollution. Acid rain problems in East Asia must also be dealt with as phenomena of transboundary air pollution. In North American and European countries, serious acid rain problems were overcome by switching to cleaner fuels, as well as by desulfurization and denitrification technologies, but intercontinental air pollution remains a major environmental issue (UNECE 2011).

The gaseous pollutants often believed to be the sources of acidic substances in the air are SO_2, NOx, and surface ozone. Photochemical oxidation of SO_2 yields sulfuric acid in the air, and this sulfuric acid forms fine particles in its native form and also upon neutralization with ammonia gas (NH_3). Photochemical reactions involving NOx bring about the formation of ozone in the troposphere, and this is the main component of photochemical smog. Photochemical oxidation of NOx also yields nitric acid in the air. The nitric acid forms fine particles upon neutralization with ammonia gas. Damage to plants occurs mostly by acids and gases producing acidic substances. Acids are usually formed as particulate matter, and they will be discussed in further detail in Chap. 2. Typical acid-producing substances are: (1) gaseous precursors of acids, such as SO_2 and NOx, and (2) oxidizing gases and radicals, such as ozone, H_2O_2, and OH radicals. The three gaseous species; namely, SO_2, NOx, and ozone, are strongly related to acid rain problems and also to damage to plants. In this chapter, we discuss these three gases and their prevalence in Asia, particularly in Japan, China, and Korea.

1.2 SO_2

In addition to being emitted by natural processes such as volcanic activity, SO_2 is emitted from anthropogenic sources such as fossil fuel combustion and the smelting of pyrites. A total of 86,273 kt of SO_2 was emitted worldwide from anthropogenic sources in 2010 (Cofala et al. 2012). Emissions from China, Japan, and Korea account for 34 %, 1 %, and 1 %, respectively, of worldwide emissions (Fig. 1.1).

SO_2 causes human health problems, such as asthma; these effects are observed in areas near large-scale SO_2 emission sources. In addition, SO_2 that is oxidized to sulfuric acid (H_2SO_4) in the air can be transported long distances on a regional or continental scale.

Oxidation of SO_2 in the atmosphere takes place in two ways. One is by means of a homogeneous gas-phase process. The main oxidizing agent in the homogeneous gas-phase is OH radicals. OH radicals react with SO_2 as follows:

$$SO_2 + OH + M \rightarrow HOSO_2 + M$$

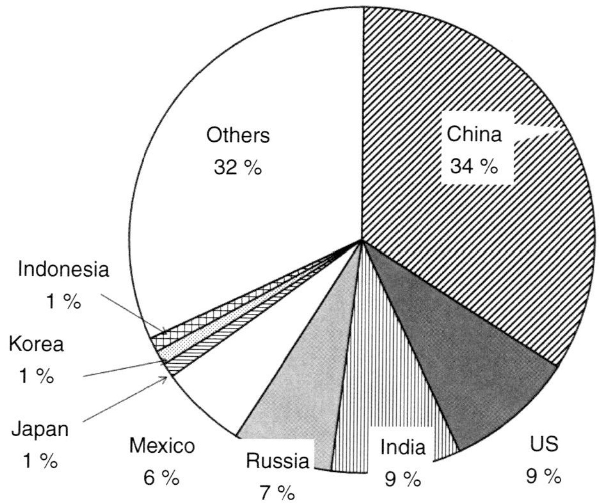

Fig. 1.1 Emissions of SO_2 by country in 2010 (Adapted from Cofala et al. 2012)

$$HOSO_2 + O_2 \rightarrow SO_3 + HO_2$$

$$SO_3 + H_2O \rightarrow H_2SO_4$$

where M is a third body whose purpose is to remove excess energy by means of adduct formation.

The other means by which SO_2 is oxidized in the atmosphere is by a heterogeneous reaction on the surfaces of cloud droplets or a homogeneous aqueous-phase process within cloud droplets. Many oxidizing agents have been suggested to promote these types of SO_2 oxidation, with the dominant oxidizing agent being H_2O_2 (Schwartz 1984). H_2O_2 reacts with HSO_3^- (the dominant chemical form of SO_2 dissolved in water at pH less than 5) as follows:

$$HSO_3^- + H_2O_2 \rightarrow A^- + H_2O$$

$$A^- + H^+ \rightarrow 2H^+ + SO_4^{2-}$$

Here A^- is speculated to be O^-—S (O)——OOH (Hoffmann and Jacob, 1984).

SO_2 and its oxidation product SO_4^{2-} can cause serious environmental problems by forming acid rain in areas far away from SO_2 emission sources. Acid rain was thought to be a major cause of forest decline in Europe and North America for a long time.

1.2.1 Japan

In Japan, air pollution by SO_2 was most serious in the 1960s and early 1970s. Particularly, around Yokkaichi City many people suffered from a serious respiratory disease that is referred to as Yokkaichi-Zensoku (Yokkaichi asthma); high concentrations of SO_2 (i.e., annual average concentrations higher than 0.05 ppm) and of its oxidation product H_2SO_4 are considered to be the causes of this disease. In 1973 the Japanese air quality standard for SO_2 was established to lower the SO_2 level. The air quality standard for SO_2 is 0.1 ppm hourly and 0.04 ppm daily, calculated as the average of all hourly averages in one day. Once these standards were established, the ambient SO_2 level in Japan decreased to 0.005 ppm; it has remained at or below this level for 20 years (Fig. 1.2; Ministry of the Environment, Japan 2012, 2015). Both air quality standards for SO_2 were achieved in nearly 100 % of residential areas in 2013.

1.2.2 China

In China, national air quality standards are classified into three grades. Grade I applies to special protected areas, such as natural conservation areas, scenic areas, and historical sites; Grade II applies to residential areas, mixed commercial/residential areas, and cultural, industrial, and rural areas; and Grade III applies to special industrial areas. Chinese standards for the three grades are listed in Table 1.1. In China, 88.2 % of cities achieved the national air quality standards in 2014 (Ministry of Environmental Protection, China 2014).

Lu et al. (2010) estimated that SO_2 emission in China increased from 21.7 Tg in 2000 to 33.2 Tg in 2006, a 53 % increase. They also suggested that emissions began to decrease after 2006 mainly due to the widespread application of flue-gas

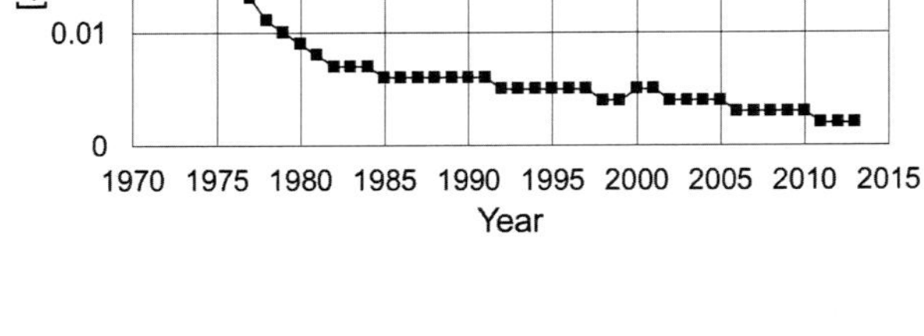

Fig. 1.2 Ambient SO_2 levels in Japan (Adapted from Ministry of the Environment, Japan 2012, 2015)

Table 1.1 National air quality standards in China for SO_2 (mg m^{-3})

	Grade I	Grade II	Grade III
Annual average	0.02	0.06	0.1
Daily average	0.05	0.15	0.25
Hourly average	0.15	0.5	0.7

0.1 mg m^{-3} of SO_2 is approximately 38.2 ppb (25 °C)

desulfurization devices in power plants in response to a new policy implemented by China's government. The emission of SO_2 in China has decreased since 2006, as can be seen in Fig. 1.3, in which the graph displays the official SO_2 emission data reported by the Ministry of Environmental Protection of China in their State of Environment report.

One major source of SO_2 emission is power plants. As shown in Fig. 1.4, ~80 % of the electric power generated in China in 2012 came from thermal power plants, and 95 % of that thermal power came from coal combustion (China Electricity Council 2012). Power generation by oil combustion accounted for only 0.14 % of the power generation of thermal power plants with capacity of 6,000 kW and higher in China during that same year. This distribution of power sources can be expected to persist for many years, because oil production in China is limited, whereas coal deposits in China are large. For this reason, the reduction of SO_2 emission remains a major challenge for China.

1.2.3 Korea

In contrast to China, Korea reduced SO_2 emissions very quickly. As shown in Fig. 1.5 (Choi 2014), annual average concentrations of SO_2 in Seoul decreased from ~50 ppb in 1990 to ~5 ppb in 2000. Such a rapid improvement in air quality was attained by the transition to cleaner fuel (Choi 2014). During the 1970s and

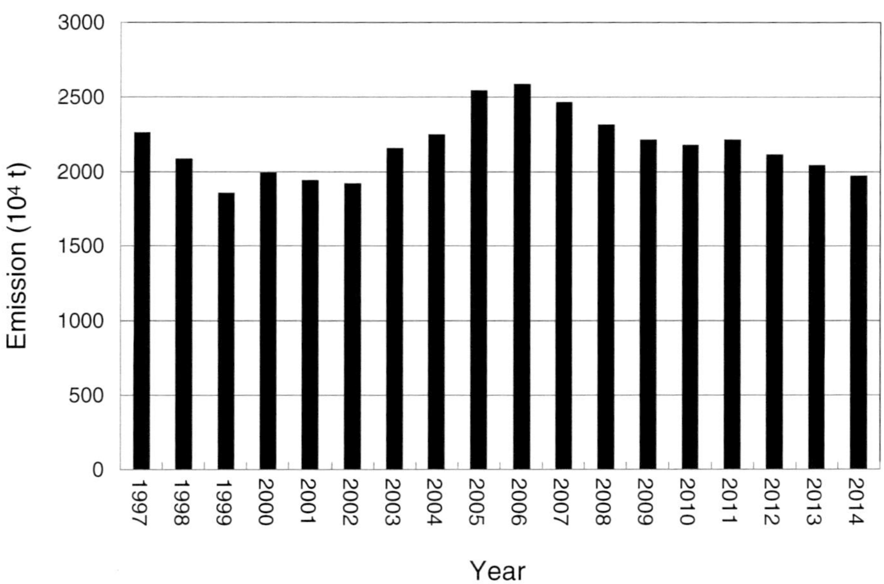

Fig. 1.3 Emissions of SO_2 in China (From the Report on the State of the Environment in China, Ministry of Environmental Protection of China)

1980s, the major SO_2 emission sources in Seoul were fossil fuels, such as coal and oils with high sulfur content. However, the use of solid fuels such as coal and wood for business facilities in the Seoul metropolitan area was prohibited beginning in 1985, and the use of liquefied natural gas (LNG), a cleaner fuel, was mandated. This switch in fuel sources proceeded very smoothly, and both businesses and households in Korea had moved from using wood and coal in the 1960s to using LNG by the 1990s.

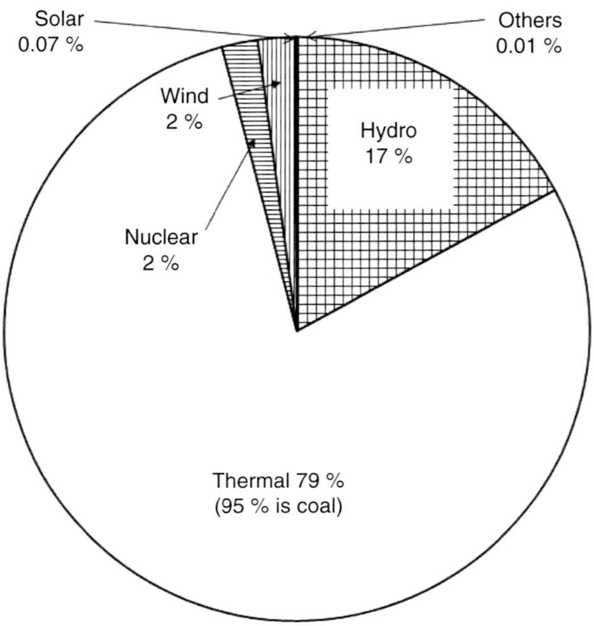

Fig. 1.4 Power generation in China in 2012 (Adapted from China Electricity Council 2012)

Fig. 1.5 Ambient annual SO_2 levels in Seoul, South Korea (Adapted from Choi 2014)

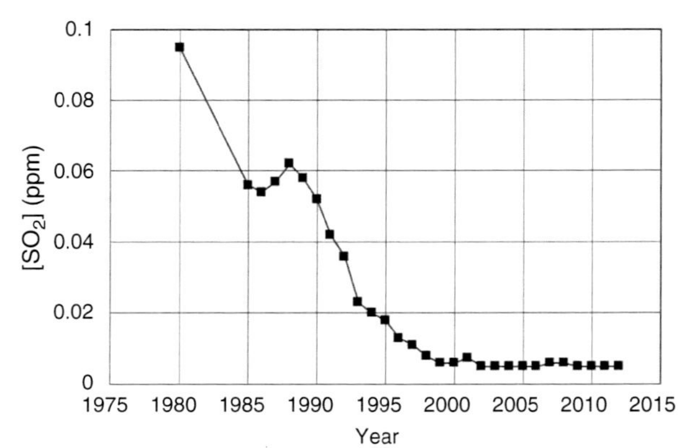

1.3 NOx

The worldwide emission of NOx was determined to be 84,696 Gg in 2010 (Cofala et al. 2012). Countries with particularly high emissions are China, the United States, and India (Figs. 1.6 and 1.7). East Asian countries such as China, Japan, and Korea emitted only 25 %, 2 %, and 1 % of worldwide NOx, respectively.

Although SO_2 is emitted from the combustion of sulfur-containing fuels, particularly coal, NOx is emitted from all types of combustion processes, because NOx is formed mainly by the reaction $N_2 + O_2 \rightarrow 2NO$ at high temperature. This reaction is endothermic, so it generally does not occur at room temperature and regular atmospheric pressure. However, this reaction is accelerated at higher temperatures,

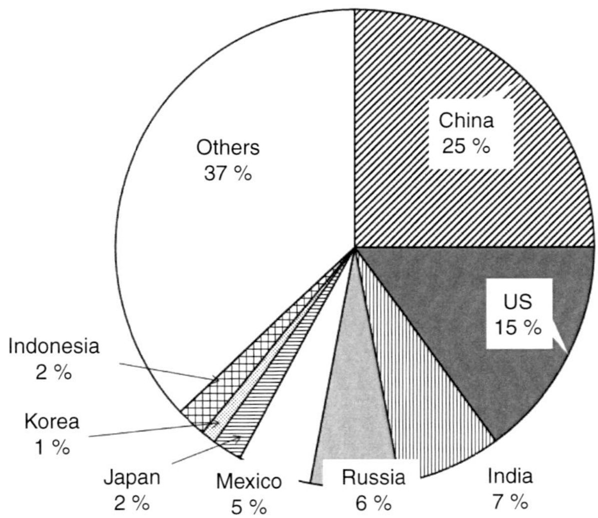

Fig. 1.6 Emissions of NOx by country in 2010 (Adapted from Cofala et al. 2012)

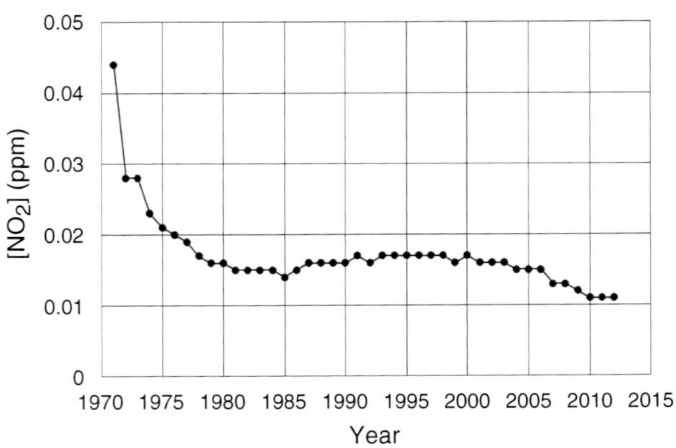

Fig. 1.7 Ambient NO_2 levels in Japan (Adapted from Ministry of the Environment, Japan 2012, 2015b)

particularly those above 1,900 °C. Such high temperatures are found inside internal combustion engines and power station boilers during the combustion of fuels in the presence of air. The emitted NO is then easily converted to NO_2 in air by oxygen molecules, ozone, and peroxy radicals such as HO_2 and RO_2 (where R is an alkyl group). The sum of NO and NO_2 is referred to collectively as NOx. Although NOx is an important atmospheric species that affects atmospheric chemistry, air quality, and climate (IPCC 2013), the most important role of NOx is to control the tropospheric ozone (O_3) budget and the formation of nitrate aerosol. Tropospheric ozone will be discussed in the next section of this chapter, and nitrate aerosol will be discussed in the next chapter. NOx is an important target of air pollution control measures worldwide.

The transport of NOx within East Asia is a substantial problem. Satellite measurements over East Asia (Lee et al. 2014) have clearly shown the high column concentration of NOx over China and the Yellow Sea, suggesting the contribution of NOx transport to downwind regions, in addition to local emissions.

A distinct seasonal cycle of NOx exists in most regions of East Asia (Mijling et al. 2013), as shown by satellite measurements acquired from 2007 to 2011. Van der A et al. (2006) showed that anthropogenic sources caused NO_2 concentrations to peak in winter, whereas soil emissions peaked in summer.

1.3.1 Japan

Air pollution by NOx in Japan was at its most serious in the 1960s and early 1970s. After 1973, when strict regulations for automobile exhausts were enacted, the emission of NOx decreased, and ambient concentrations subsequently decreased, as shown in Fig. 1.6 (Ministry of the Environment of Japan 2012, 2015a). However, after 1985, NO_2 did not decrease further, most likely owing to the increase in the number of cars in use in Japan, although the emission from individual cars decreased due to strict emission regulations.

The Japanese air quality standard for NO_2 is <0.04–0.06 ppm per day, calculated as the average of all hourly averages per day. At present, this air quality standard for NO_2 is achieved in nearly 100 % of residential areas.

1.3.2 China

As with SO_2, China maintains three grades of national air quality standards for NOx, as well as for NO_2. Chinese standards for these three grades are listed in Table 1.2. In 2014, 62.7 % of cities in China achieved these national air quality standards (Ministry of Environmental Protection, China 2014).

Kurokawa et al. (2013) reported that the emission of NOx as NO_2 in China in 2008 was 26,969 Gg (89 % growth compared with emission in 2000). Zhao et al.

Table 1.2 National air quality standards in China for NO_2 (mg m^{-3})

	Grade I	Grade II	Grade III
Annual average	0.04	0.04	0.08
Daily average	0.08	0.08	0.12
Hourly average	0.12	0.12	0.24

0.1 mg m^{-3} of NO_2 is approximately 53.1 ppb (25 °C)

(2013) also estimated NOx emission in China, which was 26,100 Gg in 2010, a major increase from the reported 11,000 Gg in 1995. Quite recently, the emission of NOx in China was reported to have decreased, from 24,043 Gg in 2011 to 20,780 Gg in 2014 (Ministry of Environmental Protection, China 2014). As suggested by Zhang et al. (2007), the reason for this observed decrease could be the installation of low-NOx boilers, especially in new large power plants. Nevertheless, because it has also been reported that the number of privately owned automobiles in China has increased to 100 million (China Daily 2011-09-17), and that this number continues to increase (Statista 2015), trends in NOx emission should be monitored more carefully.

The increase in NOx column abundance from 1996 to 2004 was observed clearly through satellite observations. In particular, the increment of NOx column abundance over the Beijing, Jinan, and Shanghai areas is as high as 8×10^{14} molecules cm^{-2} $year^{-1}$ (van der A 2006).

Wang et al. (2012) estimated the contributions of six major sources (power plants; biomass burning; soils and fertilizers; lightning; aircraft; and other anthropogenic sources, including industry, transportation, and biofuels) of NOx emission in 2007 to be 9.58, 0.23, 1.77, 0.59, 0.05, and 13.77 Tg NO_2, respectively, based on the Goddard Earth Observing System (GEOS)-Chemistry model. Biomass burning is an unimportant source of NOx in East China, contributing less than 1 % to total NOx emissions.

1.3.3 Korea

Korean air quality standards for NO_2 are 0.03, 0.06, and 0.10 ppm for annual, daily, and hourly average emissions, respectively. The national annual average air pollution level of NO_2 has been maintained below the air quality standard of 0.03 ppm (which was enacted in 2007). However, NO_2 emissions have improved only slowly in Seoul, where the annual average exceeded 0.03 ppm between 1999 and 2011 before the air quality standard was achieved in 2012 (Choi 2014). Key measures responsible for this improvement included switching to clean heating fuel (coal → oil → natural gas), introducing and strengthening emission standards for motor vehicles since 1991, and promoting the use of low-emission vehicles.

Nguyen et al. (2015) analyzed the long-term trends of NO_2 emission in seven major Korean cities over two decades (1989–2010). Because notable environmental policies were initiated in June 2000, these investigators divided the observation period into period I (1989–1999) and period II (2000–2010). They found that the mean concentrations of NO_2 in five cities were 1–26 % higher in period II than in period I. For example, the annual average concentrations of NO_2 in period I and period II were 31.2 and 36.1 ppb, respectively, in Seoul; 24.2 and 25.3 ppb, respectively, in Daegu; and 17.0 and 21.5 ppb, respectively, in Gwangju. This increase is thought to be due to the increasing consumption of petroleum and LNG. In Busan and Ulsan, lower concentrations of NO_2 were observed in period II than in period I.

As briefly mentioned under the heading 'NOx' above, the transport of NO_2 in East Asia is a very important phenomenon for Korea and Japan. Monthly variations in NO_2 columns, as measured by the Ozone Monitoring Instrument (OMI) satellite instrument over China and Eastern China clearly show a strong winter maximum, whereas over the Yellow Sea, Korea, the Sea of Japan, and Japan, an NO_2 peak is observed in spring and early summer (March–June; Lee et al. 2014). Surface observations of NO_2 in Korea show a similar spring peak. Because the springtime NO_2 enhancement agrees with the maximum CO observed at remote sites, despite the lifetime of CO being much longer than that of NO_2, Lee et al. (2014) suggested that NOx could be transported by a means similar to that of CO.

1.4 Tropospheric Ozone

Ozone (O_3) is a major air pollutant that is known to exist both in the stratosphere and in the troposphere. In the stratosphere, ozone is formed mainly by the Chapman mechanism, as follows:

$$O_2 + h\nu \rightarrow 2O \quad (1.1)$$

$$O + O_2 + M \rightarrow O_3 + M \quad (1.2)$$

$$O_3 + O \rightarrow 2O_2 \quad (1.3)$$

$$O_3 + h\nu \rightarrow O + O_3 \quad (1.4)$$

where h is Plank's constant, ν is the frequency of light, and M is a third body that removes excess energy from O_3.

As long as ozone is present only in the stratosphere, it is valuable not only for human beings but also for all the living things on the surface of the earth, because it blocks high-energy ultraviolet (UV) light, i.e., that with wavelengths shorter than 300 nm. However, ozone is a toxic gas, and so it is not desirable for ozone to exist at high concentrations in the lower troposphere. High ozone concentrations are harmful to respiration and lung function in humans (Bernard et al. 2001; Bell et al. 2007; Amman et al. 2008). Ozone also has detrimental effects on plants, the details of which will be discussed in later chapters.

High concentrations of ozone are often produced as the result of photochemical smog formation processes. Therefore, ozone is considered to be a major air pollutant.

Moreover, ozone is now recognized as a strong greenhouse gas. The atmospheric lifetime of ozone is much shorter than that of CO_2, but ozone can still induce regional warming and climate change (IPCC 2013).

In the lower troposphere, which is not penetrated by high-energy UV light, the only source of O_3 is the photolysis of NO_2 and subsequent reactions, as follows:

$$NO_2 + h\nu \rightarrow NO + O \quad (1.5)$$

$$O + O_2 + M \rightarrow O_3 + M \quad (1.2)$$

$$O_3 + NO \rightarrow NO_2 + O_2 \quad (1.6)$$

Since NO_2 is brown, it can absorb visible light within the wavelength region of 450–700 nm and subsequently decompose to form NO and an oxygen atom, as shown in Reaction 1.5 above. An oxygen atom then reacts with an oxygen molecule to form O_3 in the same manner as in the stratosphere. Thus, NO_2 alone can form ozone in the troposphere. However, NO_2 alone cannot produce high concentrations of O_3, such as those seen in photochemical smog, because NO can destroy O_3 by means of Reaction 1.6. This reaction is referred to as a titration reaction of NO for ozone.

In the atmosphere, there are many kinds of volatile organic compounds (VOCs), most of which are hydrocarbons. Under photochemical smog conditions, gas-phase chemical reactions take place mainly as OH-radical-initiated reactions, and these reactions produce the peroxy radicals HO_2 and RO_2 (where R is usually an alkyl group). HO_2 or RO_2 can readily react with NO to form NO_2 without destroying O_3. A schematic diagram of the formation of ozone in the absence and in the presence of VOCs is shown in Fig. 1.8.

Mechanisms of reactions influencing atmospheric ozone are described in detail in a recent publication by Calvert et al. (2015).

1.4.1 *Japan*

In Japan, the air quality standard for ozone is included in that for general photochemical oxidants, since the initial official measurement method was based on a wet chemical method utilizing iodometric analyses, which can be used to detect both

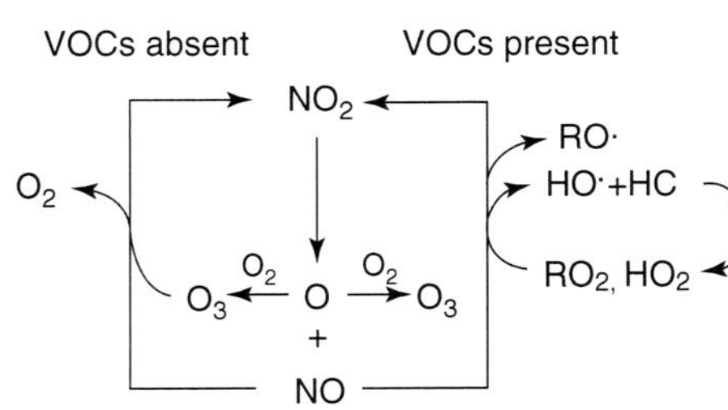

Fig. 1.8 Formation of ozone in the absence and the presence of volatile organic compounds (VOCs)

ozone and other gaseous oxidative compounds. However, UV absorption and chemiluminescence methods are now the official means for monitoring photochemical oxidants in Japan, and most monitoring is now achieved by these dry methods. Therefore, we may consider the officially reported concentrations of photochemical oxidants in Japan to be equivalent to the concentrations of ambient ozone. The air quality standard for photochemical oxidants (ozone) is 0.06 ppm (hourly average).

Photochemical smog in Japan was a very serious problem in the late 1960s and early 1970s. After the strict regulation of NOx emission from automobiles started in 1973, as described above Sect. 1.3.1, the concentration of photochemical oxidants decreased remarkably by 1980 (Fig. 1.9; Ministry of the Environment, Japan 2015a). However, the photochemical oxidant concentration gradually increased after 1980, although the concentration of an important ozone precursor, NO_2, decreased over the same time period. In addition, the concentration of VOCs has also been decreasing. Therefore, the increase in the photochemical oxidant concentration is mysterious. The reduction of NOx from automobile exhausts is assumed to be one cause of this increase: such reduction is believed to decrease the efficiency of the titration reaction by NO. Another important cause of the gradual increase of photochemical oxidants is the transport of ozone or its precursors from China (e.g., Yoshitomi et al. 2011).

The low rate of achievement of the air quality standard for photochemical oxidants is another big problem in Japan. The percentages of monitoring stations in Japan that showed concentrations lower than the air quality standard in 2012, 2013, and 2014 were 0.5 %, 0.3 %, and 0.3 %, respectively (Ministry of the Environment, Japan 2015b). One important factor is that the Japanese air quality standard for photochemical oxidants is very strict: Japan's hourly average standard is only 0.06 ppm, whereas those for the United States and Korea are 0.1 and 0.12 ppm,

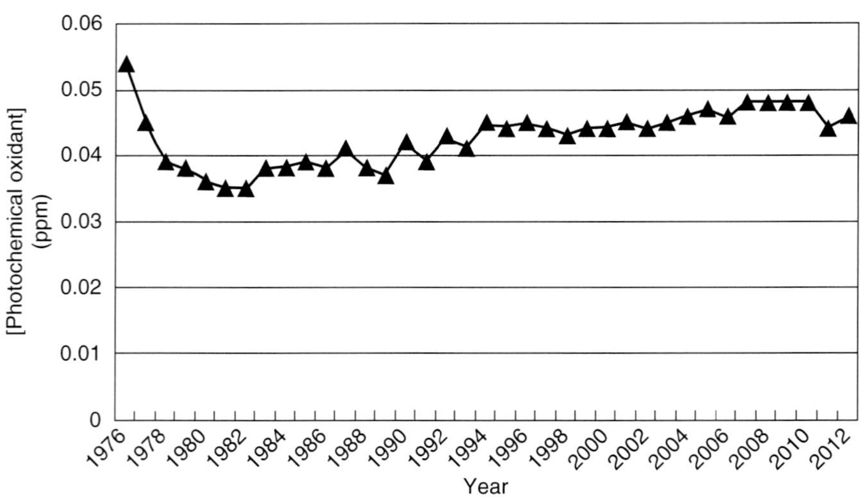

Fig. 1.9 Ambient photochemical oxidant levels in Japan (adapted from Ministry of the Environment, Japan 2015b)

respectively. China has three grades of standards, as described in the next section. Moreover, if the hourly average concentration of photochemical oxidants at a given monitoring site in Japan exceeded 0.06 ppm even once over the course of a year, that monitoring site was reported as not having achieved the standard. In contrast, among the 1,152 monitoring stations in residential areas in Japan, a total of 93.2 % reported daytime photochemical oxidant levels lower than 0.06 ppm in 2013.

1.4.2 China

As with SO_2 and NO_2, national air quality standards for ozone are classified into three grades in China. The Chinese standards for these three grades are listed in Table 1.3. In 2014, 78.2 % of cities in China achieved these national air quality standards (Ministry of Environmental Protection, China 2015).

Recently, Verstraeten et al. (2015) showed that tropospheric ozone concentrations over China had increased by about 7 % between 2005 and 2010 in response to two factors: an approximately 21 % increase in Chinese emissions and the increased downward transport of stratospheric ozone. The same report also stated that ozone from China affects air quality in the United States.

Annual average concentrations of ozone in China in 2013 and 2014 were 139 and 140 $\mu g\ m^{-3}$, or ~70.8 and ~71.3 ppb, respectively, at 25 °C (Ministry of Environmental Protection, China 2014, 2015).

Yoshitomi et al. (2011) estimated that the Chinese contribution to surface ozone over Japan in the spring season was 4.0 ± 2.8 ppb, which was larger than the European and North American contributions (3.5 ± 1.1 and 2.8 ± 0.5 ppb, respectively). This Chinese contribution could be expected for a closer source region, and levels are generally highest near cold fronts preceding the influence of more distant sources. Local sources over Japan and Korea have been reported to have a relatively small impact on the mean ozone level over Japan, at 2.4 ± 7.6 ppb, respectively.

1.4.3 Korea

Korean air quality standards for ozone are 0.06 and 0.1 ppm for daily and hourly averages, respectively. Annual averages of ozone for Korea were 0.020, 0.020, 0.022, 0.023, and 0.026 ppm in 1998, 2000, 2005, 2010, and 2013, respectively (Ministry of Environment of Korea 2015). Ozone is said to have been maintained

Table 1.3 National air quality standards in China for O_3 ($mg\ m^{-3}$)

	Grade I	Grade II	Grade III
Hourly average	0.12	0.16	0.20

0.1 $mg\ m^{-3}$ of O_3 is approximately 50.9 ppb (25 °C)

below the air quality standards for the past two decades. In the Seoul metropolitan area, the average concentration of ozone is also low (Choi et al. 2014). Seo et al. (2014) pointed out that O_3 levels for coastal cities are high due to the dynamic effects of the sea breezes, while the levels for inland cities and the Seoul metropolitan area are low due to NOx titration by local precursor emissions. However, the annual average concentration of O_3 in Seoul is on the rise (Choi et al. 2014).

1.4.4 Trends in Ozone Concentrations Above the East China Sea Based on Aerial Observations

In this section, trends in ozone concentrations over the East China Sea are discussed based on 20 years of aerial observations carried out by the authors. Data used for the analysis are those obtained in the Perturbation of East Asian Continent Air Mass to Pacific Ocean Troposphere (PEACAMPOT) campaign in October 1991, November 1992, March and December 1994, January and December 1997, February 1999, and March 2001; in the Lagrangian Experiment on Long-Range Transported Aerosols (LEXTRA) campaign in March-April 2008; and in the Impact of Aerosols in East Asia on Plants and Human Health (ASEPH) campaign in October 2009 and December 2010, for a total of 11 datasets (Hatakeyama 2000; Hatakeyama et al. 1995a, 1995b, 1997, 2001, 2004, 2011, 2014). Aerial observation is an event-based experiment, and thus the period, season, and area of observation are different every time, which can present challenges in evaluating long-term trends. However, aerial observations are seldom carried out over almost the same area for 20 years, as was done here. Recently, Clarke et al. (2010) reported many aerial observations carried out above oceans all over the world; but those observations were not carried out for the duration of the observations described here. Thus, we can say that our observations provide a unique and valuable dataset. The trends of ionic species observed in these measurements will be discussed in the next chapter; in this section, only the trends in ozone are discussed.

Ozone data were divided into two groups; namely, data from 1991 to 1999 and data obtained after 2000. These two periods, i.e., before and after the year 2000, were analyzed. Before 2000, the concentration of ozone was most frequently in the range of $40 < O_3 < 45$ ppb, whereas after 2000, it was most frequently in the range of $65 < O_3 < 70$ ppb. The results suggest that high concentrations of ozone were present more frequently after 2000, even after accounting for the fact that the number of springtime observations was larger after 2000. The reason for such a shift in the

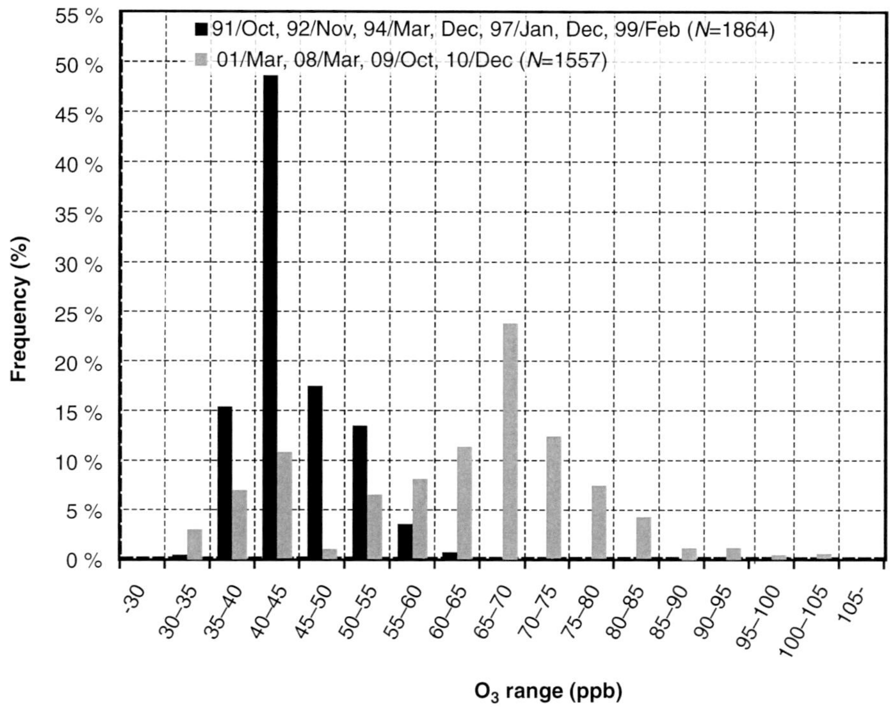

Fig. 1.10 Frequency distribution of ozone concentrations for two time periods (before and after 2000) at altitudes below 1,500 m above sea level

frequency of high-concentration ozone observations is likely an increase in NOx emissions from the Asian continent. In May 2007, high concentrations of ozone, up to 120 ppb in Kyushu and the area facing the Sea of Japan, were observed. Hayasaki et al. (2008) reported that the contribution of transboundary transport to this ozone measurement was large.

Data in the boundary layer below 1,500 m and those in the free troposphere above 1,500 m are shown as histograms in Figs. 1.10 and 1.11, respectively. The above-mentioned trend can also be seen in both atmospheric layers. However, the difference between the two periods is more prominent in the boundary layer. In the free troposphere, the mode appeared at 40 $<O_3$ <45 ppb for both periods, suggesting that the ozone being transported from the upper layer to the boundary layer remains relatively stable, whereas the ozone produced in the boundary layer by human activity is growing.

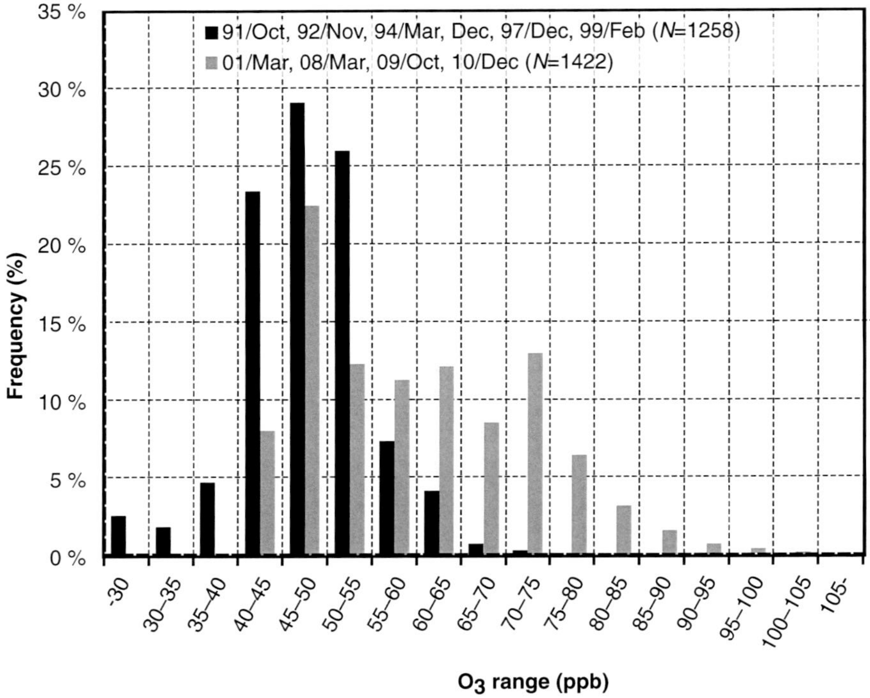

Fig. 1.11 Frequency distribution of ozone concentrations for two time periods (before and after 2000) at altitudes between 1,500 and 3,000 m a.s.l

References

Akimoto H (2003) Global Air Quality and Pollution. Science 302: 1716-1719. doi: 10.1126/science.1092666

Amann M, Derwent D, Forsberg B, Hänninen O, Hurley F, Krzyzanowski M, de Leeuw F, Liu SJ, Mandin C, Schneider J, Schwarze P, Simpson D (2008) Health risks of ozone from long-range transboundary air pollution. WHO Report, pp.93. (http://www.euro.who.int/__data/assets/pdf_file/0005/78647/E91843.pdf#search=%27health+effect+ozone%27. Accessed 27 Nov 2015)

Bell ML, Goldberg R, Hogrefe C, Kinney PL, Knowlton K, Lynn B, Rosenthal J, Rosenzweig C, Patz JA (2007) Climate change, ambient ozone, and health in 50 US cities. Clim Change 82:61–76. doi:10.1007/s10584-006-9166-7

Bernard SM, Samet JM, Grambsch A, Ebi KL, Romieu I (2001) The potential impacts of climate variability and change on air pollution-related health effects in the United States. Environ Health Persp 109(Suppl 2):199–209

Calvert JG, Orland JJ, Stockwell WR, Wallington TJ (2015) The mechanisms of reactions influencing atmospheric ozone. Oxford University Press, New York, p 590

China Daily (2011) Number of cars in China hits 100m, Updated on 2011-09-17. (http://www.chinadaily.com.cn/bizchina/2011-09/17/content_13725715.htm. Accessed on 24 Nov 2015)

China Electricity Council (2012) Annual statistics of China power industry 2012. (http://english.cec.org.cn/No.110.1481.htm. Accessed 28 Oct 2015)

Choi Y.-J Seoul's efforts to tackle air pollution *Performance and Challenges* 11th metropolis world congress 2014. (https://seoulsolution.kr/sites/default/files/notice/Seoul's%20Efforts%20to%20

Tackle%20Air%20Pollution_Ph.d%20Choi.pdf#search=%27So2+Seoul+trend%27. Accessed 28 Oct 2015)

Clarke A, Kapsutin V (2010) Hemispheric aerosol vertical profiles anthropogenic impacts on optical depth and cloud nuclei. Science 329:1488–1492

Cofala J, Bertok I, Borken-Kleefeld J, Heyes C, Klimont Z, Rafaj P, Sander R, Schöpp W, Amann M (2012) Emissions of air pollutants for the world energy outlook 2012 energy scenarios, draft final report submitted to International Energy Agency, Paris, France under Contract for Services between IEA and IIASA. (http://www.worldenergyoutlook.org/media/weowebsite/energy-model/documentation/IIASA_WEO2012_air_pollution.pdf#search=%27world+SO2+emission%27. Accessed 29 Oct 2015)

Hatakeyama S (2000) PEACAMPOT and PEACAMPOT II campaigns. IGACtivities (Newsletter of the International Global Atmospheric Chemistry project), No.20: 11–14

Hatakeyama S, Murano K, Bandow H, Mukai H, Akimoto H (1995a) High Concentration of SO_2 Observed over the Sea of Japan. Terre Atmos Oceanic Sci 6:403–408

Hatakeyama S, Murano K, Bandow H, Sakamaki F, Yamato M, Tanaka S, Akimoto H (1995b) '91 PEACAMPOT Aircraft Observation of Ozone, NOx, and SO_2 over the East China Sea, the Yellow Sea, and the Sea of Japan. J Geophys Res 100:23143–23151

Hatakeyama S, Murano K, Mukai H, Sakamaki F, Bandow H, Watanabe I, Yamato M, Tanaka S, Akimoto H (1997) SO2 and Sulfate Aerosols over the Seas between Japan and the Asian Continent. EarozoruKenkyu (J. Aerosol Res.) 12: 91-95

Hatakeyama S, Murano K, Sakamaki F, Mukai H, Bandow H, Komazaki Y (2001) Transport of atmospheric pollutants from East Asia. Water Air Soil Pollution 130:373–378

Hatakeyama S, Takami A, Sakamaki F, Mukai H, Sugimoto N, Shimizu A, Bandow H (2004) Aerial measurement of air pollutants and aerosols during March 20–22, 2001, over the East China Sea. J Geophys Res 109:D13304. doi:10.1029/2003JD004271

Hatakeyama S, Hanaoka S, Ikeda K, Watanabe I, Arakaki T, Sadanaga Y, Bandow H, Kato S, Kajii Y, Sato K, Shimizu A, Takami A (2011) Aerial observation of aerosols transported from East Asia —chemical composition of aerosols and layered structure of an air mass over the East China sea. Aerosol Air Qual Res 11:497–507

Hayasaki M, Ohara T, Kurokawa J, Uno I, Shimizu A (2008) Episodic pollution of photochemical ozone during 8–9 May 2007 over Japan: observational data analyses. J Jpn Soc Atmos Environ 43:225–237. (in Japanese)

Hoffmann MR, Jacob DJ (1984) Kinetics and mechanisms of the catalytic oxidation of dissolved sulfur dioxide in aqueous solution, in in SO_2, NO and NO_2 oxidation mechanism: atmospheric considerations. In: Calvert JG (ed) Acid precipitation series, vol 3, Butterworth Publications, Boston

IPCC: climate change 2013, the physical science basis. (https://www.ipcc.ch/pdf/assessment-report/ar5/wg1/WGIAR5_SPM_brochure_en.pdf#search=%27IPCC+2013%27. Accessed 1 Nov 2015)

Kurokawa J, Ohara T, Morikawa T, Hanayama S, Janssens-Maenhout G, Fukui T, Kawashima K, Akimoto H (2013) Emissions of air pollutants and greenhouse gases over Asian regions during 2000–2008: Regional Emission inventory in ASia (REAS) version 2. Atmos Chem Phys 13:11019–11058. doi:10.5194/acp-13-11019-2013

Lee H-J, Kim S-W, Brioude J, Cooper OR, Frost GJ, Kim C-H, Park RJ, Trainer M, Woo J-H (2014) Transport of NOx in East Asia identified by satellite and in situ measurements and Lagrangian particle dispersion model simulations. J Geophys Res Atmos 119:2574–2596. doi:10.1002/2013JD021185

Lu Z, Streets DG, Zhang Q, Wang S, Carmichael GR, Cheng YF, Wei C, Chin M, Diehl T, Tan Q (2010) Sulfur dioxide emissions in China and sulfur trends in East Asia since 2000. Atmos Chem Phys 10:6311–6331. doi:10.5194/acp-10-6311-2010

Mijling B, van der ARJ, Zhang Q (2013) Regional nitrogen oxides emission trends in East Asia observed from space. Atmos Chem Phys 13:12003–12012

Ministry of Environment of Korea (2015) ECOREA – environmental review 2015, Korea. (http://www.me.go.kr/home/file/readDownloadFile.do?fileId=94238&fileSeq=3. Accessed 27 Nov 2015)

Ministry of Environmental Protection, China (2014) Report on the state of the environment in China 2013. (http://jcs.mep.gov.cn/hjzl/zkgb/2013zkgb/201406/t20140605_276521.htm. Accessed on 26 Oct 2015) (in Chinese)
Ministry of Environmental Protection, China (2015) Report on the state of the environment in China 2014. (http://jcs.mep.gov.cn/hjzl/zkgb/2014zkgb/201506/t20150608_303142.htm. Accessed 26 Oct 2014) (in Chinese)
Ministry of the Environment, Japan (2012) The status of air pollution in 2010. (http://www.env.go.jp/air/osen/jokyo_h22/index.html. Accessed 28 Oct 2015) (in Japanese)
Ministry of the Environment, Japan (2015) The status of air pollution in 2013. (http://www.env.go.jp/press/files/jp/24227.pdf. Accessed 28 Oct 2015) (in Japanese)
Ministry of the Environment, Japan (2015b) 2015 White paper of the ministry of Environment. (http://www.env.go.jp/policy/hakusyo/h27/html/hj15020401.html#n2_4_1_1. Accessed on 27 Nov 2015) (in Japanese)
Nguyen HT, Kim K-H, Park C (2015) Long-term trend of NO_2 in major urban areas of Korea and possible consequences for health. Atmos Environ 106:347–357
Ohara T, Akimoto H, Kurokawa J, Horii N, Yamaji K, Yan X, HayasakaT (2007) An Asian emission inventory of anthropogenic emission sources for the period 1980–2020. AtmosChem Phys7: 4419–4444. doi:10.5194/acp-7-4419-2007n
Schwartz SE (1984) Gas-aqueous reactions of sulfur and nitrogen oxides in liquid water clouds, in SO_2, NO and NO_2 oxidation mechanism: atmospheric considerations. In: Calvert JG (ed) Acid precipitation series, vol 3. Butterworth Publications, Boston
Seo J, Youn D, Kim JY, Lee H (2014) Extensive spatiotemporal analyses of surface ozone and related meteorological variables in South Korea for the period 1999–2010. Atmos Chem Phys 14:6395–6415. doi:10.5194/acp-14-6395-2014
Statista: The Statistics Portal (2015) Number of owned automobiles in China from 2007 to 2014 (in millions). (http://www.statista.com/statistics/285306/number-of-car-owners-in-china/. Accessed on 24 Nov 2015)
UNECE (2011) Hemispheric transport of air pollution 2010. UNECE, New York
van der A RJ, Peters DHMU, Eskes H, Boersma KF, Van Roozendael M, De Smedt I, Kelder HM (2006) Detection of the trend and seasonal variation in tropospheric NO_2 over China. J Geophys Res 111: D12317. doi:10.1029/2005JD006594
Verstraeten WW, Neu JL, Williams JE, Bowman KW, Worden JR, Boersma KF (2015) Rapid increases in tropospheric ozone production and export from China. Nat Geosci 8:690–695. doi:10.1038/ngeo2493
Wang SW, Zhang Q, Streets DG, He KB, Martin RV, Lamsal LN, Chen D, Lei Y, Lu Z (2012) Growth in NOx emissions from power plants in China: bottom-up estimates and satellite observations. Atmos Chem Phys 12:4429–4447. doi:10.5194/acp-12-4429-2012
Yoshitomi M, Wild O, Akimoto H (2011) Contributions of regional and intercontinental transport to surface ozone in the Tokyo area. Atmos Chem Phys 11:7583–7599
Zhang Q, Streets DG, He KB, Wang Y, Richter A, Burrows JP, Uno I, Jang CJ, Chen D, Yao Z, Lei Y (2007) NOx emission trends for China, 1995–2004: the view from the ground and the view from space. J Geophys Res 112:D22306. doi:10.1029/2007JD008684
Zhao B, Wang SX, Liu H, Xu JY, Fu K, Klimont Z, Hao JM, He KB, Cofala J, Amann M (2013) NOx emissions in China: historical trends and future perspectives. Atmos Chem Phys 13:9869–9897. doi:10.5194/acp-13-9869-2013

Chapter 2
Aerosols

Shiro Hatakeyama

Abstract Aerosols in East Asia, in particular the chemical constituents of PM2.5 (particulate matter 2.5 μm or less in diameter), are described in this section. Aerosols are colloidal systems in which small (<100 μm in diameter) solid or liquid particulate matter is suspended in air. These small particles are themselves also called aerosols. PM2.5 and *kosa* (Asian yellow sand dust) are important aerosol constituents, and their chemical components; namely, elemental and organic carbon, polycyclic aromatic hydrocarbons (PAHs), metallic elements, and inorganic ionic species, are discussed. The long-range transport of the aerosol components is the major topic in this section. The chemical transformation of aerosol components was evaluated and source apportionment was performed on the basis of model simulations and statistical analyses.

Keywords PM2.5 • *Kosa* • Elemental carbon (EC) • Organic carbon (OC) • Polycyclic aromatic hydrocarbon (PAH) • Metallic elements • Ionic species

2.1 Introduction

Because of rapid economic growth in Asia, particularly in East Asia, emissions of atmospheric pollutants are growing quite rapidly. Besides gaseous air pollutants such as O_3, nitrogen oxides (NOx), SO_2, and volatile organic compounds (VOCs), many kinds of particulate air pollutants, consisting of sulfates, nitrates, and carbonaceous particles, as well as soil dust particles such as *kosa* (Asian yellow sand dust), are also emitted directly or produced secondarily by precursor gases during long-range transport. Such aerosols (see aerosol definition below) are transported downwind and they even cross the Pacific Ocean to North America, where they contribute to pollution. Clarification of the chemical change processes that occur during aerosol transport, and determination of the impacts of aerosols after

---

S. Hatakeyama
Center for Environmental Science in Saitama, Kazo, Saitama 347-0115, Japan
e-mail: hatashir@cc.tuat.ac.jp

T. Izuta (ed.), *Air Pollution Impacts on Plants in East Asia*,
DOI 10.1007/978-4-431-56438-6_2

transport on animals and plants, as well as on human health, are important environmental issues.

In January 2013, very high concentrations of PM2.5 (particulate matter with a diameter of less than 2.5 μm); described in detail in a later section) were observed in China, particularly in Beijing, where the mass concentrations of fine particles reported as PM2.5 were up to 1000 μg m^{-3}. The reported concentrations were far exceeded the daily air quality standards of not only Japan and USA (35 μg m^{-3}) but also even China (75 μg m^{-3}). The high levels of PM2.5 then became a topic of public concern in Japan and many people began to pay attention to the PM2.5 forecasts on television. The term "PM2.5" was not so familiar at that time, yet people were concerned that something very dangerous was being transported from East Asia.

The International Agency for Research on Cancer (IARC, an agency of the World Health Organization) Working Group has unanimously classified particulate matter and outdoor air pollution as carcinogenic to humans (IARC Group 1), based on sufficient evidence of carcinogenicity in humans and experimental animals and strong mechanistic evidence (Loomisa et al. 2013). The IARC describes an agent in this group as follows: "The agent is carcinogenic to humans. This category is used when there is sufficient evidence of carcinogenicity in humans. Exceptionally, an agent may be placed in this category when evidence of carcinogenicity in humans is less than sufficient but there is sufficient evidence of carcinogenicity in experimental animals and strong evidence in exposed humans that the agent acts through a relevant mechanism of carcinogenicity." At present, 117 agents are classified in this group (IARC 2015).

2.2 Aerosols and PM2.5

Aerosols are colloidal systems in which small (<100 μm in diameter) solid or liquid particulate matter is suspended in air. These small particles are themselves also called aerosols. Atmospheric pollutants transported from Asia are not limited to fine particles classified as PM2.5, but, as previously mentioned, PM2.5 aerosols have recently attracted much attention. Aerosols, including PM2.5, are not a single compound, but are composed of many different chemical species. Aerosols include sulfuric acid mist, photochemical smog, asbestos particles, and diesel exhaust soot. Figure 2.1 shows the average chemical composition of PM2.5 aerosols in urban areas of Japan (Ministry of the Environment of Japan 2009). These aerosols affect not only the local environment but also contribute to global environmental issues, such as global warming, stratospheric ozone depletion, and acid rain. Sources of aerosols are not only anthropogenic but also natural. Major anthropogenic sources are photochemical smog, automobiles, and certain industries, while major natural sources are plants, soils, and seawater. Plant pollen (the pollen of Japanese cedar trees is a notorious allergen in Japan) and *kosa* transported from China are examples of natural-origin aerosols. Aerosol particles span a range of sizes from very fine (less than 1 μm in diameter) to coarse (up to 100 μm). In addition to the variety in

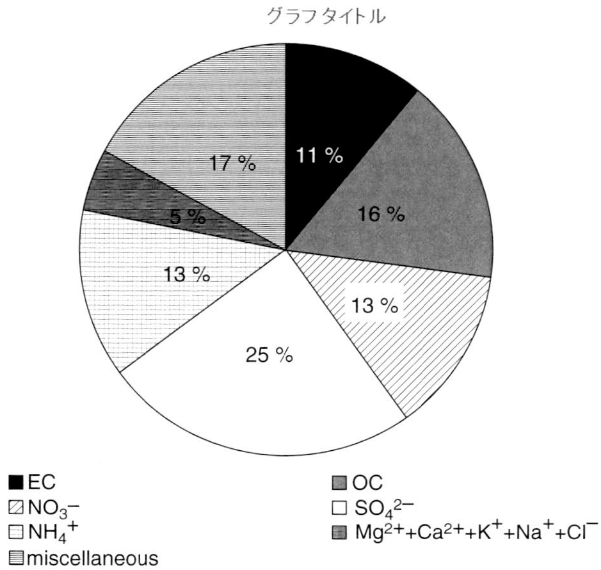

Fig. 2.1 Chemical composition of aerosols in urban areas of Japan (average of 2004–2008). *EC*, elemental carbon, *OC* organic carbon

size, they have a wide variety of chemical characteristics (different kinds of chemical compounds) and physical characteristics (liquid and solid; and a variety of forms, including both spherical and other symmetrical shapes [e.g., snow] and rugged, angular shapes). As we have been obtaining more information from recent progress in the chemical and physical analyses of aerosols, it has become clear that atmospheric aerosols play an important role in the environment.

Aerosols can be classified into four classes based on their origin. Particles that are particulate in origin are classified as primary and those that were originally gases but have been converted to particulate species by some chemical reactions in the air following emission are classified as secondary. Both primary and secondary particles have natural and anthropogenic sources. Some examples of each class are listed below.

(i) Primary natural: Soil particles, such as *kosa*; sea salt; pollen; mold; and mushroom spores
(ii) Primary anthropogenic: Soot and fly ash (a coal combustion product), dust from studded tires, and polycyclic aromatic hydrocarbons (PAHs) emitted during fuel combustion
(iii) Secondary natural: Products of the atmospheric reaction of ozone and OH radicals with natural hydrocarbons emitted by plants (isoprene and terpenes) and stratospheric aerosols that make up the so-called Junge layer (fine sulfuric acid particles produced in the stratosphere by the oxidation of SO_2 that is emitted by volcanic activity)
(iv) Secondary anthropogenic: H_2SO_4 (formed by the oxidation of SO_2 emitted from, for example, industrial sources, i.e., not from volcanoes) in air; ammonium salts (formed by the neutralization reactions of H_2SO_4 and HNO_3 produced by the oxidation of NOx); and organic particles created by the

reaction of ozone and OH radicals in air with anthropogenic hydrocarbons, such as cyclic olefins and aromatic hydrocarbons

International Organization for Standardization (ISO) PM2.5 particles are defined as those that can pass through a size-selective inlet with a 50 % efficiency cut-off at 2.5 μm aerodynamic diameter (ISO 2012). PM2.5 is not a new species, but its levels have recently increased and it is believed to pose a great health risk. Because of their small size, such fine particles can lodge deeply in the lungs.

The term PM2.5 refers only to the size of the particles, not to their chemical constituents, which can include tobacco smoke and benzo[a]pyrene, a well known carcinogenic PAH. Therefore, we need to pay attention not only to the mass concentration of PM2.5 particles but also to their chemical composition. The main components of ordinary long-range-transported PM2.5 in East Asia are sulfate and nitrate salts, as well as organic compounds.

Why is the PM2.5 problem an important environmental issue? In Japan an environmental standard for particles smaller than 10 μm (SPM; suspended particulate matter) was established in 1973, and monitoring and regulation have been performed according to this standard. However, particles smaller than 2.5 μm can cause more serious damage to human health because they can penetrate deeply into the lungs and their concentrations show high correlation with human mortality (Dockery et al. 1993). In the United States, National Ambient Air Quality Standards for PM2.5 were established in 1997 and and have been revised twice (2006 and 2013). The present standards are 35 μg m^{-3} for the daily average and 12 μg m^{-3} for the annual average. The World Health Organization (WHO 2005) suggested stricter limits for PM2.5 (25 and 10 μg m^{-3} for daily and annual averages, respectively). A Japanese Air Quality Standard for PM2.5 was established in 2009 (35 and 15 μg m^{-3} for daily and annual averages, respectively).

Another reason to regulate fine particles is that they consist mostly of anthropogenic species, whereas coarse particles are mainly of natural origin (Fig. 2.2). We cannot regulate the emissions of natural-origin aerosols, but we can and should

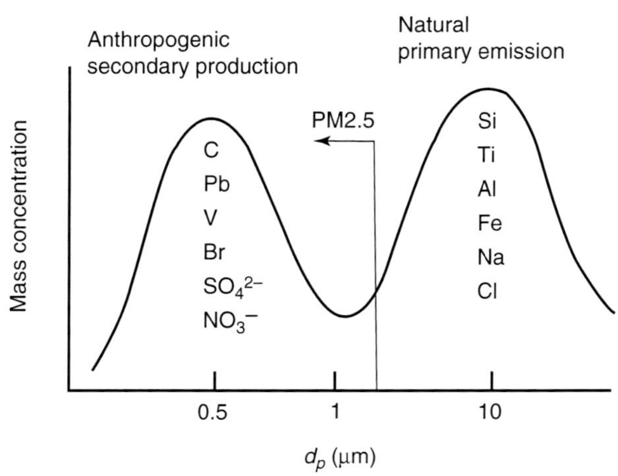

Fig. 2.2 General size distribution of aerosols and their main components

regulate anthropogenic emissions of aerosols and their precursors. We can expect the regulation of fine particles to have an effect. Moreover, the mass of a particle is proportional to the third power of its diameter, so one particle of 10-μm diameter corresponds to 1,000 particles of 1-μm diameter. If we regulate only aerosols containing coarse particles, it is difficult to control the concentration of fine particles.

2.3 *Kosa* and Chemical Species Transported with *kosa*

When very fine sand particles are blown up by strong winds in the Taklimakan and Gobi Deserts in inland China, the westerly wind transports them long distances. Figure 2.3 shows such a sandstorm, observed on 7 April 2001 (NASA 2008), that was caused by a very strong cyclone. In China, as well as in Korea, such a strong dust storm is regarded as a natural disaster.

Such strong dust storms can transport fine sand particles toward the Pacific Ocean, even reaching North America (NASA 2005). When an air mass containing dust particles passes over the coastal cities and industrial areas of China, anthropogenic air pollutants are adsorbed onto the surfaces of the aerosols and chemical transformation of the adsorbed compounds can take place. For example, SO_2 can be oxidized to sulfuric acid on the surface of such an aerosol. Such pollutants transported from Asia to America cause recognized global environmental problems. In contrast, alkaline compounds, such as $CaCO_3$ contained in dust particles, can contribute to the neutralization of sulfuric and nitric acids. Thus, we can say that dust particles have pluses and minuses that can offset each other.

Clear characteristics of the ionic composition of aerosols transported with *kosa* were identified from aerial observations of atmospheric pollutants over the East China Sea carried out by the author and colleagues (Hatakeyama et al. 2004). A dust storm broke out on 21 March 2001 in the Loess Plateau area in China and moved

Fig. 2.3 A very strong sand storm over China observed from NASA satellite on 7 April 2001

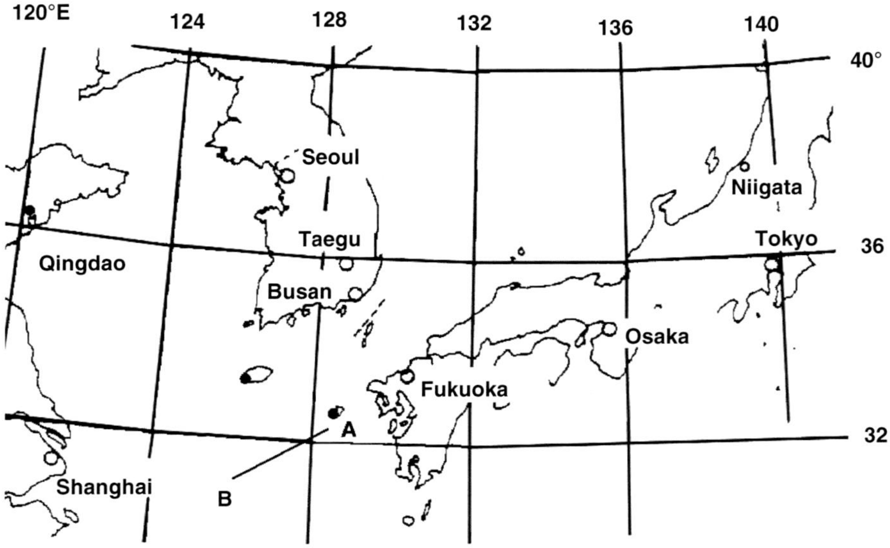

Fig. 2.4 Flight track of aircraft, between points a and b, of aerial observations carried out on 20, 21, and 22 March 2001

eastward (Darmenova et al. 2005). Our aerial measurements of atmospheric pollutants were carried out on 20, 21, and 22 March 2001 over the East China Sea (Fig. 2.4), so we were able to detect *kosa* during our experiments. Figures 2.5, 2.6, and 2.7 show the variations in the aerosol mass concentrations and the concentrations of sulfate and calcium ions in the aerosols on 20, 21, and 22 March 2001, respectively.

The concentrations of particulate matter were generally low on 20 March (Fig. 2.5); PM10 was less than 0.5 mg m^{-3} and PM2.5 was less than 12 μg m^{-3}. The sulfate and calcium ion concentrations were less than 10 and 1 μg m^{-3}, respectively. In contrast, following the arrival of *kosa*, a very high mass concentration peak of PM10 (~2.5 mg m^{-3}) and a broad peak of PM2.5 (60 μg m^{-3}) were observed on 21 March (Fig. 2.6). It is interesting to note that PM10 showed a sharp peak south of Jeju Island, Korea, whereas PM2.5 showed a broad peak over the East China Sea between Jeju Island and Fukue Island, Japan, at low altitude (~500 m). The calcium ion concentration showed a maximum (~11.5 μg m^{-3}) close to the PM10 peak (point a in (Fig. 2.6)). In contrast, the distribution of SO_4^{2-} ions was similar to the distribution of PM2.5 (point b in (Fig. 2.6)). On 22 March, the peak PM10 concentration (~3.4 mg m^{-3}) was recorded off Kyushu Island, Japan (Fig. 2.7). The peak PM2.5 concentration (~60 μg m^{-3}) also reached Kyushu Island.

On 20 March, when the concentrations of both PM2.5 and PM10 were low, the SO_4^{2-}/Ca^{2+} ratios of total suspended particles (TSP) were very large at both high and low altitudes (average, 27.8 for altitudes higher than 2,500 m and 15.3 for those lower than 1,000 m). This means that the contribution of anthropogenic sulfate was high on 20 March. On this day, a cold front passed quickly over the Korean Peninsula

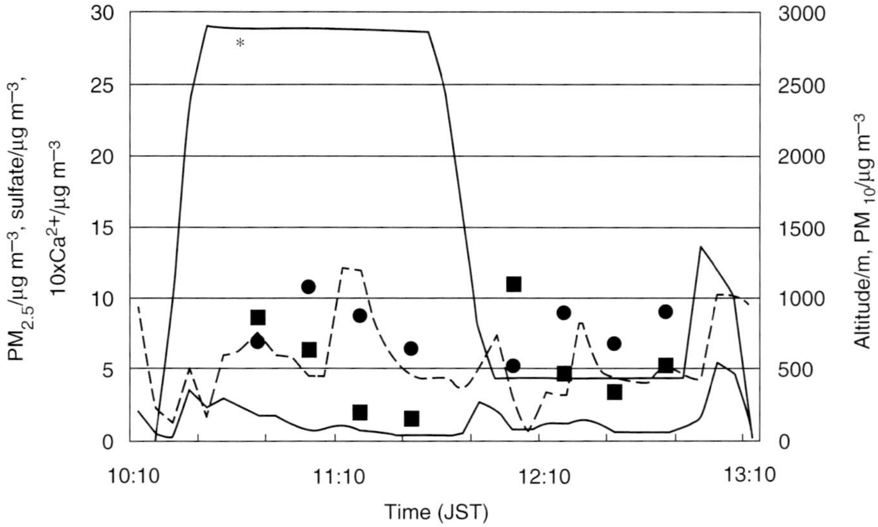

Fig. 2.5 Variations in aerosol mass concentrations (*black solid line* PM10, *dashed line* PM2.5; *black solid line marked with* * shows altitude) and concentrations of sulfate (*circles*) and calcium (*squares*) ions in the aerosols on 20 March 2001

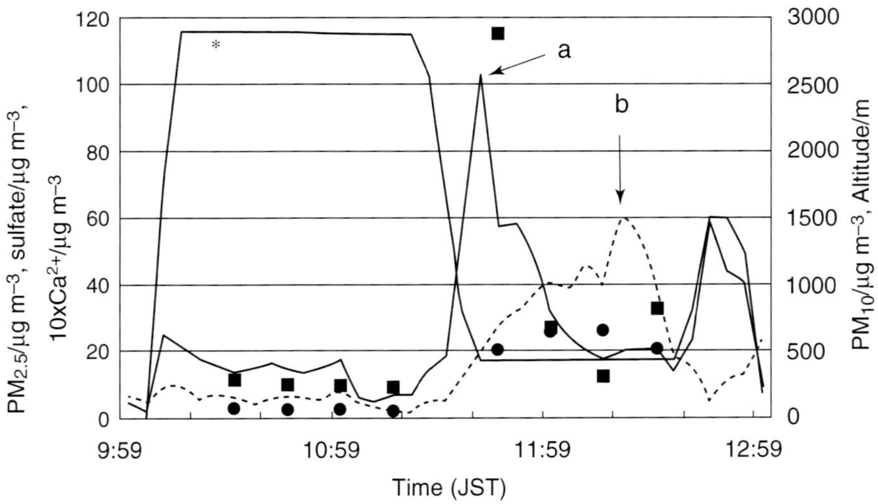

Fig. 2.6 Same items as those shown in Fig. 2.5, but on 21 March 2001

and the observation area, after which a high-pressure system moved into the observation area from central east China. The ratio of Σ(anionic species)/Σ(cationic species) (taking nitrate, sulfate, and chloride as representative anionic species; and ammonium, calcium, and sodium as representative cationic species) was between 95 and 121 %. Σ(anionic species) was always greater than Σ(cationic species).

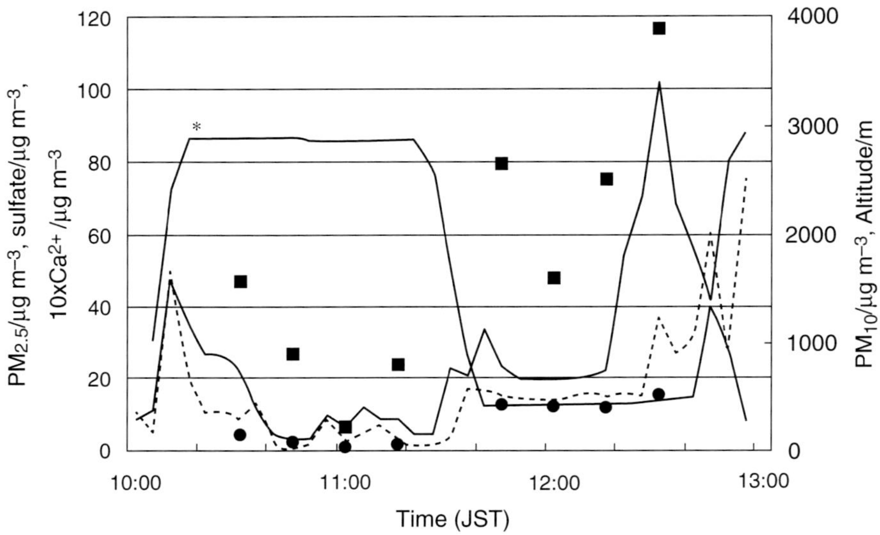

Fig. 2.7 Same items as those shown in Fig. 2.5, but on 22 March 2001

Therefore, the anthropogenic contribution to the chemical composition of the aerosols was very large on 20 March.

In contrast, on 22 March, when the concentration of PM10 was very high, the SO_4^{2-}/Ca^{2+} ratios were very small (average, 1.0 for altitudes higher than 2,500 m and 1.8 for those lower than 1,000 m). It should be noted that the entire observation area was covered by the dust storm. The satellite image taken on 22 March 2001 (NASA 2001) shows the dust continuing to blow across the Korean Peninsula and Japan. This is the reason that the SO_4^{2-}/Ca^{2+} ratios were so small. Moreover, Σ(cationic species) was greater than Σ(anionic species) on this day. This result also suggests that soil dust particles existed in excess on 22 March.

The mass concentration of coarse particles (PM10) on 21 March was high near Jeju Island (Fig. 2.6, point a), whereas that of fine particles (PM2.5) was high near Kyushu Island (point b). The concentration of Ca^{2+} showed a sharp peak following the distribution of PM10, whereas the concentration of sulfate showed a broad peak following the distribution of PM2.5. This means that the coarse particles were mainly *kosa* and that the fine particles contained a large amount of anthropogenic pollutants. Judging from the high concentration of PM2.5, we conclude that the lower atmosphere over the western Pacific was probably polluted most of the time in spring owing to continental outflow. On 21 March the SO_4^{2-}/Ca^{2+} ratios were small at high altitude (~2,800 m) and near Jeju Island at low altitude (average 2.1), whereas the ratios near the fine-particle peak depicted in Fig. 2.6 were large (average 12.3). The Σ(anionic species)/Σ(cationic species) ratio was small (63–69 %) for the samples with small SO_4^{2-}/Ca^{2+} ratios, whereas it was large (106–129 %) for the samples at lower altitude. As mentioned above, the air mass containing *kosa* arrived near Jeju Island on 21 March, when it had not yet reached the Kyushu area.

Nishikawa et al. (1991) reported that the SO_4^{2-}/Ca^{2+} ratio in soil from the Gobi Desert was 0.004. They also reported that the SO_4^{2-}/Ca^{2+} ratios of aerosols collected on Yakushima Island, Japan, during *kosa* and non-*kosa* periods were distinctly different. During the *kosa* period in April 1988, the ratios varied from 0.7 to 0.9, whereas in the corresponding non-*kosa* period the ratios ranged from 13.5 to 22.5. On the basis of these data, we can conclude that the air mass observed on 20 March was affected by anthropogenic pollutants. In contrast, the air mass observed on 22 March was affected by the Asian dust storm. Although the atmosphere near Jeju Island on 21 March was already affected by the Asian dust storm, the atmosphere near Fukue Island had not yet been affected by *kosa* particles.

Although sandstorms appear to be natural phenomena, increases in the frequency of sandstorms and subsequent deterioration of the environment arise from human activity. It has been reported that dust storms struck northwestern China on average once every 31 years from 300 CE to 1949, whereas since 1990 there has been one almost every year (Liu and Diamond 2005). This increased frequency is believed to be the result of overgrazing, deforestation, soil degradation, and desertification in Central Asia and across desert areas of China.

2.4 Elemental Carbon (EC) and Organic Carbon (OC)

Carbonaceous material is a major component of PM2.5 in the megacities of China. Many studies have reported the chemical composition of PM2.5 in Beijing; Chan and Yao (2008) reported that carbonaceous species (EC + OC) accounted for 27–42 % of the PM2.5 mass in that city, and concentrations of EC + OC varied from 10.8 to 51.9 $\mu g\ m^{-3}$. In Shanghai, carbonaceous material accounted for 41.4 % of the PM2.5 mass (Ye et al. 2003) and OC represented 73 % of the carbonaceous material. The annual average concentrations of EC and OC were 6.77 and 15.43 $\mu g\ m^{-3}$, respectively, at a monitoring site close to downtown Shanghai.

Elemental carbon (EC), also known as black carbon (BC), originates from the incomplete combustion of fuel during anthropogenic activities such as fossil fuel combustion, biomass burning, and biofuel burning for cooking and heating. EC is considered to be one of the most important aerosol components, because it is likely the second largest contributor to global warming. As one of the short-lived climate pollutants (SLCPs; which also include methane and tropospheric ozone), EC has become a target in the effort to lower global warming (Climate and Clean Air Coalition 2015). These SLCPs are also dangerous air pollutants, with various detrimental impacts on human health, agriculture, and ecosystems. It is easier to obtain a consensus for their reduction, compared with reaching a consensus for CO_2 reduction; demands for CO_2 reduction often meet opposition because this requires the reduction of fossil fuel energy use, which is widely considered to lead to the lowering of living standards.

The author's group has utilized the Cape Hedo Atmosphere and Aerosol Monitoring Station (CHAAMS) (Takami et al. 2007) at Cape Hedo, Okinawa,

Japan, to monitor the chemical composition of aerosols transported from East Asia. Among the many kinds of aerosols collected at Cape Hedo, EC and OC were the most important chemical components. Data obtained from March 2004 to March 2008 were used for the analyses of EC and OC, as well as other air pollutants (Shimada et al. 2011). Since EC and OC at Cape Hedo are mainly transported from China, the concentrations of EC and OC were high during the spring and winter, when most air masses come from East Asia (EC: 0.14–0.19 μg C m^{-3}, OC: 0.62–0.82 μg C m^{-3}). Concentrations of other air pollutants were also high at this time, e.g., PM2.5: 14.8–19.9 μg m^{-3}, O_3: 43–55 ppbv, and CO: 183–221 ppbv. In contrast, the concentrations of these pollutants were low during summer, when a Pacific high-pressure zone covered all of Okinawa (EC: 0.03–0.08 μg C m^{-3}, OC: 0.28–0.44 μg C m^{-3}). Concentrations of other pollutants were also low, e.g., PM2.5: 9.85–14.8 μg m^{-3}, O_3: 14–23 ppbv, CO (carbon monoxide): 68–93 ppbv. Thus, there is observational evidence that most air pollutants at Cape Hedo, Okinawa, come from East Asia, including China.

The source of carbonaceous aerosols was evaluated by determining the OC/EC ratios and the correlation between EC and OC. The ratio was low in spring and winter (OC/EC: 5.7–8.0) but high in summer (OC/EC: 10.2–18.9). These results imply a conclusion similar to that noted above. The anthropogenic pollutants were transported to CHAAMS from Asia in spring and winter, whereas in summer, monsoon winds transported clean oceanic air masses. The high OC/EC ratio also enabled us to identify the photochemical oxidation reaction.

In Eastern Asia (from Pakistan, China, and Mongolia to Japan, and also India, Southeast Asia, Philippines, and Korea), EC and OC emissions from China accounted for 61 % and 49 %, respectively, of the total emissions from the entire area (Zhang et al. 2009). Concentrations of EC and OC in the major cities in China were high (Yang et al. 2005). Weekly carbonaceous concentrations varied over wide ranges in, for example, Beijing (8.6–59 μg m^{-3} for OC and 1.5–25.4 μg m^{-3} for EC) and Shanghai (5.1–38.4 μg m^{-3} for OC and 2.3–13.0 μg m^{-3} for EC).

Generally speaking, air pollutants are transported from China toward Japan in association with the north or northwest monsoon in winter. High levels of air pollutants in China in winter are due to higher emission and poorer dispersion owing to low boundary layers and low wind speed. In spring, large-scale transport of air pollutants is caused by the movement of a low-pressure system or a migratory high-pressure system (Uno et al. 1998; Shimada et al. 2011).

2.5 PAHs

In many aerosols, organic (carbon-containing) particles are very small (around 0.1–1 μm in diameter) and these particles are one of the main components of PM2.5. Organic aerosols can be divided into two kinds: primary and secondary. Primary organic aerosols (POAs) are mainly emitted by fossil fuel combustion.

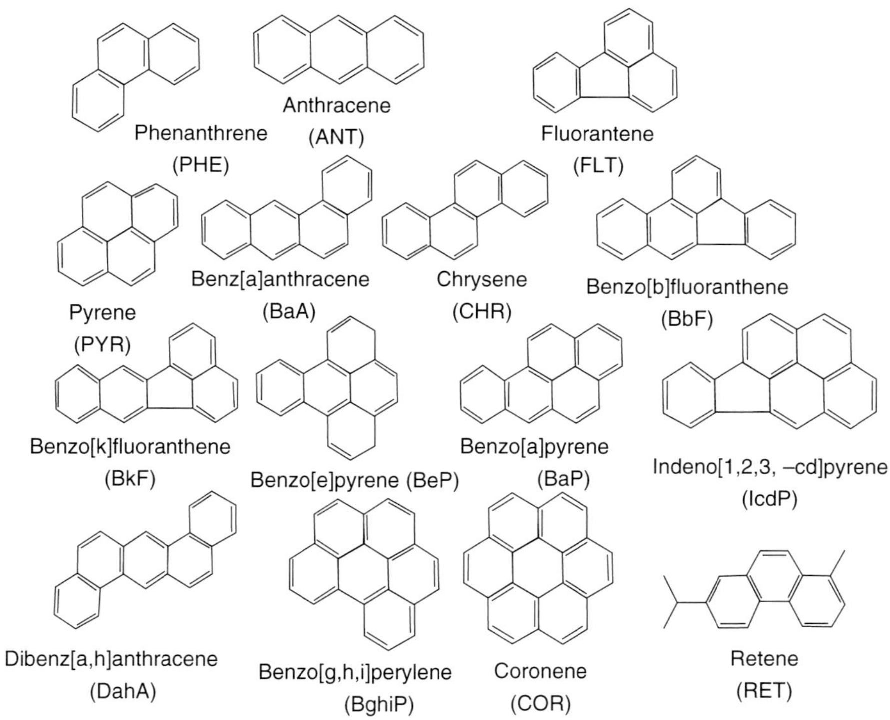

Fig. 2.8 Typical polycyclic aromatic hydrocarbons (PAHs) measured in this study

Cooling of high-temperature emission gas produces POA compounds, owing to the coagulation of organic compounds with dozens or more carbon atoms. Such compounds are emitted into the air as fine particles (0.1–0.5 μm in diameter). Secondary organic aerosols (SOAs) are produced in the atmosphere by the photo-oxidation of semi-volatile organic compounds forming very fine particles. Organic aerosols contain compounds such as fatty acids, low-molecular-weight di-carboxylic acids, long-chain alkanes, and PAHs. Since these organic compounds have origins of their own, they can be used as tracers to obtain information about the origins of aerosols.

PAHs (typical examples are shown in Fig. 2.8) are formed by the aromatization of parts of low-molecular-weight organic compounds that are produced by the thermal decomposition of organic compounds in the combustion process. PAHs are emitted mainly by the incomplete combustion of fossil fuels and bio-fuels, and often contain harmful compounds that are carcinogenic or mutagenic. For example, benzo[a]pyrene is classified as IARC Group 1 (carcinogenic to humans). PAHs represent 35–82 % of the carcinogenicity of aerosols (Pedersen et al. 2004). They acquire higher toxicity or carcinogenicity by transformation to nitro-substituted PAHs or by transformation to quinones by atmospheric reaction with ozone, OH radicals, and NO_2.

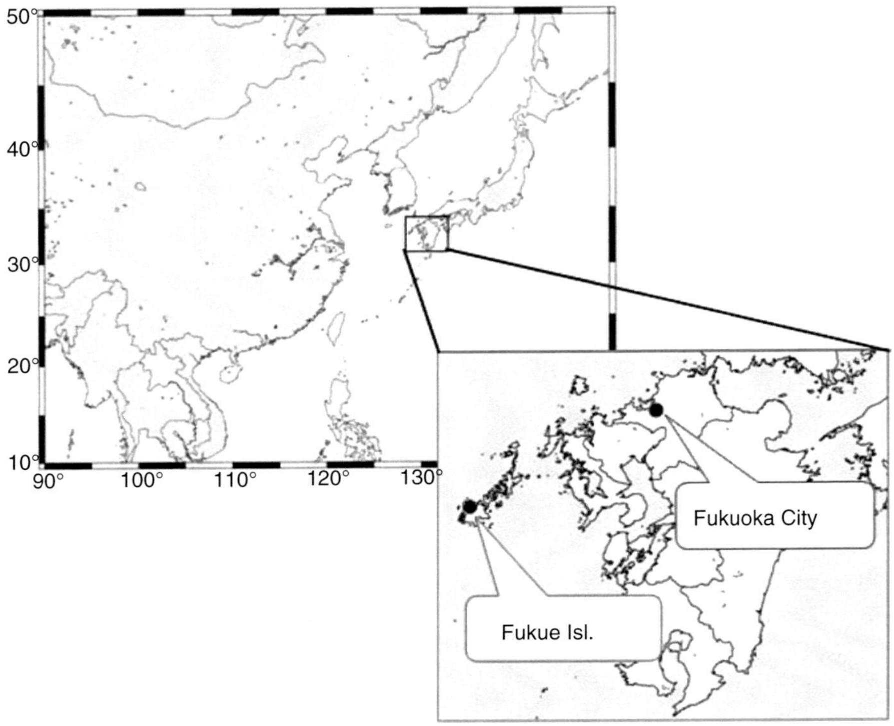

Fig. 2.9 Locations of Fukue Island and Fukuoka

The emission of PAHs in China accounts for 20 % or more of world emission (Zhang and Tao 2009). The annual average concentrations of benzo[a]pyrene in Beijing and Xi'an are 11.6 and 15.6 ng m^{-3}, respectively, levels which exceed Chinese air quality standards (10 ng m^{-3}; the WHO guideline is 1 ng m^{-3}) (Ma et al. 2011; Okuda et al. 2010).

Since Japan is located downwind of East Asia, aerosols emitted in East Asia can be easily transported to Japan. It is expected that domestic emission in Japan will decrease owing to the air quality regulations, but, nonetheless, an increase in PAH concentrations is expected because of their long-range transport. In western areas of Japan, such as Fukue Island, Nagasaki, and Fukuoka, the monthly average concentration levels and variations of PM2.5 are quite similar and short-term increases in concentration are mainly due to PM2.5 transport from East Asia (Kaneyasu et al. 2010, 2011). The same situation can be expected for PAHs.

From this viewpoint, the author and co-workers continually observed 15 PAHs (as shown in Fig. 2.8) on Fukue Island and in Fukuoka (Fig. 2.9) from 2009 to 2011 (Ogawa et al. 2012; Sato et al. 2007, 2009; Yoshino et al. 2011).

The seasonal average concentrations of 14 of these PAHs, except for retene (Fig. 2.8) on Fukue Island were found to be in the following order: summer (0.12 ng m^{-3}) < fall (1.68 ng m^{-3}) < spring (1.96 ng m^{-3}) < winter (3.26 ng m^{-3}) (Fig. 2.10).

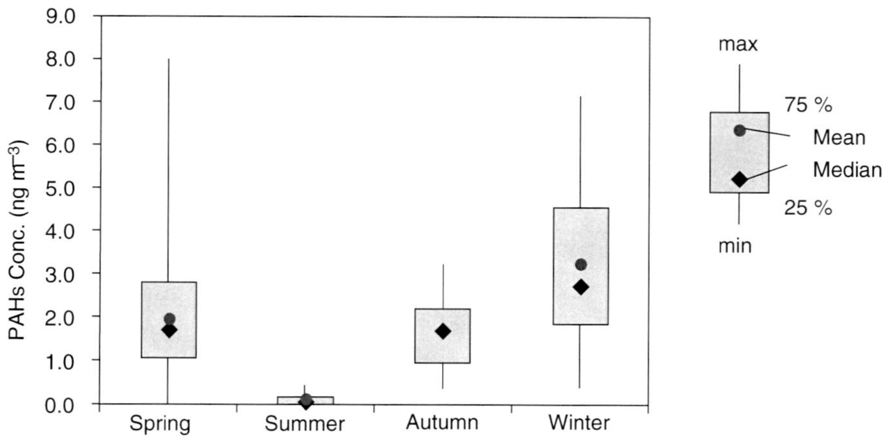

Fig. 2.10 Seasonal average concentrations of the PAHs shown in Fig. 2.8 (with the exception of retene) on Fukue Island

In winter, compared with the concentrations in spring and fall, the concentrations of CHR (chrysene), BbF (Benzo[b]fluoranthene), BkF (Benzo[k]fluoranthene), and BeP (Benzo[e]pyrene) (acronyms are shown in Fig. 2.8) were high, which suggests an increase in coal combustion. This is reasonable because in winter the use of coal for heating increases in northern China.

In contrast, in the Japanese megacity of Fukuoka, in addition to the PAHs arising from long-range transport, local-origin PAHs originating from automobiles have been shown to contribute to high PAH concentrations. If we assume that all PAHs measured on Fukue Island arrive by long-range transport, we can use the ratio of PAHs in Fukue to those in Fukuoka to estimate the fraction of transported PAHs in Fukuoka. The ratios in spring and winter are 0.66 ± 0.40 and 0.75 ± 0.20, respectively, which means that the effect of long-range PAH transport in Fukuoka is at a level that cannot be disregarded.

2.6 Metallic Elements and Source Apportionment

Although the proportion of metallic elements in aerosols is not very large (less than 17 %, as shown in Fig. 2.1), these elements are among the most important components of PM2.5. The fine particles contain many elements. Some of these elements have been identified as being carcinogenic or mutagenic, such as Cr, Ni, Cd, and As (IARC 2015). In addition, metallic elements can serve as indicators of the origin of aerosols, because they do not participate in chemical reactions during long-range transport.

The concentrations of some metallic elements in Beijing (Yu et al. 2013) are listed in Table 2.1, where they are compared with those in Fukue Island, Nagasaki, and Fukuoka (Suzuki et al. 2014). The concentrations in Beijing are from 10 to more than 50 times higher than those in Fukuoka.

Table 2.1 Concentrations of metallic elements in Beijing, China, and Fukue Island and Fukuoka, Japan

Elements	Beijing[a] (ng m^{-3})	Fukue[b] (ng m^{-3})	Fukuoka[b] (ng m^{-3})	Beijing[c] Fukuoka
V	15.3	1.48	1.97	10.3
Cr	22.4	0.54	0.84	41.5
Mn	62	4.74	6.83	13.1
Fe	1089	96.5	152	11.3
Ni	28.1	1.21	1.18	23.2
Cu	36.8	1.85	2.13	19.9
Zn	313	26.7	29.3	11.7
As	37.9	1.23	1.35	30.8
Se	38.1	0.67	0.67	56.9
Ba	88.2	1.53	4.73	57.6
Pb	117.3	8.41	10.1	13.9

[a]Average for four seasons
[b]Average for three seasons (spring, summer, and winter)
[c]Ratio of the concentration at Beijing against that at Fukuoka

Recently, Yu et al. (2013) evaluated the origin of aerosols by the positive matrix factorization (PMF) method (Paatero 1997). Their analysis included not only metallic elements but also non-metallic elements such as S, Si, P, Cl, and Br. They optimized the aerosol origin factors and categorized them into seven types: secondary sulfur (13.8 μg m^{-3}, 26.5 %), vehicle exhaust (8.9 μg m^{-3}, 17.1 %), fossil fuel combustion (8.3 μg m^{-3}, 16 %), road dust (6.6 μg m^{-3}, 12.7 %), biomass burning (5.8 μg m^{-3}, 11.2 %), soil dust (5.4 μg m^{-3}, 10.4 %), and metal processing (3.1 μg m^{-3}, 6.0 %). They also reported that: "Fugitive dusts (including soil dust and road dust) showed the highest contribution of 20.7 μg m^{-3} in the spring, doubling those in other seasons. On the contrary, contributions of the combustion source types (including biomass burning and fossil fuel combustion) were significantly higher in the fall (14.2 μg m^{-3}) and in the winter (24.5 μg m^{-3}) compared to those in the spring and summer (9.6 and 8.0 μg m^{-3}, respectively)."

In contrast, Suzuki et al. (2014) did not include non-metallic elements, and they determined five aerosol source factors for Fukue Island and four factors for Fukuoka. The five aerosol source factors for Fukue Island were *kosa*, sea salt, heavy oil combustion, coal combustion, and road dust, and the four factors for Fukuoka were *kosa*, heavy oil combustion, coal combustion, and road dust. Suzuki et al. (2014) also analyzed the source area of aerosols by using the total potential source contribution function (TPSCF) method (Hopke et al. 1995). At both sites, they found that *kosa* and emissions from coal combustion, both of which were evaluated on the basis of PMF analyses, originated in continental Asia, especially in China. A TPSCF plot of the road dust factor shows distributions over large cities such as Shanghai and Seoul. TPSCF plots of the oil combustion factor indicate that emissions from marine ships are also important at both sites.

2.7 Sulfate, Nitrate, and Ammonium Aerosols

The most common aerosols in the atmosphere consist of sulfate (SO_4^{2-}) and nitrate (NO_3^-). These are not primary aerosols; they are secondary aerosols that form after precursor gases (SO_2 and NOx) emitted into the air are oxidized by atmospheric reactions. In the atmosphere, there are both natural aerosols produced from gases emitted by, e.g., volcanos, and anthropogenic aerosols produced from gases produced by fossil fuel combustion.

Gaseous SO_2 is mainly oxidized in the air by OH radicals as follows:

$$SO_2 + OH \rightarrow HOSO_2$$

$$HOSO_2 + O_2 \rightarrow SO_3 + HO_2$$

$$SO_3 + H_2O \rightarrow H_2SO_4$$

The resulting sulfuric acid (H_2SO_4) has such low volatility that it forms very fine particles. Moreover, it readily reacts with surrounding ammonia gas to form fine particles of ammonium sulfate [$(NH_4)_2SO_4)$], which are carried across wide areas by winds.

Similarly, NOx is oxidized in air to form nitric acid (HNO_3) mainly by OH radical reactions, as follows:

$$NO_2 + OH \rightarrow HNO_3$$

Nitric acid, however, is a gas under normal atmospheric conditions. Fine particles are created by its reaction with ammonia gas in the surrounding environment.

$$HNO_3 + NH_3 \rightarrow NH_4NO_3$$

$$NH_4NO_3 \rightarrow HNO_3 + NH_3$$

Unlike ammonium sulfate, ammonium nitrate is not very stable. It thermally decomposes during transport to form nitric acid gas and ammonia gas. The resulting nitric acid gas is adsorbed on the surfaces of coarse particles such as soil and sea-salt particles, and nitrate salts such as calcium nitrate [$Ca(NO_3)_2$]. Thus, nitrate is contained in coarse particles after long-range transport. As shown in Fig. 2.11, most nitrates are initially contained in fine particles, but they are primarily contained in coarse particles in Okinawa after long-range transport (Takiguchi et al. 2008). These findings show that air pollutants and atmospheric aerosols can be transformed during long-range transport.

More detailed size distributions of ionic species in aerosols were obtained by size-segregated collections and analyses of aerosols (Yumoto et al. 2015). Samples were collected by using a cascade impactor sampler with five stages. Aerosols in the size ranges of >10, 2.5–10, 1–2.5, and 0.5–1 μm were collected on Teflon filters,

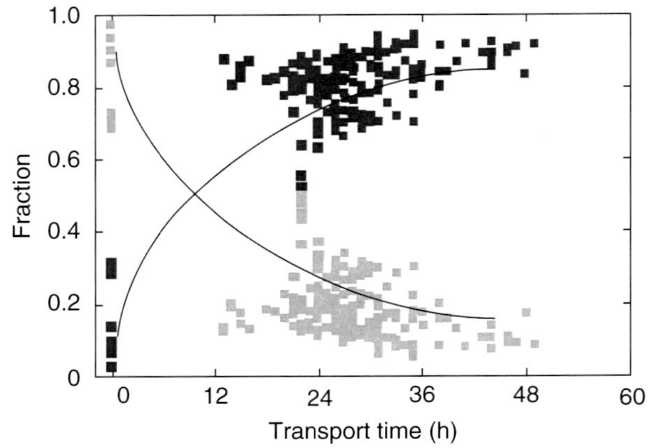

Fig. 2.11 Variations in nitrate fractions of particles with transport time. *Gray squares*: fraction of fine particulate nitrate in PM2.5. *Black squares*: fraction of coarse particulate nitrate in PM10. The transport time was calculated based on back trajectory analyses, by assuming that time 0 was when the trajectory left mainland China

and aerosols in the size range of 0.1–0.5 μm were collected with a stainless steel fiber filter (Otani et al. 2007).

During the sampling periods of April, October, and December 2012 and April 2013 at Cape Hedo, species from anthropogenic sources, such as NH_4^+ and SO_4^{2-}, were dominant among fine-mode particles (0.5–1, 0.1–0.5 μm). In contrast, species from natural sources, such as Ca^{2+} and Na^+, were dominant among coarse-mode particles (> 10, 2.5–10 μm). SO_4^{2-} was found in nearly all the samples. SO_4^{2-} in fine mode was non-sea salt, whereas in coarse mode it was sea salt. The size distribution of NO_3^- indicated that NH_4NO_3 had decomposed to gaseous NH_3 and HNO_3 during transport, and the resulting HNO_3 had been adsorbed on coarse particles by reacting with sea salt or crustal particles.

2.8 Trends of Ionic Species in Aerosols Transported from East Asia Measured by Aircraft

Anthropogenic emission in East Asia has been increasing because of rapid economic growth in this area. Atmospheric pollutants emitted in East Asia are observed even in North America, and they affect the entire northern hemisphere climate (UNECE 2011). Not only are the global aspects of long-range transport of pollutants very important, but so also are the regional aspects. From this perspective, we have continued aerial observations over the sea between continental Asia and Japan for 20 years in order to analyze the transport and transformation of long-range-transported atmospheric pollutants. The areas covered are the East China Sea, Sea of Japan, and the Yellow Sea. During this period, there was remarkable economic growth in East Asia, particularly in China, and as a result, the emission of atmospheric pollutants increased tremendously. Here, the results of our analyses of the long-term trends in the ionic components of aerosols are presented.

For the analyses we used 11 data sets, which were obtained in October 1991, November 1992, March and December 1994, January and December 1997, February

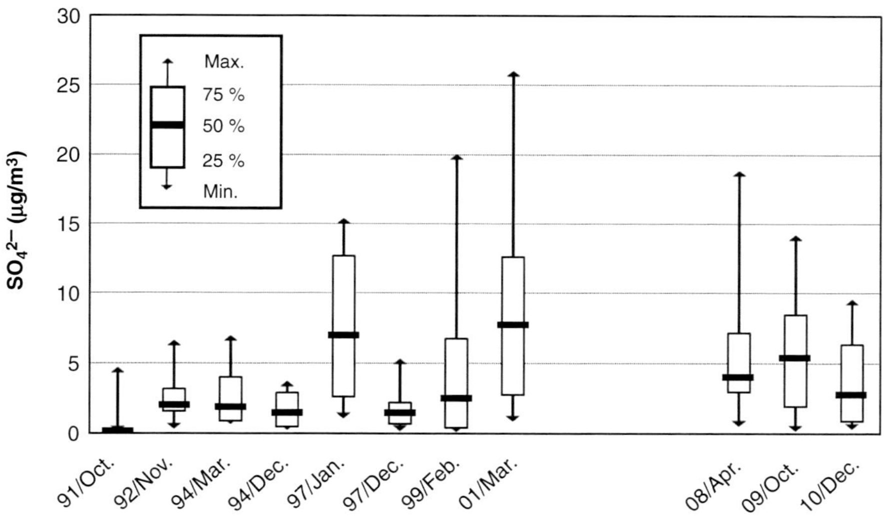

Fig. 2.12 Trend of sulfate in aerosols collected by aircraft

1999, March 2001, March–April 2008, October 2009, and December 2010 (Hatakeyama et al. 1995a, b, 1997, 2000, 2001, 2004, 2011, 2014).

Aerial observation is usually event-based and is difficult to continue over a long time. Many large projects have measured atmospheric aerosols all over the world (Clarke and Kapstin 2010). Although a lot of good data are available, long-term trends are not the usual target of such aerial observations. The greatest advantage of ground-based observations is that long-term trends and seasonal variations of air pollutants are usually the main targets of such studies. However, we continued aerial observations for about 20 years in almost the same area. Such observations have seldom been made elsewhere in the world, and we obtained very rare data.

The aircraft employed were a Cessna 404 and a Fairchild Swearingen Merlin IV, chartered from Showa Aviation Company (Yao, Japan), and a Beechcraft King Air 200 T, chartered from Diamond Air Service Incorporation (Toyoyama, Japan). The target area of the observations was the lower troposphere (below 3,000 m).

The common observations during 20 years of experiments were of gaseous species, such as ozone, SO_2, and NOx or NOy (total reactive nitrogenous compounds: gaseous HNO_3, organic nitrates, and inorganic nitrate salts in addition to NOx), as well as ionic species in aerosols, collected with a high-volume tape sampler. Sulfate, nitrate, ammonium, and calcium were the main ionic species analyzed. We determined the trends of anthropogenic species in the aerosol phase (SO_4^{2-}, NO_3^-, and NH_4^+).

Figures 2.12, 2.13, and 2.14 present the trends of sulfate, ammonium, and nitrate ion concentrations, respectively, in aerosols collected on board the aircraft. Sulfate and ammonium ions showed quite similar trends, which indicates that these compounds existed in the form of $(NH_4)_2$ SO_4 or (NH_4) HSO_4. Sulfate seems to have decreased from 2001, but unfortunately, since we lack data from 2002 to 2007

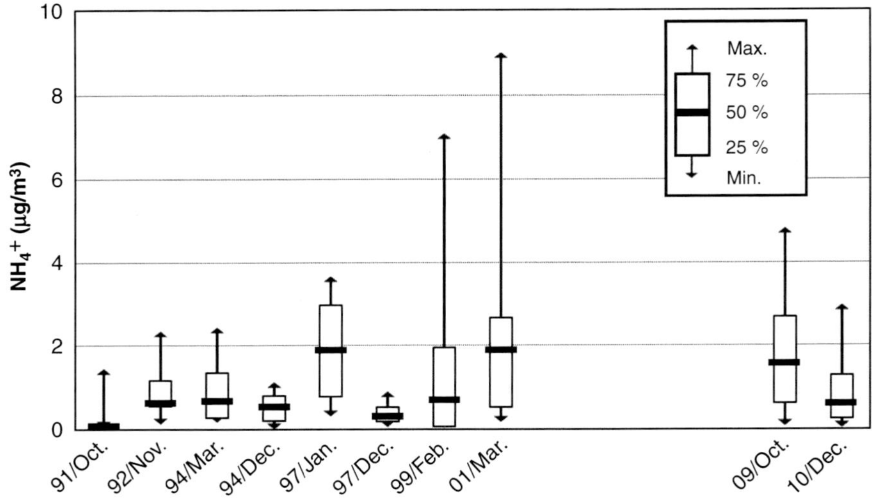

Fig. 2.13 Same as Fig. 2.12, for ammonium

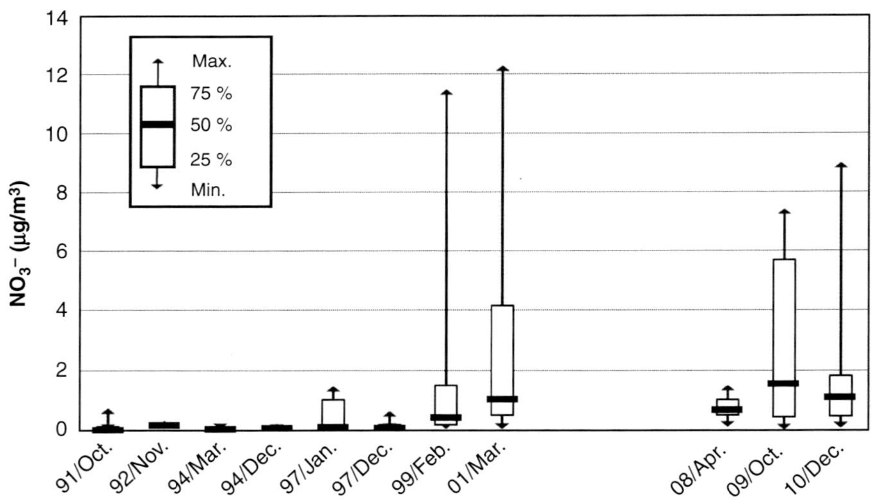

Fig. 2.14 Same as Fig. 2.12, for nitrate

(we carried out aerial observations over mainland China during that period), the trend is not clearly demonstrated. However, it has been reported that the emission of SO_2 in China has been decreasing since 2006 (Lu et al. 2010). Our results appear to agree with this finding, although a decreasing trend of SO_2 is not entirely evident in their work.

In contrast, nitrate was low before December 1997 (maximum 1.5 μg m^{-3}), but it increased after 1999, with maximum concentrations exceeding 7 μg m^{-3}. This seems to show an increasing trend. The emission of NOx, which is a precursor of NO_3^-, is

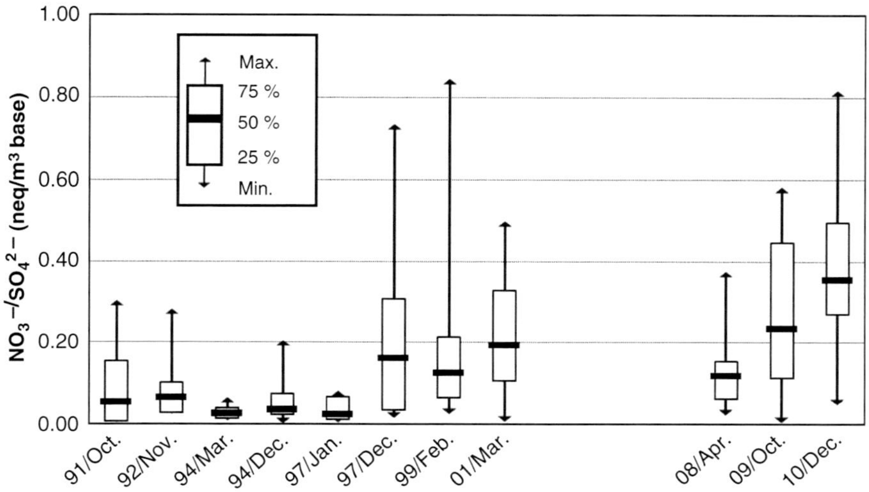

Fig. 2.15 Same as Fig. 2.12, for the ratio of nitrate/sulfate

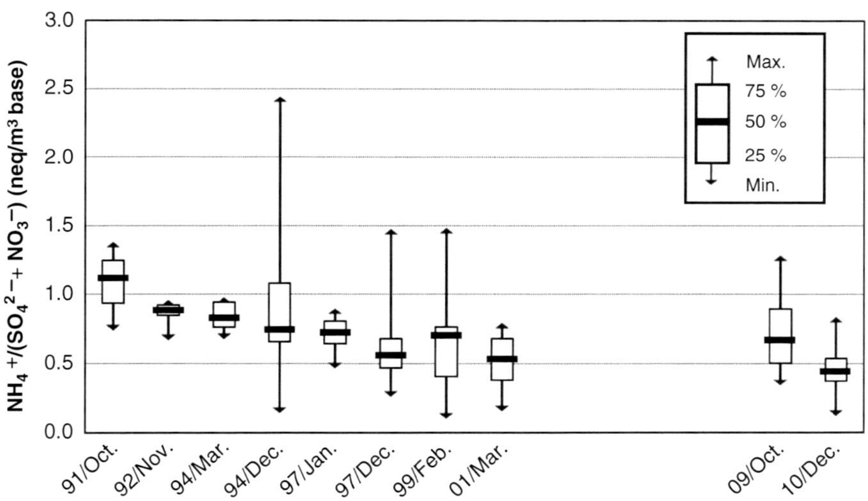

Fig. 2.16 Same as Fig. 2.12 for the ratio of ammonium/(sulfate + nitrate)

increasing in China (Kurokawa et al. 2013), and this could cause NO_3^- to increase over the East China Sea. More clearly, the ratio of NO_3^-/SO_4^{2-} has increased, and this reflects a decreasing trend of sulfate and an increasing trend of nitrate (Fig. 2.15). Furthermore, the ratio of NH_4^+ to the sum of SO_4^{2-} and NO_3^- is steadily coming down (Fig. 2.16). This means that the acidity of aerosols over the sea between continental Asia and Japan is still growing, and this could lead to more serious acid deposition in this area.

References

Chan CK, Yao X (2008) Air pollution in mega cities in China. Atmos Environ 42:1–42. doi:10.1016/j.atmosenv.2007.09.003

Clarke A, Kapstin V (2010) Hemispheric aerosol vertical properties: anthropogenic impacts on optical depth and cloud nuclei. Science 329:1488–1492

Climate and Clean Air Coalition (2015) Short-lived climate pollutants. (http://www.unep.org/ccac/Short-LivedClimatePollutants/Definitions/tabid/130285/Default.aspx. Accessed 21 Oct. 2015)

Darmenova K, Sokolik IN, Darmenov A (2005) Characterization of East Asian dust outbreaks in the spring of 2001 using ground-based and satellite data. J Geophys Res 110:D02204. doi:10.1029/2004JD004842

Dockery DW, Pope CA, Xu X, Spengler JD, Ware JH, Fay ME, Ferris BG, Speizer FE (1993) An association between air pollution and mortality in six U.S. Cities. N Engl J Med 329: 1753–1759

Hatakeyama S (2000) PEACAMPOT and PEACAMPOT II campaigns. IGACtivities (Newsletter of the International Global Atmospheric Chemistry Project) No.20:11–14

Hatakeyama S, Murano K, Bandow H, Sakamaki F, Yamato M, Tanaka S, Akimoto H (1995a) '91 PEACAMPOT aircraft observation of ozone, NOx, and SO_2 over the East China Sea, the Yellow Sea, and the Sea of Japan. J Geophys Res 100:23143–23151

Hatakeyama S, Murano K, Bandow H, Mukai H, Akimoto H (1995b) High concentration of SO_2 observed over the Sea of Japan. Terre Atmos Oceanic Sci 6:403–408

Hatakeyama S, Murano K, Mukai H, Sakamaki F, Bandow H, Watanabe I, Yamato M, Tanaka S, Akimoto H (1997) SO_2 and sulfate aerosols over the seas between Japan and the Asian continent. Earozoru Kenkyu 12:91–95

Hatakeyama S, Murano K, Sakamaki F, Mukai H, Bandow H, Komazaki Y (2001) Transport of atmospheric pollutants from East Asia. Water Air Soil Pollut 130:373–378

Hatakeyama S, Takami A, Sakamaki F, Mukai H, Sugimoto N, Shimizu A, Bandow H (2004) Aerial measurement of air pollutants and aerosols during 20–22 March 2001 over the East China Sea. J Geophys Res 109:D13304. doi:10.1029/2003JD004271

Hatakeyama S, Hanaoka S, Ikeda K, Watanabe I, Arakaki T, Sadanaga Y, Bandow H, Kato S, Kajii Y, Sato K, Shimizu A, Takami A (2011) Aerial observation of aerosols transported from East Asia—chemical composition of aerosols and layered structure of an air mass over the East China Sea. Aerosol Air Qual Res 11:497–507

Hatakeyama S, Ikeda K, Hanaoka S, Watanabe I, Arakaki T, Bandow H, Sadanaga Y, Kato S, Kajii Y, Zhang D, Okuyama K, Ogi T, Fujimoto T, Seto T, Shimizu A, Sugimoto N, Takami A (2014) Aerial observations of air masses transported from east Asia to the western Pacific: vertical structure of polluted air masses. Atmos Environ 97:456–461, http://dx.doi.org/10.1016/j.atmosenv. 2014.02.040

Hopke PK, Barrie LA, Li S-M, Cheng M, Li C, Xie Y (1995) Possible sources and preferred pathways for biogenic and non-sea-salt sulfur for the high Arctic. J Geophys Res 100: 16595–16603

IARC (2015) IARC monographs on the evaluation of carcinogenic risks to humans. (http://monographsiarcfr.ENG.Monographs.vol100C.mono100Cpdf. Accessed 14 Sept 2015)

ISO (2012) ISO 13271:2012(en) Stationary source emissions – determination of PM10/PM2.5 mass concentration in flue gas – measurement at higher concentrations by use of virtual impactors. (https://www.iso.org/obp/ui/#iso:std:iso:13271:ed-1:v1:en. Accessed on 31 Aug 2015)

Kaneyasu N, Takami A, Sato K, Hatakeyama S, Hayashi M, Hara K, Chang LS, Ahn JY (2010) Long-range transport of PM2.5 in Northern Kyushu area in spring. J Jpn Soc Atmos Environ 45:227–234

Kaneyasu N, Takami A, Sato K, Hatakeyama S, Hayashi M, Hara K, Kawamoto K, Yamamoto S (2011) Year-round behavior of PM2.5 in remote island and urban sites in the Northern Kyushu area Japan. J Jpn Soc Atmos Environ 46:111–118

Kurokawa J, Ohara T, Morikawa T, Hanayama S, Janssens-Maenhout G, Fukui T, Kawashima K, Akimoto H (2013) Emissions of air pollutants and greenhouse gases over Asian regions during 2000–2008: Regional Emission inventory in ASia (REAS) version 2. Atmos Chem Phys 13:11019–11058

Liu J, Diamond J (2005) China's environment in a globalizing world. Nature (London) 435: 1179–1186

Loomisa D, Grossea Y, Lauby-Secretana B, El Ghissassia F, Bouvarda V, Benbrahim-Tallaaa L, Guhaa N, Baana R, Mattocka H, Straifa K (2013) The carcinogenicity of outdoor air pollution. Lancet Oncol 14(13):1262–1263

Lu Z, Streets DG, Zhang Q, Wang S, Carmichael GR, Cheng YF, Wei C, Chin M, Diehl T, Tan Q (2010) Sulfur dioxide emissions in China and sulfur trends in East Asia since 2000. Atmos Chem Phys 10:6311–6331, doi:105194.acp-10-6311-2010

Ma W-L, Sun D-Z, Shen W-G, Yang M, Qi H, Liu L-Y, Shen J-M (2011) Atmospheric concentrations sources and gas-particle partitioning of PAHs in Beijing after the 29th Olympic games. Environ Pollut 159(7):1794–1801

Ministry of the Environment of Japan (2009) Report of the special committee for particulate matter in the Central Environmental Council of Ministry of the Environment of Japan

NASA (2001) SeaWiFS: east Asian dust continues. (http://visibleearth.nasa.gov/view.php?id=55610. Accessed 18 Oct 2015)

NASA (2005) Dust over the Pacific. (http://earthobservatory.nasa.gov/IOTD/view.php?id=5486&eocn=image&eoci=related_image. Accessed 1 Sept 2015)

NASA (2008) A perfect storm. (http://www.nasa.gov/multimedia/imagegallery/image_feature_989.html. Accessed 1 Sept 2015)

Nishikawa M, Kanamori S, Kanamori N, Mizoguchi T (1991) Kosa aerosol as eolian carrier of anthropogenic material. Sci Total Environ 107:13–27

Ogawa Y, Kaneyasu N, Sato K, Takami A, Hayashi M, Hara K, Hatakeyama S (2012) PAHs and n-alkanes transported long-range – from the observation at cape hedo fukue island and Fukuoka city in spring and autumn 2009. J Japan Soc Atmos Environ 47(1):18–25 (in Japanese with English abstract)

Okuda T, Okamoto K, Tanaka S, Shen Z, Han Y, Huo Z (2010) Measurement and source identification of polycyclic aromatic hydrocarbons (PAHs) in the aerosol in Xi'an China by using automated column chromatography and applying positive matrix factorization (PMF). Sci Total Environ 408(8):1909–1914

Otani Y, Eryu K, Furuuchi M, Tajima N, Tekasakul P (2007) Inertial classification of nanoparticles with fibrous filters. Aerosol Air Qual Res 7:343–352

Paatero P (1997) Least squares formulation of robust non-negative factor analysis. Chemom Intell Lab Syst 37:23–35

Pedersen DU, Durant JL, Penman BW, Crespi CL, Hemond HF, Lafleur AL, Cass GR (2004) Human-cell mutagens in respirable airborne particles in the northeastern United States 1 mutagenicity of fractionated samples. Environ Sci Technol 38:682–689

Sato K, Tanaka Y, Li H, Ogawa S, Hatakeyama S (2007) Distributions and seasonal changes of organic aerosols at Cape Hedo Okinawa: polycyclic aromatic hydrocarbons observed during 2005–2006. Chikyukagaku (Geochemistry) 41:145–153

Sato K, Li H, Tanaka Y, Ogawa S, Iwasaki Y, Takami A, Hatakeyama S (2009) Long-range transport of particulate polycyclic aromatic hydrocarbons at Cape Hedo remote island site in the East China Sea between 2005 and 2008. J Atmos Chem 61(3):243–257

Sato K, Takami A, Irei S, Miyoshi T, Ogawa Y, Yoshino A, Nakayama H, Maeda M, Hatakeyama S, Hara K, Hayashi M, Kaneyasu N (2013) Transported and local organic aerosols over Fukuoka Japan. Aerosol Air Qual Res 13:1263–1272

Shimada K, Takami A, Kato S, Kajii Y, Hatakeyama S (2011) Variation of carbonaceous aerosol in polluted air mass transported from East Asia and evaluation of their source origin. J Jpn Soc Atmos Environ 46:1–9 (in Japanese with English abstract)

Suzuki R, Yoshino A, Kaneyasu N, Takami A, Hayashi M, Hara K, Watanabe I, Hatakeyama S (2014) Characteristics and source apportionment of trace metals in the aerosols at Fukue Island and Fukuoka City. J Jpn Soc Atmos Environ 49(1):15–25 (In Japanese with English abstract)

Takami A, Miyoshi T, Shimono A, Kaneyasu N, Kato S, Kajii Y, Hatakeyama S (2007) Transport of anthropogenic aerosols from Asia and subsequent chemical transformation. J Geophys Res 112:D22S31. doi:10.1029/2006JD008120

Takiguchi Y, Takami A, Sadanaga Y, Lun X, Shimizu A, Matsui I, Sugimoto N, Wang W, Bandow H, Hatakeyama S (2008) Transport and transformation of total reactive nitrogen over the East China Sea. J Geophys Res 113:D10306. doi:10.1029/2007JD009462

UNECE (2011) Hemispheric transport of air pollution 2010. UNECE, New York

Uno I, Ohara T, Murano K (1998) Simulated acidic aerosol long-range transport and deposition over east Asia—role of synoptic scale weather systems. In: Grying SE, Chaumerliac N (eds) Air pollution modeling and its application, vol 22. Plenum, New York, pp 185–193

WHO (2005) WHO Air quality guidelines for particulate matter, ozone, nitrogen dioxide and sulfur dioxide. (http://apps.who.int/iris/bitstream/10665/69477/1/ WHO_SDE_PHE_OEH_06.02_eng.pdf#search=%27PM2.5+WHO+standard%27. Accessed 3 Sept 2015)

Yang F, He K, Ye B, Chen X, Cha L, Cadle SH, Chan T, Mulawa PA (2005) One-year record of organic and elemental carbon in fine particles in downtown Beijing and Shanghai. Atmos Chem Phys 5:1449–1457

Ye B, Ji X, Yang H, Yao X, Chan CK, Cadle SH, Chan T, Mulawa PA (2003) Concentration and chemical composition of PM2.5 in Shanghai for a 1-year period. Atmos Environ 37:499–510

Yoshino A, Nakayama H, Ogawa Y, Sato K, Takami A, Hatakeyama S (2011) Long-range transport of polycyclic aromatic hydrocarbons from east Asia at Cape Hedo in 2010. Earozoru Kenkyu 26(4):307–314 (in Japanese with English abstract)

Yu L, Wang G, Zhang R, Zhang L, Song Y, Wu B, Li X, An K, Chu J (2013) Characterization and source apportionment of PM2.5 in an urban environment in Beijing. Aerosol Air Qual Res 13:574–583. doi:10.4209/aaqr.2012.07.0192

Yumoto Y, Shimada K, Araki Y, Yoshino A, Takami A, Hatakayema S (2015) Size-segregated chemical analyses of particles transported from East Asia to Cape Hedo, Okinawa and their transformation mechanisms during the transport. Earozoru Kenkyu 30:115–125. doi:10.11203/jar30.115, in Japanese with English abstract

Zhang Y, Tao S (2009) Global atmospheric emission inventory of polycyclic aromatic hydrocarbons (PAHs) for 2004. Atmos Environ 43(4):812–819

Zhang Q, Streets DG, Carmichael GR, He KB, Huo H, Kannari A, Klimont Z, Park IS, Reddy S, Fu J, Chen S, Duan DL, Lei Y, Wang LT, Yao ZL (2009) Asian emissions in 2006 for the NASA INTEX-B mission. Atmos Chem Phys 9:5131–5153

Chapter 3
Acid Deposition

Hiroyuki Sase

Abstract Acid deposition is a regional issue in Asia. Emission levels of acidic substances, such as SO_2 and nitrogen oxides (NO_X), are still very high, although SO_2 emissions in China have started declining. The Acid Deposition Monitoring Network in East Asia (EANET), the regional network covering Northeast and Southeast Asia, has been monitoring acid deposition and its effects on forests and inland waters. Wet deposition levels of sulfur and nitrogen in EANET countries are significantly higher than those in Europe and the United States, although total deposition, as well as dry deposition, has not been sufficiently evaluated in the region, particularly in forested areas. Soil acidification is a potential risk for plant growth in Asia. Soil acidification has been observed in areas receiving high acid deposition in China and Japan. Observational data on acid deposition and soil chemical properties should be accumulated for forest areas to ascertain the risk for plants in the region.

Keywords Acid Deposition Monitoring Network in East Asia (EANET) • Wet deposition • Dry deposition • Soil acidification

3.1 Introduction

Due to rapid economic growth in recent decades, emissions of acidic substances, such as SO_2 and nitrogen oxides (NO_X), have been increasing in Asian countries. Anthropogenic SO_2 emissions in East Asia (Japan, China, and the Republic of Korea) have been increasing significantly since the 1950s, from approximately 3,000 Gg SO_2 y^{-1} in the 1950s to 35,000 Gg SO_2 y^{-1} in 2005. Emissions in Southeast Asia have also been gradually increasing, while emissions in Europe and North America peaked in the 1970s/1980s and rapidly decreased thereafter (Smith et al. 2011). According to Ohara et al. (2007), from 1980 to 2003, the Asian emissions of SO_2 and NO_X increased by 119 % and 176 %, respectively. In particular, NO_X

H. Sase
Asia Center for Air Pollution Research, 1182 Sowa, Nishi-ku, Niigata City 950-2144, Japan
e-mail: sase@acap.asia

T. Izuta (ed.), *Air Pollution Impacts on Plants in East Asia*,
DOI 10.1007/978-4-431-56438-6_3

emissions in China increased by 280 % from 1980 to 2003 (Ohara et al. 2007). Updated emission inventories in Asia (Kurokawa et al. 2013) continue to show high growth rates from 2000 to 2008, with rates of 34 % for SO_2 and 54 % for NO_X. SO_2 emissions in China peaked in 2006 and started decreasing gradually thereafter (Lu et al. 2011), while NO_X emissions were still increasing as of 2010 (Zhang et al. 2012). The emission levels of both SO_2 and NO_X in Asia remain high compared with those in the 1950s, and therefore the actual emission trend of SO_2 and the growth rate of NO_X emissions should be carefully monitored henceforth.

Atmospheric deposition of acidic substances, hereafter referred to as "acid deposition", is a concern for the general public as "acid rain", which has been blamed for causing soil acidification (e.g., Hallbäcken and Tamm 1986), forest decline (e.g., Schulze 1989), and the decline of freshwater fish populations (Wright et al. 1976) in Europe and North America. Therefore, the monitoring of acid deposition and its effects on terrestrial ecosystems has also been conducted at the national level in Japan and China since the 1980s (Zhao et al. 1988; Tamaki et al. 2000) and at the regional level by the Acid Deposition Monitoring Network in East Asia (EANET) since 2001 (EANET 2014). Relevant studies on soil and inland water acidification have been conducted, particularly in Japan and China. In this chapter, we review the current status of acid deposition and soil acidification, as a potential inhibitor of plant growth, in Northeast and Southeast Asia, based on EANET data and other observational studies.

3.2 Acid Deposition Monitored by a Regional Network in East Asia

The EANET is an intergovernmental regional network that was formed by 13 countries in Northeast and Southeast Asia. The EANET started its preparatory-phase activities in 1998 and regular-phase activities in 2001. It has been monitoring wet deposition (precipitation), dry deposition (gaseous and particulate matter), soil chemical properties, forest vegetation (tree growth and species composition of understory vegetation), inland water chemistry, and biogeochemical processes in forest catchments. Monitoring data from 2000 onwards are disclosed to the public on the EANET website (www.eanet.asia).

3.2.1 Wet Deposition

From 2000 to 2009, the annual mean pH values of rainwater from 57 sites in 13 countries ranged from 4.18 to 6.94, with an arithmetic mean of 5.07 and a median of 4.94, according to the second EANET periodic report (EANET 2011). During this period, pH values below 4.30 were recorded at two sites (Jinyunshan and Haifu) in Chongqing, China, one site (Petaling Jaya) in Selangor, Malaysia, and one site

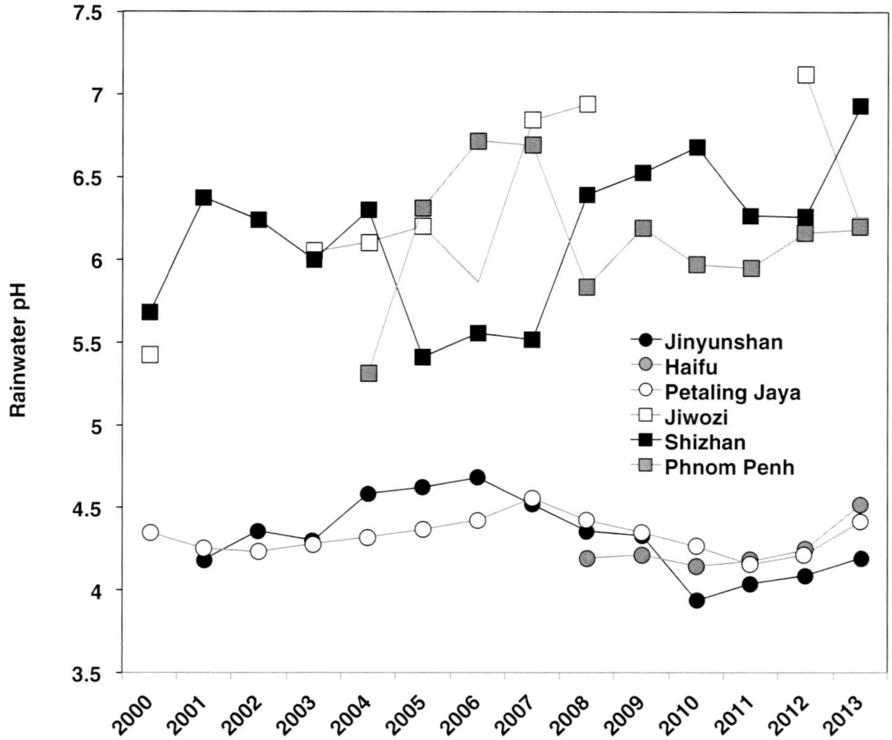

Fig. 3.1 Annual changes in rainwater pH at the EANET (Acid Deposition Monitoring Network in East Asia) sites. Sites with significantly high or low pH values are shown here – Jinyunshan and Haifu, in Chongqing, China; Petaling Jaya, in Selangor, Malaysia; Jiwozi and Shizhan in Xi'an, China; and Phnom Penh, Cambodia

(Kanghwa) in the Republic of Korea. In contrast, pH values above 6.5 were observed in two sites in Xi'an, China (Jiwozi and Shizhan) and one site (Phnom Penh) in Cambodia (EANET 2014). The annual variation of pH at these sites for the past decade is shown in Fig. 3.1 (after EANET 2014). Although the acidity of rainwater may fluctuate year by year depending on its chemical composition and precipitation amount, rainwater in China and Malaysia appears to have been significantly acidic for the past decade. However, at two sites in Xi'an, China, the rainwater was strongly neutral.

Wet deposition levels of non-sea-salt (nss) SO_4^{2-}, dissolved inorganic nitrogen (DIN), and nss-Ca^{2+} at the EANET monitoring sites from 2009 to 2013 are summarized in Table 3.1. Two sites (Jinyunshan and Haifu) in Chongqing, and the Petaling Jaya site, had significant depositions of nss-SO_4^{2-} and DIN. This may contribute to the low pH values at these sites. The Shizhan site in Xi'an also had high deposition of nss-SO_4^{2-} and relatively high deposition of DIN (94 mmol m^{-2} y^{-1}; not shown in Table 3.1), although the rainwater pH was significantly high. The high level of wet nss-Ca^{2+} deposition may have contributed to neutralization of the acids.

Table 3.1 Wet deposition amounts of non-sea-salt SO_4^{2-}, dissolved inorganic nitrogen (DIN), and non-sea-salt Ca^{2+} from 2009 to 2013 in the EANET sites

	Percentile, mmol m^{-2} y^{-1} and site name with the mean value		
Parameter	10 %	50 %	90 %
Non-sea-salt SO_4^{2-}	7.5	20	53
Site name (mean), "≤10 %" or "90 % ≤"	Listvyanka, RU (7.0)		Jinyunshan, CN (121)
	Ogasawara, JP (6.1)		Haifu, CN (113)
	Khanchanaburi, TH (5.3)		Bandung, ID (63)
	Chiang Mai, TH (4.7)		Shizhan, CN (60)
	Mondy, RU (0.74)		Petaling Jaya, MY (60)
DIN, $NO_3^- + NH_4^+$	19	56	133
Site name (mean), "≤10 %" or "90 % ≤"	Ochiishi, JP (19)		Haifu, CN (196)
	Ogasawara, JP (13)		Jinyunshan, CN (179)
	Listvyanka, RU (14)		Petaling Jaya, MY (178)
	Danum Valley, MY (10)		Bandung, ID (155)
	Mondy, RU (2.4)		Serpong, ID (133)
Non-sea-salt Ca^{2+}	3.7	12	42
Site name (mean), "≤10 %" or "90 % ≤"	Rishiri, JP (3.3)		Shizhan, CN (76)
	Nakhon Ratchasima, TH (3.2)		Haifu, CN (60)
	Ochiishi, JP (2.3)		Jinyunshan, CN (58)
	Ogasawara, JP (1.9)		Cuc Phuong, VN (57)
	Mondy, RU (0.97)		Mt. Sto. Tomas, PH (52)

After EANET 2014
Note: Total of 47 sites (whose annual data were available for at least 3 years)
CN China, *ID* Indonesia, *JP* Japan, *MY* Malaysia, *PH* Philippines, *RU* Russia, *TH* Thailand

At the sites in Chongqing, wet deposition levels of nss-Ca^{2+} were similar to those at the Shizhan site. However, as shown in Table 3.1, the deposition levels of nss-SO_4^{2-} at these Chongqing sites were almost double those at the Shizhan site, while the NO_3^- levels were more than double (70, 66, and 20 mmol m^{-2} y^{-1} at Haifu, and Jinyunshan, and Shizhan, respectively). Therefore, it seems that acidity derived from the strong acid anions in rainwater may not be sufficiently neutralized at these sites. In the sites in China, the deposition of strong acid anions and basic cations is significant, and therefore the balance between these ions is very important for evaluating the effects of acid deposition on ecosystems. The importance of cation deposition has already been suggested by previous studies (e.g., Dawei et al. 2001; Xu et al. 2001). Zhao et al. (2001) suggested that a high concentration of Ca^{2+} in the soil solution, which was derived from the high rates of Ca^{2+} deposition, controlled the toxicity of Al in the acidified soil of Chongqing. On the other hand, Larssen and Carmichael (2000) suggested that the effects of acid deposition on soil and vegetation would be manifested if the anthropogenic emission of calcium was reduced considerably faster than that of sulfur emission. Since sulfur emissions in China started declining after 2006, the emission trend of basic cations should also be monitored carefully.

The deposition levels of S and N are quite high in EANET countries. Matsubara et al. (2009) compared the wet deposition of S and N between the monitoring networks in Japan, Europe (determined by the European Monitoring and Evaluation Programme; EMEP), and the United States (determined by the National Atmospheric Deposition Program; NADP) to assess the impacts of acid deposition on river water pH. According to the comparison, the 50th percentiles for the wet deposition of nss-S in 2003 were 8.01, 2.57, and 2.85 kg S ha^{-1} y^{-1} in Japan, Europe, and the United States, respectively. Those of DIN were 7.86, 5.19, and 3.61 kg N ha^{-1} y^{-1}, respectively. Compared with deposition levels in Europe and the United States, deposition levels in Japan were significantly greater. The 50th percentiles for wet deposition across all EANET countries corresponded to 6.21 kg S ha^{-1} y^{-1} and 7.84 kg N ha^{-1} y^{-1} (calculated based on the data in Table 3.1). Although the assessment year was slightly different, the wet deposition of nss-S and N in the EANET countries was significantly greater than that in Europe and the United States. Moreover, the 90th percentiles of wet DIN deposition were 16.4, 9.20, 6.51, and 18.6 kg N ha^{-1} y^{-1} in Japan, Europe, the United States, and the EANET countries, respectively. Monitoring sites in the EANET countries and Japan significantly exceeded the threshold values of N deposition, whereby NO_3^- concentration had increased in stream water: 10 kg N ha^{-1} y^{-1} (Wright et al. 1995) and 8 kg N ha^{-1} y^{-1} (MacDonald et al. 2002) in Europe and 7 kg N ha^{-1} y^{-1} (Aber et al. 2003) in the United States. Measured values also significantly exceeded the critical N loads for epiphytic lichens: 2.4 kg N ha^{-1} y^{-1} (Giordani et al. 2014) in Europe and 3–9 kg N ha^{-1} y^{-1} (Geiser et al. 2010) in the United States. Therefore, the effects of N deposition along with acid deposition should be carefully monitored in the EANET countries, taking into account forest ecosystem sensitivities.

3.2.2 *Dry and Total Deposition*

Dry deposition of gaseous and particulate matter is also an important deposition process in terrestrial ecosystems. To evaluate acid deposition in these ecosystems, total deposition should be estimated as the sum of dry deposition and wet deposition. In the EANET countries, it was recommended that dry deposition flux should be evaluated by the inferential method (EANET 2010), in which the flux can be calculated by multiplying the air concentrations of the pollutants by their deposition velocities. However, detailed meteorological data is necessary to estimate dry deposition velocities, and sufficient meteorological data has not been compiled for the EANET sites, except for those in Japan. Therefore, as of 2015, regional assessment of dry deposition and total deposition has not yet been conducted in the EANET countries. According to the EANET second periodic report (EANET 2011), the mean dry deposition amount and mean wet deposition amount of S compounds in the Japanese EANET sites for the period from 2003 to 2007 were approximately 20 and 35 mmol m^{-2} y^{-1}, and those of N compounds were approximately 28 and 48 mmol m^{-2} y^{-1}, respectively. Compared with levels in Europe (determined by EMEP) and the United States (determined by the Clean Air Status and Trends Network; CASTNET), the total deposition levels of S and N compounds were significantly greater in Japan (EANET 2011). In particular, total S deposition was three times greater than that in Europe and the United States, due to high levels of precipitation and nss-SO_4^{2-}. Although dry deposition has not been evaluated, total deposition levels at some of the EANET sites, such as the sites in Chongqing and Malaysia, may be significantly greater than those at Japanese sites; this can be explained by the extremely high wet deposition levels (see Table 3.1). Therefore, in the future, regional assessment of total deposition in Northeast Asia and Southeast Asia should be promoted to evaluate the effects of acid and N deposition on terrestrial ecosystems.

3.2.3 Observational Data in Forested Areas Obtained by the EANET: Relevant Studies

Acid deposition monitoring in the EANET has been conducted in open spaces, away from forested areas, to obtain representative data over a relatively large area. Therefore, to gain observational data in forested areas, such data should be gathered in the region. Atmospheric deposition and its effects on forest catchments have been studied by EANET scientists in different types of forests; namely, a temperate coniferous forest in Japan from 2002 (e.g. Kamisako 2008; Sase et al. 2008), a tropical dry evergreen forest in Thailand from 2005 (e.g., Yamashita et al. 2010; Sase et al. 2012), a tropical rainforest in Sabah, Malaysia, from 2008 (Yamashita et al. 2014), and a tropical rehabilitated rainforest in Sarawak, Malaysia, from 2012 (Sase et al. 2015). Although DIN may be subject to strong effects of canopy interactions, such as uptake or consumption in the forest canopy (Sase et al. 2008), the canopy interactions of S may be negligible. Therefore, the levels of S deposition collected by throughfall (TF) and stemflow (SF), which also include dry deposition amounts, can be considered as the total deposition in forested areas.

Atmospheric deposition amounts of S and DIN collected by TF and SF in the study forests are summarized in Table 3.2 (Sase et al. 2015). The deposition amounts

Table 3.2 Atmospheric deposition amounts of S and DIN in four different forests in Asia

Site	Kajikawa	Sakaerat	Danum Valley	Bintulu
Country and province	Niigata, Japan	Nakhon Ratchasima, Thailand	Sabah, Malaysia	Sarawak, Malaysia
Forest type	Japanese cedar	Dry evergreen forest	Tropical rainforest	Rehabilitated forest
Annual precipitation (mm)	2,281	1,488	2,700	3,500
S (kg S ha^{-1})	28.5	5.76	3.6	19
DIN (kg N ha^{-1})	16.6	7.9	6.2	11.8

After Sase et al. 2015

Note: Deposition amounts under the forest canopy, determined by throughfall (+ stemflow) method, are shown in the Table of S and DIN were significant at the Kajikawa site in Japan, likely caused by the long-range transport of air pollutants from the Asian continent (Sase et al. 2012). Deposition levels at the Bintulu monitoring site, located in one of the most industrialized cities in Malaysia, were also high. Deposition levels measured in Sakaerat, Thailand, and the Danum Valley, Malaysia, were relatively low. However, the deposition pattern at the Sakaerat site showed a distinct seasonality as a result of well defined dry and wet seasons, with deposition peaks observed at the beginning of the wet season (Sase et al. 2012). At the Kajikawa site, the highest deposition levels were observed during periods of high snowfall in winter, due to strong seasonal westerly winds from the Sea of Japan (Kamisako 2008; Sase et al. 2008, 2012). Such seasonality of deposition and precipitation patterns may be closely related to phenology and biogeochemistry in ecosystems. Therefore, not only annual deposition levels but also their seasonal patterns should be carefully monitored to quantify the effects of acid deposition on forest ecosystems.

3.3 Soil Acidification in Asian Countries

The current level of acid deposition in EANET countries may not affect plant physiology directly. However, soil acidification is still a potential risk for plant growth in Asian countries. According to a model simulation by Hicks et al. (2008), acid-sensitive soil types were found in some areas of Northeast and Southeast Asia, where base saturation of the soil would decrease to less than 20 % with acidification within 50 years, although the projection varied depending on future deposition scenarios. This finding strongly indicates that long-term observational data of acid deposition and soil chemical properties should be accumulated.

Soil has buffer systems; therefore, the addition of a strong acid does not directly result in a decrease in pH (Ulrich 1991). The amount of strong acid required to reduce the pH of a system to a reference pH value is termed the "acid neutralizing capacity (ANC)"; thus, soil acidification should be defined as a decrease in the ANC of the soil (Van Breemen et al. 1983). Decreases in the ANC due to inputs of strong acids derived from external sources are not necessarily reflected in changes to indicators of soil acidification, such as pH and exchangeable H^+ and Al^{3+}. Therefore, it takes a long time to detect soil acidification phenomena from field observations. In Europe (e.g., Hallbäcken and Tamm 1986; Nilssen1986; Schulze 1989) and the United States (Drohan and Sharpe 1997), soil acidification phenomena, such as decreases in pH and in Ca/Al and Mg/Al ratios in soil solutions, were observed by re-sampling forest soils after 20–50 years. Based on the data summarized in Central Europe and Scandinavia by Nilssen (1986) and Ulrich (1991) and data from the United States (Drohan and Sharpe 1997), the reduced pH values and the initial soil pH are plotted in Fig. 3.2a (after Sase 2015). Greater pH decline can be observed in the soils with higher initial pH. This may be because buffering systems function differently in soils with different pH values, as suggested by Ulrich (1991).

As described above, re-sampling of pre-established plots after 20−50 years helped to quantify soil acidification, as proven in the 1980s in Europe and the United

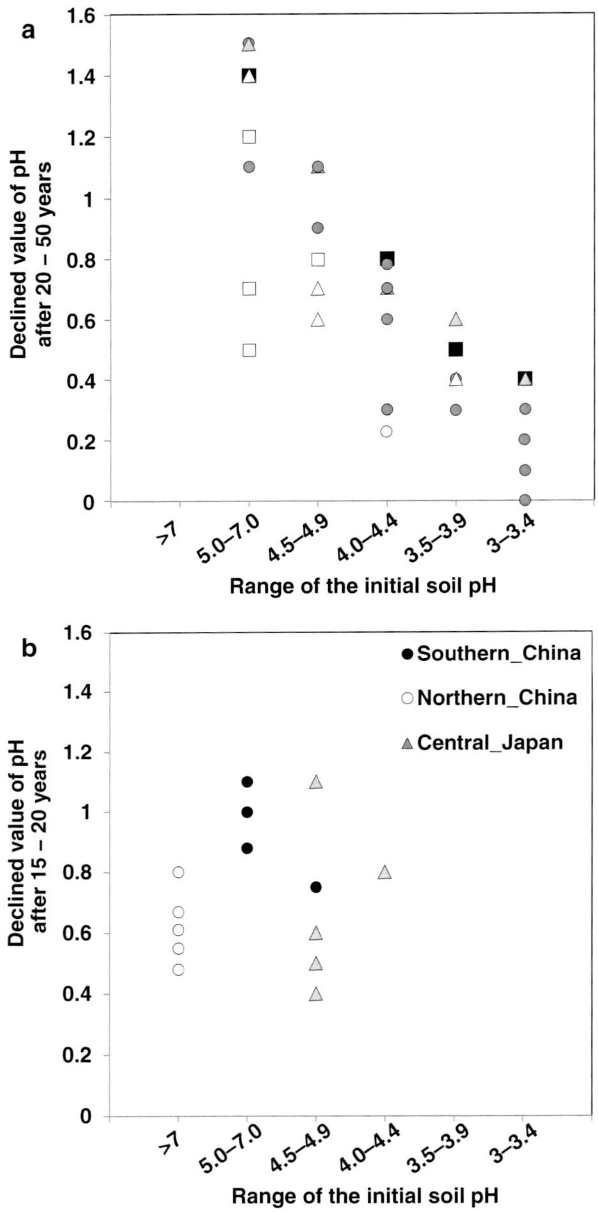

Fig. 3.2 Initial soil pH and declined pH values in (**a**) Central Europe, Scandinavia, and the United States for 20–50 years and (**b**) China and Japan for 15–20 years (After Nilssen 1986; Ulrich 1991; Drohan and Sharpe 1997; Dai et al. 1998; Liu et al. 2010; Nakahara et al. 2010; Yang et al. 2012; Sase 2015). *Closed circles* humus layers, *open circles* A-horizons, *open triangles* B-horizons, *closed triangles* silt, *open squares* C-horizons, *closed squares* sand

States. Unfortunately, in East Asia, this approach has not been conducted to a sufficient extent. Soil acidification was reported by field observation only in those areas of China and Japan that received high acid deposition. In China, this occurred in the mountainous areas of Hunan Province and the Guangxi Zhuang Autonomous Region from 1959/1960 to 1994 (Dai et al. 1998); in the Dinghushan Biosphere Reserve, Guangdong, from the 1980s to 2005 (Liu et al. 2010); and in northern China's grasslands from the 1980s to 2000s (Yang et al. 2012). In central Japan, this occurred in the Lake Ijira catchment, Gifu Prefecture, from 1990 to 2004 (Nakahara et al. 2010). The observational data from China and Japan are plotted in Fig. 3.2b (after Sase 2015). Similar to the findings noted above, in China and Japan, when initial soil pH was lower than 7.0, greater pH decline seemed to be observed in the soils with higher initial pH, although the amount of supporting data was limited. Liu et al. (2010) also showed that forest soils with originally higher pH values were acidified more rapidly than those with originally lower pH values. In soils with high pH (>7) in northern China, soil carbonates functioned as the main buffering systems (Yang et al. 2012), and therefore pH declines in these soils may be relatively insignificant. Dai et al. (1998) showed that soil acidification was not apparent in areas covered by red and yellow soils, such as acrisols, alisols, and ferralsols. These soils are highly weathered acidic soils and are rich in aluminum, and are theoretically within the aluminum buffer range. Therefore, pH values in these soils may not readily decline, even with high acid deposition.

3.4 Summary

Emission levels of S and N remain high in Asia. Accordingly, deposition levels are significantly higher in Northeast and Southeast Asia than in Europe and the United States. Wet deposition monitoring data have been accumulated owing to the efforts of the EANET, the regional network in East Asia, although total deposition and dry deposition have not yet been sufficiently evaluated. Observational data on acid deposition in forested areas are also lacking. Atmospheric deposition, including wet deposition and dry deposition, should be monitored in Asian forested areas. Soil acidification caused by acid deposition poses a potential risk for plant growth. Although the number of observations is limited, soil acidification phenomena have been observed in China and Japan. These phenomena corresponded to the soil-buffering mechanism, a finding that was similar to observations in Europe and the United States. Long-term observational data on soil chemical properties should be gathered to determine the potential risk for plant growth with greater precision.

Acknowledgments Part of the manuscript was cited from the report of the research project "ARCP2013-13CMY-Sase", which was supported by the Asia-Pacific Network for Global Change Research (APN). The author thanks APN and the project members; namely, Tsuyoshi Ohizumi, Naoyuki Yamashita, Thiti Visaratana, Bopit Kietvuttinon, Hathairatana Garivait, Nik Muhamad Majd, Ahmed Osumanu Haruna, Seca Gandaseca, Tsuyoshi Saito, and Yayoi Inomata.

References

Aber JD et al (2003) Is nitrogen deposition altering the nitrogen status of northeastern forests? BioScience 53:375–389
Dai Z et al (1998) Changes in pH, CEC and exchangeable acidity of some forest soils in Southern China during the last 32–35 years. Water Air Soil Pollut 108:377–390
Dawei Z et al (2001) Acid deposition and acidification of soil and water in the Tie Shan Ping Area, Chongqing, China. Water Air Soil Pollut 130:1733–1738
Drohan JR, Sharpe WE (1997) Long-term changes in forest soil acidity in Pennsylvania, USA. Water Air Soil Pollut 95:299–311
EANET (2010) Technical manual on dry deposition flux estimation in east Asia. Network Center for the EANET, Asia Center for Air Pollution Research, Niigata
EANET (2011) Second periodic report on the state of acid deposition in east Asia. Secretariat for the EANET, UNEP/RRC.AP, Pathumthani, Thailand, and Network Center for the EANET, Asia Center for Air Pollution Research, Niigata
EANET (2014) Data report 2013. Network Center for EANET, Asia Center for Air Pollution Research, Niigata
Geiser L et al (2010) Lichen based critical loads for atmospheric nitrogen deposition in Western Oregon and Washington Forests, USA. Environ Pollut 158:2412–2421
Giordani P et al (2014) Detecting the nitrogen critical loads on European forests by means of epiphytic lichens, a signal-to-noise evaluation. For Ecol Manage 311:29–40
Hallbäcken L, Tamm CO (1986) Changes in soil acidity from 1927 to 1982–1984 in a forest area of south-west Sweden. Scand J For Res 1:219–232
Hicks WK et al (2008) Soil Sensitivity to Acidification in Asia: Status and Prospects. AMBIO: A Journal of the Human Environment 37:295–303.
Kamisako M (2008) Seasonal and annual fluxes of inorganic constituents in a small catchment of a Japanese cedar forest near the Sea of Japan. Water Air Soil Pollut 195:51–61
Kurokawa et al (2013) Emissions of air pollutants and greenhouse gases over Asian regions during 2000–2008: Regional Emission inventory in ASia (REAS) version 2. Atmos Chem Phys 13:11019–11058
Larssen T, Carmichael GR (2000) Acid rain and acidification in China: the importance of base cation deposition. Environ Pollut 110:89–102
Liu KH et al (2010) Soil acidification in response to acid deposition in three subtropical forests of subtropical China. Pedosphere 20:399–408
Lu et al (2011) Sulfur dioxide and primary carbonaceous aerosol emissions in China and India, 1996–2010. Atmos Chem Phys 11:9839–9864
MacDonald JA et al (2002) Nitrogen input together with ecosystem nitrogen enrichment predict nitrate leaching from European forests. Glob Chang Biol 8:1028–1033
Matsubara H et al (2009) Long-term declining trends of river water pH in central Japan. Water Air Soil Pollut 200:253–265
Nakahara O et al (2010) Soil and stream water acidification in a forested catchment in central Japan. Biogeochemistry 97:141–158
Nilssen I (1986) Critical deposition limits for forest soils. Nord Ministerråd Miljø Rapp 1986(11):37–69
Ohara T et al (2007) An Asian emission inventory of anthropogenic emission sources for the period 1980–2020. Atmos Chem Phys 7:4419–4444
Sase H (2015) Soil acidification in forests of East Asia (4.3.1.1), Acidification and eutrophication (chapter 4). In: Review on the state of air pollution in East Asia p 154–162. By Task Force on Research Coordination, Science Advisory Committee, EANET, Network Center for EANET, Asia Center for Air Pollution Research (ACAP), Niigata
Sase H et al (2008) Seasonal variation in the atmospheric deposition of inorganic constituents and canopy interactions in a Japanese cedar forest. Environ Pollut 152:1–10

Sase H et al (2012) Deposition process of sulfate and elemental carbon in Japanese and Thai forests. Asian J Atmos Environ 6:246–258
Sase H et al (2015) Dynamics of sulphur derived from atmospheric deposition and its possible impacts on the east Asian forests, final report submitted to APN (Project reference number, ARCP2013-13CMY-Sase). Asia-Pacific Network for Global Change Research (APN), Kobe
Schulze ED (1989) Air pollution and forest decline in a spruce (*Picea abies*) forest. Science 244:776–783
Smith SJ et al (2011) Anthropogenic sulfur dioxide emissions: 1850–2005. Atmos Chem Phys 11:1101–1116
Tamaki M et al (2000) Progress in acid deposition monitoring technology in Japan. Glob Environ Res 4:25–38
Ulrich B (1991) An ecosystem approach to soil acidification. In: Ulrich B, Sumner ME (eds) Soil acidity. Springer, Berlin, pp 28–79
Van Breemen N et al (1983) Acidification and alkalinization of soils. Plant Soil 75:283–308
Wright RF et al (1976) Impacts of acid precipitation on freshwater ecosystems in Norway. Water Air Soil Pollut 6:483–499
Wright RF et al (1995) NITREX: responses of coniferous forest ecosystems to experimentally changed deposition of nitrogen. For Ecol Manage 71:163–169
Xu YG et al (2001) Chemical composition of precipitation, throughfall and soil solutions at two forested sites in Guangzhou, South China. Water Air Soil Pollut 130:1079–1084
Yamashita N et al (2010) Seasonal and spatial variation of nitrogen dynamics in the litter and surface soil layers on a tropical dry evergreen forest slope. For Ecol Manage 259:1502–1512
Yamashita N et al (2014) Atmospheric deposition versus rock weathering in the control of streamwater chemistry in a tropical rain-forest catchment in Malaysian Borneo. J Trop Ecol 30:481–492
Yang Y et al (2012) Significant soil acidification across northern China's grassland during 1980s–2000s. Glob Chang Biol 18:2292–2300.
Zhao D et al (1988) Acid rain in southwestern China. Atmos Environ 22:349–358
Zhang Q et al (2012) Satellite remote sensing of changes in NO*x* emissions over China during 1996–2010. Chinese Sci Bull 57:2857–2864

Part II
Effects of Gaseous Air Pollutants on Plants in Japan

Chapter 4
Effects of Ozone on Japanese Agricultural Crops

Tetsushi Yonekura and Takeshi Izuta

Abstract Field surveys and experimental studies of the effects of ozone on Japanese agricultural crops have been conducted since the early 1970s. We review these research studies comprehensively, in chronological order. In the 1970s, most studies of the effects of ozone on agricultural crops, and the evaluation of ozone-induced foliar injuries, were field surveys. Since the 1980s, there has also been considerable research using various facilities, such as open-top chambers and controlled-environment growth cabinets; these studies are concerned with the response to and mechanism of ozone influence on various crops. In addition, since the last half of the 1990s, informative studies have been carried out to assess and map critical levels of ozone in relation to Japanese agricultural crops.

Keywords Ozone • Japanese agricultural crops

4.1 Introduction

Visible foliar injuries of unknown origin have been observed in tobacco (*Nicotiana tabacum* L.) plants in the Kinki, Chugoku, and Shikoku districts of Japan since around 1965 (Kuroda et al. 1970; Shinohara et al. 1973). In 1969–1970, similar foliar injuries appeared in tobacco and rice (*Oryza sativa* L.) plants in the Kanto district, including the Tokyo metropolitan area and Chiba and Saitama Prefectures, and it soon became clear that the leaf damage was related to photochemical oxidants (Sawada et al. 1971; Ohdaira 1973; Shinohara et al. 1973; Kadota and Ohta (1972);

T. Yonekura (✉)
Center for Environmental Science in Saitama, Kazo, Saitama 347-0115, Japan
e-mail: yonekura.tetsushi@pref.saitama.lg.jp

T. Izuta
Institute of Agriculture, Tokyo University of Agriculture and Technology, Fuchu, Tokyo 183-8509, Japan
e-mail: izuta@cc.tuat.ac.jp

T. Izuta (ed.), *Air Pollution Impacts on Plants in East Asia*,
DOI 10.1007/978-4-431-56438-6_4

Matsuoka 1976). In Japan, the negative impacts of photochemical oxidant pollution on plants were first recognized around big cities. Photochemical oxidants are the products of reactions between nitrogen oxides (NOx) and a wide variety of volatile organic compounds (VOCs). These oxidants are formed with substances such as ozone (O_3), peroxyacetyl nitrate (PAN), and hydrogen peroxide (H_2O_2). The main impact on the natural environment is mostly due to elevated levels of ozone (which accounts for about 90% of photochemical oxidants). Since the 1970s, Japanese scientists have conducted field surveys and experimental studies of the effects of ozone on Japanese agricultural crops. We review these studies comprehensively, in chronological order.

4.2 Studies in the 1970s

In the 1970s, most studies of the effects of ozone on agricultural crops, and the evaluation of foliar ozone injuries, were field surveys. In these field surveys, visible leaf injuries were observed in many crop plants nationwide (Table 4.1). Visible signs on the upper surfaces of leaves affected by ozone included numerous bleached spots, relatively large bleached areas, and large bifacial necrotic bleached areas in interveinal regions (Nouchi 2002). Foliar injuries were observed in Welsh onion (*Allium fistulosum*), komatsuna (Japanese mustard spinach; *Brassica rapa*), santosai (a cabbage-lettuce hybrid; *Brassica campestris*), Swiss chard (*Beta vulgaris*), crown daisy (*Chrysanthemum coronarium*), taro (*Colocasia esculenta*), and corn (*Zea mays*) plants in many parts of the Kanto district (Sawada et al. 1971; Matsuoka et al. 1971). Leaf damage in these crops, all relatively sensitive to ozone, appeared when the daily maximum ozone concentration in the field was 60–70 ppb (Nouchi et al. 1988).

Table 4.1 List of agricultural crop plants injured by ozone in Japan in the 1970s (Sawada et al. 1971; Matsuoka et al. 1971; Matsuoka 1976; Yamazoe 1987; Nouchi et al. 1988)

Leafy vegetables
Spinach (*Spinacia oleracea*), radish (*Raphanus sativus* var. *sativus*), Swiss chard (*Beta vulgaris* var. *cicla*), crown daisy (*Chrysanthemum coronarium*), Welsh onion (*Allium fistulosum*), types of Chinese cabbage: Komatsuna (*Brassica rapa* var. *perviridis*) and Santosai (*Brassica campestris*)
Root vegetables
Potato (*Solanum tuberosum* L.), sweet potato (*Ipomoea batatas*), taro (*Colocasia esculenta* (L.) Schott), turnip (*Brassica rapa* L.), burdock (*Arctium lappa* L.), carrot (*Daucus carota* L.)
Fruit vegetables and flowers
Tomato (*Solanum lycopersicum*), cucumber (*Cucumis sativus* L.), morning glory (*Ipomoea nil* L.), petunia (*Petunia hybrida*)
Grains and others
Rice (*Oryza sativa*), soybean (*Glycine max*), sweet corn (*Zea mays*), kidney bean (*Phaseolus vulgaris*), peanut (*Arachis hypogaea*), cowpea (*Vigna unguiculata*), pea (*Pisum sativum*), tobacco (*Nicotiana tabacum* L.), alfalfa (*Medicago sativa*), red clover (*Trifolium pratense*)

Nouchi et al. (1973) carried out an ozone fumigation experiment with 15 herbaceous plants, including komatsuna, santosai, radish (*Raphanus sativus*), corn, soybean (*Glycine max*), kidney bean (*Phaseolus vulgaris*), Welsh onion, cucumber (*Cucumis sativus*), lettuce (*Lactuca sativa*), tomato (*Lycopersicon esculentum*), carrot (*Daucus carota*), and cauliflower (*Brassica oleracea*), and six woody plants. The plants were fumigated with 200 ppb ozone in a controlled-atmosphere greenhouse. The herbaceous plants were generally sensitive to injury, especially komatsuna and santosai. Ozone-induced injury was typically confined to the upper surfaces of leaves and was notably greater on mature leaves. Microscopic examination showed that palisade cells were much more prone to ozone-induced injury than other leaf tissues. In addition, total chlorophyll content and the chlorophyll a to chlorophyll b ratio in santosai were decreased due to the ozone exposure (Nouchi and Odaira 1973). Nouchi (1979) reported that spinach and radish also incurred foliar damage when exposed to 70–90 ppb ozone for 3 h in controlled-environment chambers.

Asakawa et al. (1978) examined the relationship between the extent of ozone leaf injury in the Welsh onion and aspects of its growth environment; namely, air temperature and soil moisture. The reduction in growth due to ozone was minimal under environmental conditions in which stomatal closure occurred.

In Japanese rice, foliar injury was observed in the Kanto district (e.g., Nakamura et al. 1975; Morikawa et al. 1975; Matsuoka 1976; Nakamura 1979). Nakamura et al. (1976) and Matsuoka et al. (1976) reported, based on filtered air chamber experiments, that these injuries in rice leaves were caused mainly by ozone. Nakamura (1979) reported that the extent of leaf ozone-induced injury varied among 28 rice cultivars, and that the cultivars most sensitive to ozone were 'Akiharu,' 'Harebare,' and 'MGS-359.' The activity of RuBisCO (ribulose-1,5-bisphosphate carboxylase/oxygenase), a key enzyme that catalyzes the first major step in photosynthetic carbon fixation, in rice leaves was reduced by ozone (Nakamura and Saka 1978).

In Japanese pear (*Pyrus pyrifolia var. culta*) trees, signs of chlorosis, such as browning and necrosis on the upper surfaces of mature leaves, have been observed in orchards in the Tokyo metropolitan area since 1972. The chlorophyll and nutrient contents of the injured leaves were decreased (Kawamata 1976).

Tobacco is known worldwide as an extremely ozone-sensitive plant (e.g., Taylor 1974; Heggestad 1991). Foliar injury symptoms caused by ambient ozone were first reported in tobacco in the United States in 1959 (Heggestad and Middleton 1959). Therefore, tobacco was recommended as a bioindicator for ozone (Manning and Feder 1980). As mentioned above, leaf injury symptoms in tobacco have also been observed in Japan, since around 1965. Suyama et al. (1973) reported that the cultivar 'Bel-W3' was the most sensitive to ozone among the 30 tobacco cultivars cultivated in Japan. However, because its free-market cultivation in Japan was illegal until 1985, tobacco was difficult to use as a bioindicator for ozone in this country. Instead, the Japanese morning glory has been used as a bioindicator for ozone in Japan.

Japanese morning glory, a traditional horticultural plant in Japan, is also highly sensitive to ozone (Hatta and Terakado 1975; Nouchi and Aoki 1979). The cultivars 'Scarlett O'Hara' and 'Heavenly Blue' are extremely sensitive and were selected for detecting ozone (Nakamura and Matsunaka 1974). A project monitoring photochemical oxidant levels using the Japanese morning glory ('Scarlett O'Hara') was carried out throughout Japan between 1974 and 1976, under the sponsorship of all 47 prefectural governments and a newspaper company (Matsunaka 1977). Foliar injury symptoms were observed in these plants in all but ten prefectures, including Hokkaido and Okinawa in summer.

In contrast, no visible ozone damage was observed in some agricultural crops (Box 4.1). Based on field observations in the Kanto district in the 1970s, it is thought that some crops, such as cabbage (*Brassica oleracea var. capitata*), cauliflower (*Brassica oleracea var. botrytis*), and squash (*Cucurbita moschata*), have relatively high tolerance for ozone before visible damage occurs (Nouchi et al. 1988).

Box 4.1 List of agricultural crop plants in the Kanto district of Japan in which no visible foliar injury was observed in the 1970s (Nouchi et al. 1988)
Cabbage (*Brassica oleracea* var. *capitata*), cauliflower (*Brassica oleracea* var. *botrytis*), broccoli (*Brassica oleracea* var. *italica*), onion (*Allium cepa*), shiso (*Perilla frutescens* var. *crispa*), ginger (*Zingiber officinale*), fuki (*Petasites japonicus*), udo (*Aralia cordata*), squash (*Cucurbita moschata*), melon (*Cucumis melo*), strawberry (*Fragaria × ananassa Duchesne ex Rozier*)

4.3 Studies from the 1980s to 1990s

In the 1980s to 1990s, field research on the effects of ozone on agricultural crops was carried out more systematically than previously (e.g., Yamada and Kuwata 1982; Asakawa 1982). To investigate crop losses due to ozone, experiments using open-top chambers have been conducted in North America (e.g., Heck et al. 1984; Heagle et al. 1982) and Europe (e.g., Skärby et al. 1993) since the 1980s. In Japan, there has also been considerable research using various facilities, such as controlled-field air chambers, open-top chambers, and controlled-environment growth cabinets; these studies are concerned with the response to and mechanism of ozone influence on various Japanese agricultural crops.

Monitoring of and research on ozone's effect on bioindicator plants – not only the Japanese morning glory (Okazaki et al. 1984; Nouch and Aoki 1979) but also the radish (Izuta et al. 1988b, 1993) – have been conducted. Long-term field surveys of the Japanese morning glory were carried out from 1974 to 1997 in seven prefectures in the Kanto district (Yamazoe 1987; Nouchi 2002). Annual changes in the ratio of the number of sites where foliar injuries appeared to the total number of observation sites remained approximately unchanged during this period, suggesting that ozone pollution had not decreased much in the Kanto district (Nouchi 2002). The radish has also been used an indicator plant for ozone because of its high sensitivity and growth response to ozone (Izuta et al. 1988b, 1991, 1993). In addition, Matsumaru and Takasaki (1989) and Matsumaru and Ookoshi (1991a) conducted a field survey to select crops, based on foliar injuries, to serve as ozone indicators. The survey indicated that 'Waseshiro' (potato; *Solanum tuberosum* L.), 'Chibahandachi,' (peanut; *Arachis hypogaea*), and 'Dodare' (taro) , were sensitive cultivars, and that they could be used as ozone indicator crops. However, varietal differences in ozone sensitivity and visible foliar injury among potato plants did not necessarily agree with varietal differences in potato yield (Matsumaru and Ookoshi 1991b).

Asakawa et al. (1981a) reported a reduction in Japanese rice yields in six of the eight cultivars, including 'Nipponbare,' 'Hounenwase,' and 'Yamadanishiki,' studied, as determined by comparing the yields of rice cultivated with either nonfiltered or charcoal-filtered treatment in Hyogo Prefecture.

An experiment using rice, including the cultivars 'Koshihiokari' and 'Nipponbare,' exposed to five levels of ozone in field exposure chambers (Kobayashi et al. 1994) for 3 years found that light-use efficiency in leaves, as well as growth and yield, were reduced by ozone, and the rates of reduction were higher with rising ozone concentrations (Kobayashi and Okada 1995; Kobayashi et al. 1995). Furthermore, the experiments led to a model of the relationship between the mean ozone concentration and the relative rice yield loss (Kobayashi et al. 1995). The impact of ozone on rice production was estimated with a simple model (Kobayashi 1997) of rice growth for the Kanto district (Kobayashi 1999). The estimation showed an approximately 10 % yield loss due to ozone, with large variability by year and location.

Asakawa et al. (1981b) suggested that the negative impact of ozone on rice yield was larger at the reproductive stage than at the vegetative growth stage. The decrease in the photosynthetic rate in rice leaves during ozone fumigation was mainly caused by stomatal closure, and the ozone absorption rate during ozone fumigation was suppressed by stomatal closure (Kouchi 1980a, b). Jeong et al. (1980, 1981) investigated the effects of abscisic acid (ABA) on ozone injury and ethylene production in rice leaves. Endogenous ABA content in the ozone-tolerant cultivar 'Tongil' was higher than that in the ozone-sensitive 'Nipponbare.' The ABA content reduced the ethylene production caused by ozone fumigation. Furthermore, ozone-induced foliar injury was reduced with increasing ABA content, of which production was increased under water stress conditions (Jeong and Ota 1981a). In contrast, ABA content decreased with increasing nitrogen content in rice leaves (Jeong and Ota 1981b). Kuno and Arai (2002) reviewed studies concerned with the effects of nutrients on visible foliar injury in Japanese plants exposed to ozone. Toyama et al. (1989) examined the ultrastructural changes in chloroplasts in the ozone-damaged leaves of young rice plants. Destruction of the chloroplast envelope was observed after fumigation with 100 ppb ozone for 4 h per day for 4 days. This ozone injury was more servere in abaxial leaf cells than in adaxial ones. Similar results were reported in the leaves of spinach (Miyake et al. 1984) and radish plants (Miyake et al. 1989).

In spinach, susceptibility to ozone-induced foliar injury varied markedly among 15 cultivars (Kuno 1989a). Spinach cultivars with leaves high in ascorbic acid content and superoxide dismutase activity tended to be resistant to ozone (Kuno 1989b, d). In addition, glutathione reductase activity in leaves increased immediately after ozone exposure in ozone-resistant spinach cultivars (Kuno 1989c).

In radish, Izuta et al. (1988a) reported that growth sensitivity to ozone was affected by the growth temperature. Ozone-induced reduction in dry-weight growth was magnified when plants were grown at relatively high temperatures. The primary factor determining the cultivar sensitivity to ozone, based on dry-weight growth, was related to inhibition of the rate of net photosynthesis per unit of ozone absorbed by leaves (Izuta et al. 1994) and the capacity for physiological detoxification of ozone in leaves (Izuta et al. 1999).

In sunflowers (*Helianthus annuus*), whole-plant growth was reduced by 11 % and 32 % of the control when the plants were exposed to 100 ppb and 200 ppb ozone, respectively, for 12 days (Shimizu et al. 1981). Root growth was markedly inhibited by ozone. A similar root-growth response was reported in other crops (Cooley and Manning 1989; Nouchi et al. 1991). According to Okano et al. (1984), ozone reduced carbon dioxide fixation in leaves and inhibited chromosomal translocation. In particular, assimilates exported from leaves to roots decreased considerably.

Studies of ozone's influence on lipid metabolism systems in leaves were conducted for spinach (Sakaki et al. 1983, 1985. 1990a, b, c), parsley (*Petroselinum crispum*; Yamada and Kuwata 1982), and morning glory and kidney bean (Nouchi and Toyama 1988). These studies investigated the mechanism of fatty acid resolution in leaves affected by ozone exposure. In addition, increases in mRNA levels in response to ozone in rice leaves were reported by Ohki et al. (1999).

4.4 Studies in the early 2000s and Beyond

Since 2000, there have been several experimental studies on the effects of ozone on Japanese crops such as Japanese rice (Yonekura et al. 2005a; Yamaguchi et al. 2008, 2014, 2015; Inada et al. 2008; Frei et al. 2008, 2010, 2011, 2012; Sawada and Kohno 2009; Tsukahara et al. 2012), wheat (Inada et al. 2012), soybean (Yonekura et al. 2000; Ahsan et al. 2010), and radish and komatsuna (Yonekura et al. 2005b).

Yonekura et al. (2005a) exposed 16 rice cultivars (9 Japanese and 7 from Southeast Asia or North America) to three levels of ozone. The magnitudes of ozone-induced reductions in yields were smaller in the Japanese rice cultivars than these yield reductions in the other cultivars (Fig. 4.1). This experiment shows that Japanese rice cultivars have a relatively low level of ozone-induced reduction in grain yield compared with findings in the rice cultivars in other countries examined, such as the United States, Vietnam, the Philippines, and India. Critical ozone levels, estimated using the AOT40 (accumulated exposure to ozone above a threshold of 40 ppb) – rice yield response relationship method (Table 4.2), in nine Japanese cultivars and seven other cultivars, above which a 5 % yield reduction is expected, were 7 ppm h and 3.5 ppm h, respectively. The critical level for Japanese rice was higher than that for wheat in Europe (Mills et al. 2007). Thus, it is considered that it is difficult to introduce the critical level value used in Europe to Japan. According to estimations of yield loss in rice affected by ozone in the Kanto district of Japan,

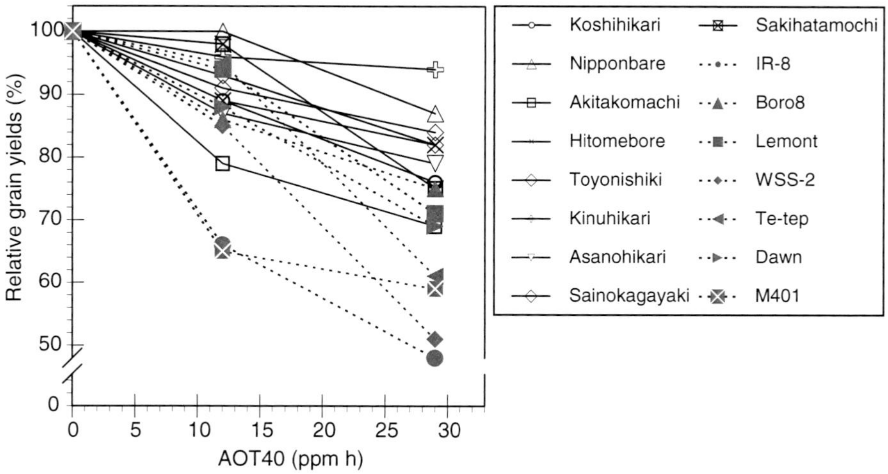

Fig. 4.1 Effects of ozone on grain yield in 16 rice cultivars relative to yield with clean air (CA) treatment. The values for accumulated exposure to ozone over a threshold of 40 ppb (AOT40) for 8 h during the growth period of 100 days in CA, ambient air (AA), and one and a half times AA treatments were 0, 12.3, and 28.9 ppm h, respectively (Revised by Yonekura et al. 2005a)

Table 4.2 Linear regression between relative grain yield in clean air (CA) (Yg) and at accumulated exposure to ozone over a threshold of 40 ppb (AOT40) (D_{AOT40})

Cultivar (origin)	Regression line	Cultivar (origin)	Regression line
Koshihikari (Japan)	$Yg = -0.82\ D_{AOT40} + 99.5$	IR-8 (Philippines)	$Yg = -1.77\ D_{AOT40} + 95.4$
Nipponbare (Japan)	$Yg = -0.49\ D_{AOT40} + 102.1$	Boro8 (India)	$Yg = -0.85\ D_{AOT40} + 98.9$
Akitakomachi (Japan)	$Yg = -1.06\ D_{AOT40} + 97.2$	WSS-2 (Vietnam)	$Yg = -1.71\ D_{AOT40} + 102.3$
Hitomebore (Japan)	$Yg = -0.61\ D_{AOT40} + 99.0$	Tetep (Vietnam)	$Yg = -1.37\ D_{AOT40} + 104.3$
Toyonishiki (Japan)	$Yg = -0.63\ D_{AOT40} + 100.4$	Lemont (USA)	$Yg = -1.03\ D_{AOT40} + 102.4$
Kinuhikari (Japan)	$Yg = -0.21\ D_{AOT40} + 99.6$	Dawn (USA)	$Yg = -1.07\ D_{AOT40} + 100.5$
Asanohikari (Japan)	$Yg = -0.72\ D_{AOT40} + 98.6$	M401 (USA)	$Yg = -1.37\ D_{AOT40} + 93.3$
Sainokagayaki (Japan)	$Yg = -0.53\ D_{AOT40} + 98.9$		
Sakihatamochi (Japan)	$Yg = -0.89\ D_{AOT40} + 103.2$		
Nine Japanese rice cultivars	$Yg = -0.67\ D_{AOT40} + 99.8$	Seven other rice cultivars	$Yg = -1.31\ D_{AOT40} + 99.6$

The data are from Fig. 4.1 (Revised by Yonekura et al. 2005a)

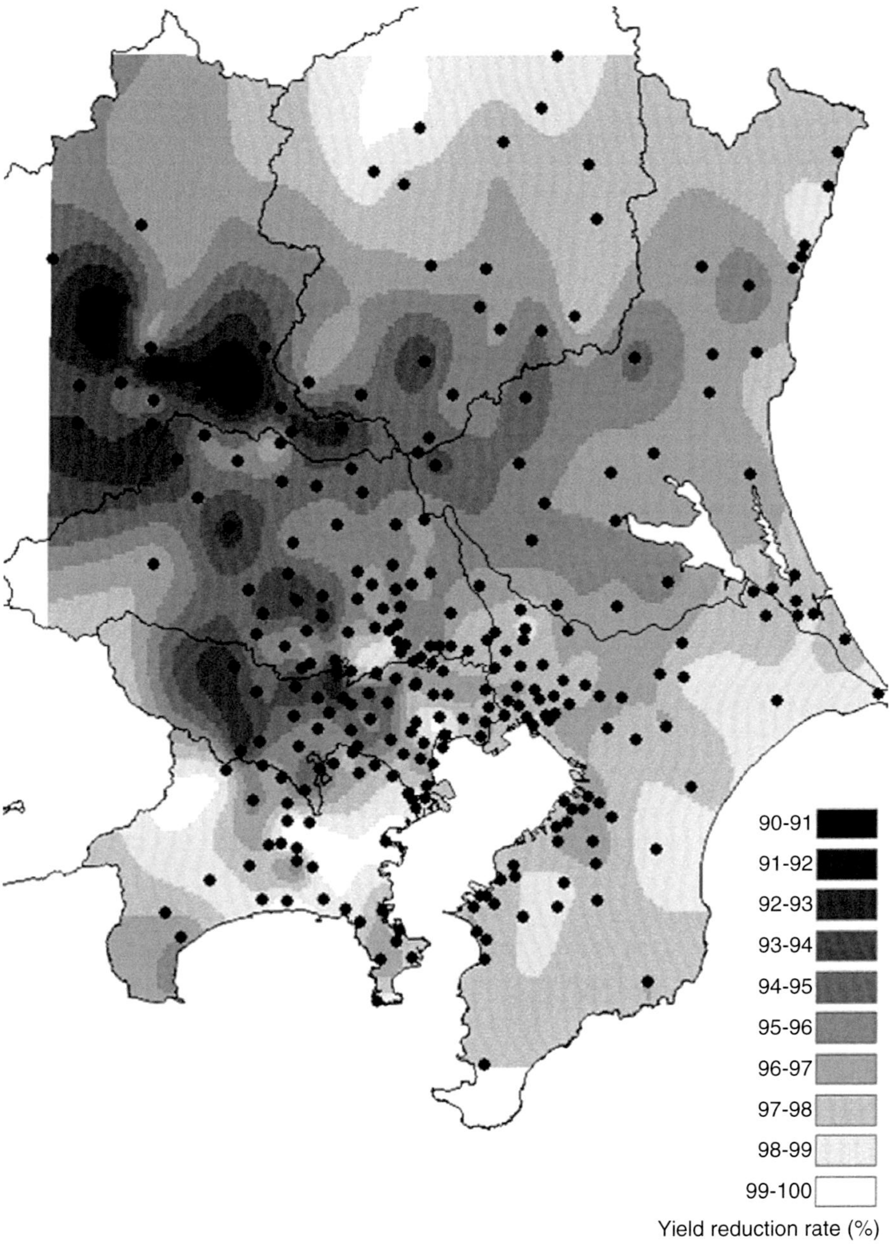

Fig. 4.2 Prediction map of yield-loss in Japanese rice cultivars in the Kanto district of Japan obtained by using the AOT40-yield response relationship method in nine Japanese cultivars (Table 4.2) under current ozone level (average AOT40 during 1990–2000) (Revised by Yonekura et al. 2005a)

obtained by using the data shown in Table 4.2, the Japanese rice yield was reduced by about 5–10 % of the expected value by the current ozone level (Fig. 4.2). Takagi and Ohara (2003) estimated, using the model of mean ozone concentration-relative rice yield loss relationship established by Kobayashi et al. (1995), that the relative yield loss for rice caused by ambient ozone was almost 3.5 % in the Kanto district. Of note, Nakanishi et al. (2009) estimated that the reduction rates of rice yields in the Kanto district, based on the AOT40 and AOT30, were 6.8 % and 9.2 %, respectively.

Yamaguchi et al. (2008) and Inada et al. (2008) reported the effects of ozone on the growth, yields, leaf gas exchange rates, photosynthetic components, and radical scavenging systems of two Japanese rice cultivars ('Koshihikari' and 'Kinuhikari'). The rates of reduction in grain yield caused by exposure to ozone at 60 and 100 ppb were 3 % and 23 %, respectively, in 'Koshihikari' and 18 % and 34 %, respectively, in 'Kinuhikari'. The exposure to ozone reduced the net photosynthetic rate and stomatal diffusive conductance, and chlorophyll concentrations in both cultivars were reduced by exposure to 100 ppb ozone. On the other hand, ozone significantly increased the activities of ascorbate peroxidase (APX), monodehydroascorbate reductase (MDAR), and glutathione reductase (GR) in leaves in both cultivars. Yamaguchi et al. (2015) evaluated the effects of ozone on the net photosynthetic rate in the flag leaves of rice, based on stomatal ozone flux and reactive oxygen species (ROS) scavenging enzyme activities. The degree of ozone-induced reduction in the net photosynthetic rate and the activities of ROS scavenging enzymes such as APX and catalase were better explained by the cumulative flux of ozone compared with concentration-based ozone indices. Furthermore, Yamaguchi et al. (2014) established a stomatal conductance model for a Japanese rice cultivar, to estimate stomatal ozone uptake in the leaves, and they evaluated the ozone effect on the yield of Japanese rice using the cumulative ozone uptake. Several researchers have also established multiplicative models to estimate stomatal ozone uptake in the leaves of rice and wheat in Japan and China (Oue et al. 2008, 2009, 2011; Feng et al. 2012). Sawada and Kohno (2009) reported differential rice cultivar sensitivities to ozone, as indicated by visible foliar injury and grain yield. The sensitivity of rice cultivars to ozone as evaluated by visible injury did not correspond to that evaluated by the reduction in grain yield. Frei et al. (2012) reported that rice grain quality was changed by ozone fumigation. The concentrations of proteins and lipids in rice grain increased significantly in response to ozone, while the starch concentration decreased. In recent years there have been several studies evaluating ozone's influence on rice and soybeans, using proteomics or quantitative trait locus (QTL) analysis (Frei et al. 2008, 2011; Ahsan et al. 2010; Tsukahara et al. 2012). In addition, there have been studies on the interactive effects of ozone and carbon dioxide on growth, yields, and leaf gas exchange rates in rice (Imai and Kobori 2008; Imai and Ookoshi 2011; Kobayakawa and Imai 2011).

In Japanese cultivars of wheat, Inada et al. (2012) investigated the effects of chronic exposure to ambient levels of ozone on gas exchange rates, the activity and concentration of RuBisCO, the activity of ROS-scavenging enzymes, and the concentration of antioxidants in the flag leaf in two cultivars ('Norin 61' and

'Shirogane-komugi'). The net photosynthetic rate, and the activity and/or concentration of the enzymes RuBisCO and MDAR were reduced by ozone only in the flag leaf of 'Shirogane-komugi'. It was concluded that the difference in ozone sensitivity between the two cultivars was mainly due to the effects of ozone on the plant's ability to detoxify ROS, which is mainly determined by the activity of ROS-scavenging enzymes such as catalase (CAT) and MDAR.

The interactive effects of ozone and carbon dioxide on the growth of komatsuna and radish plants were investigated by Yonekura et al. (2005b). The growth response, affected by ozone under elevated carbon dioxide, was different in the two plants. In radishes, the reduced growth rate caused by ozone under elevated carbon dioxide was similar to the reduced growth rate caused by ozone under ambient-level carbon dioxide. Contrarily, in komatsuna plants, the reduced growth rate caused by ozone under elevated carbon dioxide was lower than the reduced growth rate caused by ozone under ambient-level carbon dioxide.

Agricultural countermeasures taken against ozone-induced crop injury in Japan were reviewed by Nouchi (2003). This review found five types of agriculture technology being used to avoid damage by ozone: (1) inherited resistance properties of crops; (2) fertilizer management; (3) treatment with specific chemicals; (4) biotechnological activation of defense genes; and (5) in-facilities cultivation.

To summarize, in Japan, various studies have been carried out concerning the effects of ozone on agricultural crops in this country. However, it seems that research on the combined effects of ozone and other environmental factors, such as carbon dioxide concentration, temperature, and water availability, is called for, and studies to provide information for the assessment and mapping of critical levels of ozone for vegetation are also needed.

References

Ahsan N, Nanjo Y, Sawada H, Kohno Y, Komatsu S (2010) Ozone stress induced proteomic changes in leaf total soluble and chloroplast proteins of soybean reveal that carbon allocation is involved in adaptation in the early developmental stage. Proteomics 10:2605–2619

Asakawa F (1982) Evaluation of the effect of photochemical oxidants on plants: relation between ozone concentration and exposure time on the plant injury. J Jpn Soc Air Pollut 17:279–287 [in Japanese with English summary]

Asakawa F, Imai T, Kusaka S (1978) Effects of soil moisture and temperature on the susceptibility of welsh onion to ozone: on the role of stomata in ozone damage. Jpn J Soil Sci Plant Nutr 49:204–209 [in Japanese]

Asakawa F, Tanaka H, Kusaka S (1981a) Influence of air pollution, photochemical oxidants, on the growth and yield of rice plants: (1) the effects of filtered ambient air of the growth and yield of rice plants. Jpn J Soil Sci Plant Nutr 52:201–206 [in Japanese]

Asakawa F, Tanaka H, Kusaka S (1981b) Influences of air pollution, photochemical oxidants, on the growth yield of rice plants: (2) differences of the influence of photochemical oxidants on the growth and yield at the different growth stages of early or late maturing rice plants. Jpn J Soil Sci Plant Nutr 52:289–296 [in Japanese]

Cooley D, Manning WJ (1989) The impact of ozone on assimilate partition in plants: a review. Environ Pollut 47:95–113

Feng Z, Tang H, Uddling J, Pleijel H, Kobayashi K, Zhu J, Oue H, Guo W (2012) A stomatal ozone flux-response relationship to assess ozone-induced yield loss of winter wheat in subtropical China. Environ Pollut 164:16–23

Frei M, Tanaka JP, Wissuwa M (2008) Genotypic variation in tolerance to elevated ozone in rice: dissection of distinct genetic factors linked to tolerance mechanisms. J Exp Bot 59:3741–3752

Frei M, Tanaka JP, Chen CP, Wissuwa M (2010) Mechanisms of ozone tolerance in rice: characterization of two QTLs affecting leaf bronzing by gene expression profiling and biochemical analyses. J Exp Bot 61:1405–1417

Frei M, Kohno Y, Wissuwa M, Makkar HPS, Becker K (2011) Negative effects of tropospheric ozone on the feed value of rice straw are mitigated by an ozone tolerance QTL. Glob Chang Biol 17:2319–2329

Frei M, Kohno Y, Tietze S, Jekle M, Hussein MA, Becker T, Becker K (2012) The response of rice grain quality to ozone exposure during growth depends on ozone level and genotype. Environ Pollut 163:199–206

Hatta H, Terakado K (1975) Effect of photochemical oxidant on morning glory and its validity as indicator plant: (I) characteristics of oxidant damage on morning glory and correlation between leaf damage index and oxidant dose. J Jpn Soc Air Pollut 9:722–728 [in Japanese with English summary]

Heagle AS, Kress LS, Temple PJ, Kohut RJ, Miller JE, Heggestad HE (1982) Factors influencing ozone dose-yield response relationship in open-top field chamber studies. In: Heck WW, Taylor OC, Tingey DT (eds) Assessment of crop loss from air pollutions. Elsevier Science, London, pp 141–179

Heck WW, Cure WW, Rawlings JO, Zaragoza LJ, Heagle AS, Heggestad HE, Kohut RJ, Kress LW, Temple PJ (1984) Assessing impacts of ozone on agricultural crops: II. Crop yield functions and alternative exposure statistics. J Air Pollut Control Assoc 34:810–817

Heggestad HE (1991) Origin of Bel-W3, Bel-C and Bel-B tobacco varieties and their use as indicators of ozone. Environ Pollut 74:264–291

Heggestad HE, Middleton JT (1959) Ozone in high concentrations as cause of tobacco leaf injury. Science 129:208–210

Imai K, Kobori K (2008) Effects of the interaction between ozone and carbon dioxide on gas exchange, ascorbic acid content, and visible leaf symptoms in rice leaves. Photosynthetica 46:387–394

Imai K, Ookoshi T (2011) Elevated CO_2 ameliorates O_3-inhibition of growth and yield in paddy rice. Environ Control Biol 49:75–82

Inada H, Yamaguchi M, Satoh R, Hoshino D, Nagasawa A, Negishi Y, Nouchi I, Kobayashi K, Izuta T (2008) Effects of ozone on photosynthetic components and radical scavenging system in leaves of rice (*Oryza sativa* L.). J Agri Meteorol 64:243–255

Inada H, Kondo T, Akhtar N, Hoshino D, Yamaguchi M, Izuta T (2012) Relationship between cultivar difference in the sensitivity of net photosynthesis to ozone and reactive oxygen species scavenging system in Japanese winter wheat (*Triticum aestivum*). Physiol Plant 146:217–227

Izuta T, Funada S, Ohashi T, Miyake H, Totsuka T (1988a) Effects of ozone on the growth of radish plants at different temperatures. J Jpn Soc Air Pollut 23:209–217 [in Japanese with English summary]

Izuta T, Takikawa M, Horie K, Miyake H, Totsuka T (1988b) An evaluation of atmospheric environment by a plant indicator based on the growth of radish plant grown in open-top chambers. J Jpn Soc Air Pollut 23:288–292 [in Japanese with English summary]

Izuta T, Funada S, Ohashi T, Miyake H, Totsuka T (1991) Effects of low concentrations of ozone on the growth of radish plants under different light intensities. Environ Sci 1:21–33

Izuta T, Miyake H, Totsuka T (1993) Evaluation of air-polluted environment based on the growth of radish plants cultivated in small-sized open-top chambers. Environ Sci 2:25–37

Izuta T, Ohtsu G, Miyake H, Totsuka T (1994) Effects of ozone on dry weight, growth, net photosynthetic rate and leaf diffusive conductance in three cultivars of Radish plants. J Jpn Soc Air Pollut 29:1–8

Izuta T, Takahashi K, Matsumura H, Totsuka T (1999) Cultivar difference of *Brassica campestris* L. in the sensitivity to O_3 based on the dry weight growth. J Jpn Soc Air Pollut 34:137–146

Jeong YH, Ota Y (1981a) Physiological studies on photochemical oxidant injury in rice plants: (3) relationship between abscisc acid (ABA) and water metabolism in water-stressed rice plants. Jpn J Crop Sci 50:566–569 [in Japanese with English summary]
Jeong YH, Ota Y (1981b) Physiological studies on photochemical oxidant injury in rice plants: IV. Effect of nitrogen application on endogenous abscisic acid (ABA) production and ozone injury of rice plants. J Crop Sci 50:570–574 [in Japanese with English summary]
Jeong YH, Nakamura T, Ota Y (1980) Physiological studies on photochemical oxidant injury in rice plants: (1) varietal difference of abscisic acid content and its relation to the resistance to ozone. Jpn J Crop Sci 49:456–460 [in Japanese with English summary]
Jeong YH, Nakamura T, Ota Y (1981) Physiological studies on photochemical oxidant injury in rice plants: (2) effect of abscisic acid (ABA) on ozone injury and ethylene production in rice plants. Jpn J Crop Sci 50:560–565 [in Japanese with English summary]
Kadota M, Ohta K (1972) Ozone sensitivity of Japanese plant species in summer, with special reference to a tentative sensitivity grade list for applying to field survey on ozone injury. J Jpn Soc Air Pollut 7:19–26 [in Japanese with English summary]
Kawamata S (1976) Effects of air pollution on fruit trees: photochemical smog injury to Japanese pear tree. J Jpn Soc Horticult Sci 45:15–23
Kobayakawa H, Imai K (2011) Effects of the interaction between ozone and carbon dioxide on gas exchange, photosystem II and antioxidants in rice leaves. Photosynthetica 49:227–238
Kobayashi K (1997) Variation in the relationship between ozone exposure and crop yield as derived from simple models of crop growth and ozone impact. Atmos Environ 31:703–714
Kobayashi K (1999) Assessing the impacts of tropospheric ozone on agricultural production. J Jpn Soc Air Pollut 34:162–175 [in Japanese with English summary]
Kobayashi K, Okada M (1995) Effects of ozone on the light use of rice (*Oryza sativa* L.) plants. Agric Ecosyst Environ 53:1–12
Kobayashi K, Okada M, Nouchi I (1994) A chamber system for exposing rice (*Oryza sativa* L.) to ozone. New Phytol 126:317–325
Kobayashi K, Okada M, Nouchi I (1995) Effects of ozone on dry matter partitioning and yield of Japanese cultivars of rice (*Oryza sativa* L.). Agric Ecosyst Environ 53:109–122
Kouchi H (1980a) Studies on relationships between ozone absorption rate and stomatal diffusion resistance of plant leaf. J Jpn Soc Air Pollut 15:109–117 [in Japanese with English summary]
Kouchi H (1980b) Relationships between ozone absorption and leaf injury: especially, on the relationships between differences in ozone susceptibility accompanying leaf age and ozone absorption rate. J Jpn Soc Air Pollut 15:389–393 [in Japanese with English summary]
Kuno H (1989a) Characteristics of spinach leaf injury by photochemical oxidants and mechanisms of ozone resistance in spinach cultivars: (I) relationship of growth stages to leaf injury. J Jpn Soc Air Pollut 24:180–187 [in Japanese with English summary]
Kuno H (1989b) Characteristics of Spinach leaf injury by photochemical oxidants and mechanisms of ozone resistance in spinach cultivars: (II) resistance to oxidants in Spinach cultivars. J Jpn Soc Air Pollut 24:188–195 [in Japanese with English summary]
Kuno H (1989c) Characteristics of Spinach leaf injury by photochemical oxidants and mechanisms of ozone resistance in spinach cultivars: (III) effects of ozone on stomatal diffusion resistances and activities of enzymes participating in the detoxication of active oxygen. J Jpn Soc Air Pollut 24:253–258 [in Japanese with English summary]
Kuno H (1989d) Characteristics of Spinach leaf injury by photochemical oxidants and mechanisms of ozone resistance in spinach cultivars: (IV) effects of low ozone levels on enzymes and substances participating in the detoxication of active oxygen. J Jpn Soc Air Pollut 24:259–263 [in Japanese with English summary]
Kuno H, Arai K (2002) Countermeasures with fertilization to reduce oxidant-induced injury to plants. In: Omasa K, Saji H, Youssefian N, Kondo N (eds) Air pollution and plant biotechnology: prospects for phytomonitoring and phytoremediation. Springer, Tokyo, pp 269–284
Kuroda S, Shinohara T, Kunisawa K, Suyama I (1970) Weather fleck in tobacco: relation of fleck outbreaks to air polluting oxidants. Proc Crop Sci Soc Jpn 39:89–90, in Japanese, No English title
Manning WJ, Feder WA (1980) Biomonitoring air pollutants with plants. Applied Science Publishers, London, p 142

Matsumaru T, Ookoshi T (1991a) Photochemical oxidant injury on potatoes of several cultivars: (1) occurrence of brown spots on potato leaves in fields and its possible causes. J Jpn Soc Air Pollut 26:378–384

Matsumaru T, Ookoshi T (1991b) Photochemical oxidant injury on potatoes of several cultivars: (2) analysis of effects of photochemical oxidants on growth and yields of potatoes using filtered chambers. J Jpn Soc Air Pollut 26:385–391

Matsumaru T, Takasaki T (1989) Selecting potato, peanut and taro cultivars as indicator plants of photochemical oxidants. J Jpn Soc Air Pollut 24:287–289 [in Japanese with English summary]

Matsunaka S (1977) Utilization of morning glory as an indicator plant for photochemical oxidants in Japan. In: Kasuga S, Suzuki N, Yamada T, Kimura G, Inagaki K, Onoe K (eds) Proceedings of the fourth international clean air congress. Japan Union of Air Pollution Prevention Association, Tokyo, pp 91–94 [in Japanese, no English title]

Matsuoka Y (1976) Plant damage by photochemical air pollution in the Tokyo metropolitan area. J Jpn Soc Air Pollut 11:195–203 [in Japanese with English summary]

Matsuoka Y, Takasaki T, Udagawa O (1971) Evaluation of the relationship of crop plants damage and oxidants. J Jpn Soc Air Pollut 6:140 [in Japanese]

Matsuoka Y, Takasaki T, Morikwa M, Matsumaru T, Shiratori K (1976) Studies on the visible injury to rice plants caused by photochemical oxidants: (1) identification of the leaf injury caused by photochemical oxidants. Proc Crop Sci Soc Jpn 45:124–136 [in Japanese with English summary]

Mills G, Buse A, Gimeno B, Bermejo V, Holland M, Emberson L, Pleijel H (2007) A synthesis of AOT40-based response functions and critical levels of ozone for agricultural and horticultural crops. Atmos Environ 41:2630–2643

Miyake H, Furukawa A, Totsuka T, Maeda E (1984) Differential effects of ozone and sulphur dioxide on the fine structure of spinach leaf cells. New Phytol 96:215–228

Miyake H, Matsumura H, Fujinuma Y, Totsuka T (1989) Effects of low concentrations of ozone on the fine structure of radish leaves. New Phytol 111:187–195

Morikawa et al. (1975) Effects of photochemical oxidant on the growth of rice: I. Visible foliar injury responses in different growth stages of rice to ozone. Bull Chiba Prefectural Agric Exp Station 16:103–112 [in Japanese, No English title]

Nakamura T (1979) Study on the photochemical oxidants injury in rice plant. Bull Natl Inst Agric Sci D-30:1–68 [in Japanese, no English title]

Nakamura T, Matsunaka S (1974) Indicator plants for air pollutants: (1) susceptibility of morning glory to photochemical oxidants: varietal difference and effect of environmental factors. Proc Crop Sci Soc Jpn 43:517–522 [in Japanese with English summary]

Nakamura T, Saka H (1978) Photochemical oxidants injury in rice plants: (3) effects of ozone on physiological activities in rice plants. Proc Crop Sci Soc Jpn 47:707–714 [in Japanese with English summary]

Nakamura T, Hashimoto S, Ota Y, Nakamura M (1975) Photochemical oxidants injury in rice plants: (I) occurrence of photochemical oxidants injury in rice plants at Kanto area and its symptoms. Proc Crop Sci Soc Jpn 44:312–319 [in Japanese with English summary]

Nakamura H, Ota Y, Hashimoto S, Okino H (1976) Photochemical oxidants injury in rice plants: (2) the effect of filtered ambient air on the growth and yield of rice plants. Proc Crop Sci Soc Jpn 45:630–636 [in Japanese with English summary]

Nakanishi J, Shinozaki Y, Inoue K (eds) (2009) Ozone: photochemical oxidant. Maruzen, Tokyo, p 289 [in Japanese, no English title]

Nouchi I (1979) Effects of ozone and PAN concentrations and exposure duration on plant injury. J Jpn Soc Air Pollut 14:489–496 [in Japanese with English summary]

Nouchi I (2002) Responses of whole plants to air pollution. In: Omasa K, Saji H, Youssefian N, Kondo N (eds) Air pollution and plant biotechnology: prospects for phytomonitoring and phytoremediation. Springer, Tokyo, pp 3–39

Nouchi I (2003) Agricultural countermeasures for avoiding crop injury from ozone in Japan. J Agric Meteorol 59:59–67

Nouchi I, Aoki I (1979) Morning glory as a photochemical oxidant indicator. Environ Pollut 18:289–303
Nouchi I, Odaira T (1973) Influence of ozone on plant pigments. J Jpn Soc Air Pollut 8:120–125 [in Japanese with English summary]
Nouchi I, Odaira T, Sawada T, Oguchi K, Komeiji T (1973) Plant ozone injury symptoms. J Jpn Soc Air Pollut 8:113–119 [in Japanese with English summary]
Nouchi I, Takasaki T, Totsuka T (1988) Relative photochemical oxidant sensitivity of agricultural and horticultural plants. J Jpn Soc Air Pollut 23:355–370 [in Japanese with English summary]
Nouchi I, Toyama S (1988) Effects of ozone and peroxyacetyl nitrate on polar lipids and fatty acids in leaves of morning glory and kidney bean. Plant Physiol 87:638–646
Nouchi I, Ito O, Harazono Y, Kobayashi K (1991) Effects of chronic ozone exposure on growth, root respiration and nutrient uptake of rice plants. Environ Pollut 74:149–164
Ohdaira T (1973) Studies on photochemical smog phenomena and its effects on plant damage in Tokyo. Jpn J Health Hum Ecol 39:118–145 [in Japanese with English summary]
Ohki T, Matsui H, Nagasaka A, Yoshioka T, Watanabe A, Satoh S (1999) Induction by ozone of ethylene production and an ACC oxidase cDNA in rice (*Oryza sativa* L.) leaves. Plant Growth Regul 28:123–127
Okano K, Ito O, Takebe G, Shimizu A, Totsuka T (1984) Alteration of 13C-assimilate partitioning in plants of *Phaseolus vulgaris* exposed to ozone. New Phytol 97:155–163
Okazaki J, Okabe S, Shiratori T (1984) The relationship between leaf injury degree of morning glory and combined air pollutants under field conditions. J Jpn Soc Air Pollut 19:379–386
Oue H, Motohiro S, Inada K, Miyata S, Mano M, Kobayashi K, Zhu J (2008) Evaluation of ozone uptake by the rice canopy with the multi-layer model. J Agric Meteorol 64:223–232
Oue H, Feng Z, Pang J, Miyata A, Mano M, Kobayashi K, Zhu J (2009) Modeling the stomatal conductance and photosynthesis of a flag leaf of wheat under elevated O_3 concentration. J Agric Meteorol 65:239–248
Oue H, Kobayashi K, Zhu J, Guo W, Zhu X (2011) Improvements of the ozone dose response functions for predicting the yield loss of wheat due to elevated ozone. J Agric Meteorol 67:21–32
Sakaki T, Kondo N, Sugahara K (1983) Breakdown of photosynthetic pigments and lipids in Spinach leaves with ozone fumigation: role of active oxygens. Physiol Plant 59:28–34
Sakaki T, Ohnishi J, Kondo N, Yamada M (1985) Polar and neutral lipid changes in spinach leaves with ozone fumigation: triacyiglycerol synthesis from polar lipids. Plant Cell Physiol 26: 253–262
Sakaki T, Saito k, Kawaguchi A, Kondo N, Yamada M (1990a) Conversion of monogalactosyldiacylglycerols to triacylglycerols in ozone-fumigated spinach leaves. Plant Physiol 94:766–772
Sakaki T, Kondo N, Yamada M (1990b) Pathway for the synthesis of triacylglycerols from monogalactosyldiacylglycerols in ozone-fumigated spinach leaves. Plant Physiol 94:773–780
Sakaki T, Kondo N, Yamada M (1990c) Free fatty acids regulate two galactosyltransferases in chloroplast envelope membranes isolated from spinach leaves. Plant Physiol 94:781–787
Sawada H, Kohno Y (2009) Differential ozone sensitivity of rice cultivars as indicated by visible injury and grain yield. Plant Biol 11(suppl 1):70–75
Sawada T, Mori G, Nouchi I, Komeiji T (1971) Plant ozone injury symptoms affect by photochemical oxidant. J Jpn Soc Air Pollut 6:139 [in Japanese, No English title]
Shimizu H, Motohashi S, Iwaki H, Furukawa A, Totsuka T (1981) Effects of low concentrations of O_3 on the growth of sunflower plants. Environ Control Biol 19:137–147
Shinohara T, Yamamoto Y, Kitano H, Fukuda M (1973) The relation between ozone treatment and the injury in Tobacco. Proc Crop Sci Soc Jpn 42:412–417
Skärby L, Selldén G, Mortensen L, Bender J, Jones M, De Temmerman L, Wenzel A, Fuhrer J (1993) Responses of cereals exposed to air pollutants in open-top chambers. In: Jäger HJ, Unsworth MH, De Temmerman L, Mathy P (eds) Effects of air pollution on agricultural crops in Europe. Commission of the European Communities, Brussels, pp 241–259

Suyama I, Kuroda S, Shinohara T, Kimura T, Shyoda M, Miyake Y, Kurihara K (1973) Study of weather fleck in tobacco: (7) the cultivar difference in sensitivity to leaf injury. Bull Okayama Tobacco Exp Station 33:55–62 [in Japanese, No English title]

Takagi K, Ohara T (2003) Estimation of ozone impact on plants by damage functions in the Kanto area. J Jpn Soc Air Pollut 38:205–216

Taylor GS (1974) Ozone injury on tobacco seedlings can predict susceptibility in the field. Phytopathology 64:1047–1048

Toyama S, Yoshida M, Niki T, Ohashi T, Koyama I (1989) Studies on ultrastructure and function of photosynthetic apparatus in Rice cells: IV. Effect of low dose and intermittent fumigation of ozone on the ultrastructure of chloroplasts in rice leaf cells. Jpn J Crop Sci 58:664–672

Tsukahara K, Sawada H, Matsumura H, Kohno Y, Tamaoki M (2012) Quantitative trait locus analyses of ozone-induced grain yield reduction in rice. Environ Exp Bot 88:100–106

Yamada K, Kuwata K (1982) Effects of air pollution on leaf surface lipids of parsley (*Petroselinum crispum*): variation in amount of total hydrocarbon and in composition of saturated, and unsaturated hydrocarbons. J Jpn Soc Air Pollut 17:258–264 [in Japanese with English summary]

Yamaguchi M, Inada H, Satoh R, Hoshino D, Nagasawa A, Negishi Y, Sasaki H, Nouchi I, Kobayashi K, Izuta T (2008) Effects of ozone on the growth, yield and leaf gas exchange rates of two Japanese cultivars of rice (*Oryza sativa* L.). J Agric Meteorol 64:131–141

Yamaguchi M, Hoshino D, Inada H, Akhtar N, Sumioka C, Takeda K, Izuta T (2014) Evaluation of the effects of ozone on yield of Japanese rice (*Oryza sativa* L.) based on stomatal ozone uptake. Environ Pollut 184:472–480

Yamaguchi M, Hoshino D, Kondo T, Satoh R, Inada H, Izuta T (2015) Evaluation of O_3 effect on net photosynthetic rate in flag leaves of rice (Oryza sativa L.) by stomatal O_3 flux and radical scavenging enzyme activities. J Agric Meteorol 71:211–217

Yamazoe F (1987) History and aspect of researches on effect of air pollution on plants. J Jpn Soc Air Pollut 22:199–210 [in Japanese]

Yonekura T, Ohshima K, Hattori M, Izuta T (2000) Effects of ozone and soil water stress, singly and in combination, on growth, yield, seed quality and germination rate of Soybean plants. J Jpn Soc Air Pollut 35:36–50 [in Japanese with English summary]

Yonekura T, Shimada T, Miwa M, Arzate AT, Ogawa K (2005a) Impacts of tropospheric ozone on growth and yield of rice (*Oryza sativa* L.). J Agric Meteorol 60:1045–1048

Yonekura T, Kihira A, Shimada T, Miwa M, Arzate A, Izuta T, Ogawa K (2005b) Impacts of O_3 and CO_2 enrichment on growth of Komatsuna (*Brassica campestris*) and Radish (*Raphanus sativus*). Phyton (Ann Bot) 45:229–235

Chapter 5
Effects of Ozone on Japanese Trees

Makoto Watanabe, Yasutomo Hoshika, Takayoshi Koike, and Takeshi Izuta

Abstract The effects of ozone (O_3) on tree species in Japan have been studied since the 1970s. Based on the results from O_3 fumigation studies, current ambient levels of O_3 have negative impacts on the growth and physiological functions of Japanese forest tree species, although there is a big variation of O_3 sensitivity between species. Stomatal O_3 uptake is one of the key factors that can explain the differences in O_3 sensitivity between species, and modeling of this factor has been intensively studied during the past decade. Although O_3 generally induces stomatal closure, less efficient stomatal control, so-called stomatal sluggishness, is also induced by chronic exposure to O_3. These opposite phenomena result in complex responses of stomata to O_3. Detailed gas exchange analysis has revealed that O_3-induced reductions in the photosynthetic rate of Japanese forest tree species were mainly due to a biochemical limitation in chloroplasts, but not due to stomatal closure. Risk assessments of the O_3 impact on Japanese forest tree species, based on the results of experimental studies, national monitoring data of air pollutant concentrations, and vegetation surveys, indicate that the areas with high O_3-induced reduction in growth do not necessarily correspond to the areas with relatively high O_3 exposure. Free-air O_3 fumigation systems in Japan were developed in 2011. Studies with this novel technology have clarified differences in leaf O_3 sensitivities between canopy positions, and have estimated the effects of O_3 on whole-canopy carbon budgets. As future perspectives, not only we need clarification of the

M. Watanabe (✉) • T. Izuta
Institute of Agriculture, Tokyo University of Agriculture and Technology,
Fuchu, Tokyo 183-8509, Japan
e-mail: nab0602@cc.tuat.ac.jp; izuta@cc.tuat.ac.jp

Y. Hoshika
Institute for Sustainable Plant Protection, National Research Council of Italy,
Via Madonna del Piano, Sesto Fiorentino 50019, Italy
e-mail: hoshika0803@gmail.com

T. Koike
Silviculture and Forest Ecological Studies, Hokkaido University,
Sapporo, Hokkaido 060-8589, Japan
e-mail: tkoike@for.agr.hokudai.ac.jp

T. Izuta (ed.), *Air Pollution Impacts on Plants in East Asia*,
DOI 10.1007/978-4-431-56438-6_5

physiological mechanisms of O_3 impact, but we also need clarification of the effects of interactions between trees and other biotic factors such as diseases, herbivores, and symbiotic microbes.

Keywords Ozone • Japanese forest tree species • Growth • Physiological functions • Stomatal function

5.1 Introduction

The Japanese archipelago is located on the eastern edge of the Eurasian continent and extends from southwest to northeast in an arcuate shape. The length of the island chain is approximately 3,000 km, from the subtropics to the subarctic, covering a broad range of functional types of trees. Forests cover 67 % of Japan's total land area. In general, the number of tree species in the East Asian temperate biome is higher than that in the European and North American temperate biomes (Fine and Ree 2006).

Japan was the first country in East Asia to be recognized as suffering from serious air pollution, from the 1960s to the early 1970s (e.g., Izuta 2003, 2006; Nouchi 2001). As mentioned in Chap. 1, although the concentration of O_3 in Japan fell in the late 1970s, it has increased again from the 1980s. Furthermore, a continuous increase of O_3 concentration is predicted in this century. Therefore, the effects of O_3 on forest tree species will be serious in the near future. The understanding of tree responses to elevated O_3 is important to predict future forest contributions to our life through timber production, carbon sequestration, disaster prevention, and biodiversity conservation.

In Japan, studies on the effects of O_3 on plants have been published since the 1970s (Nouchi et al. 1973; Furukawa et al. 1983). These studies revealed the acute effects of O_3 on trees in roadsides and parks, such as *Prunus yedoensis*, *Ginkgo biloba*, and *Zelkova serrata* (Fujinuma et al. 1987; Furukawa 1991; Izuta 2006). Experimental studies on the effects of chronic O_3 exposure on the growth and physiology of forest trees commenced in the 1990s. These studies indicated that exposure to ambient levels of O_3 below 100 ppb for several months was sufficient to inhibit growth and physiological functions in sensitive Japanese forest tree species, such as *Fagus crenata*, *Z. serrata*, and *Pinus densiflora* (Izuta et al. 1996, 1998; Matsumura et al. 1996, 1998; Shan et al. 1997). In this chapter, we summarize the effects of O_3 on the growth and physiological characteristics of Japanese forest tree species, based on the experimental studies conducted mainly since the 1990s. In addition, advanced free-air O_3 fumigation studies of representative Japanese deciduous tree species, showing the effects of O_3 on defense capacity against biotic stresses, and symbiosis with ectomycorrhizal fungi are described. Nomenclature follows that described by Ohwi (1983).

5.2 Effects of O_3 on Growth and Physiological Characteristics of Japanese Trees

5.2.1 Effects of O_3 on Growth

5.2.1.1 Dry-Matter Growth

When plants are exposed to elevated O_3, their dry-matter growth is generally decreased. There are many studies concerning the effects of O_3 on the growth of Japanese forest tree species, using O_3 fumigation chambers (Fig. 5.1; e.g. Kohno et al. 2005; Yamaguchi et al. 2011). For example, Matsumura et al. (1996) reported that the whole-plant dry mass of *Cryptomeria japonica*, *Chamaecyparis obtusa*, and *Z. serrata* seedlings was reduced by increasing the exposure to O_3 from 0.4 to 3.0 times ambient O_3 concentration for 24 weeks (Fig. 5.2). Ozone-induced growth reductions have been reported for all functional types of tall tree species, such as deciduous broad-leaved (Matsumura 2001; Matsumura et al. 1996, 1998, 2005: Yonekura et al. 2001a, b; Watanabe et al. 2007; Yamaguchi et al. 2007), deciduous conifer (Matsumura 2001; Watanabe et al. 2006; Koike et al. 2012;

Fig. 5.1 (**a**) Naturally lit environmental control chambers, (**b**) greenhouse-type chambers, (**c**) open-top chambers, and (**d**) growth measurement of the seedlings in the open-top chamber

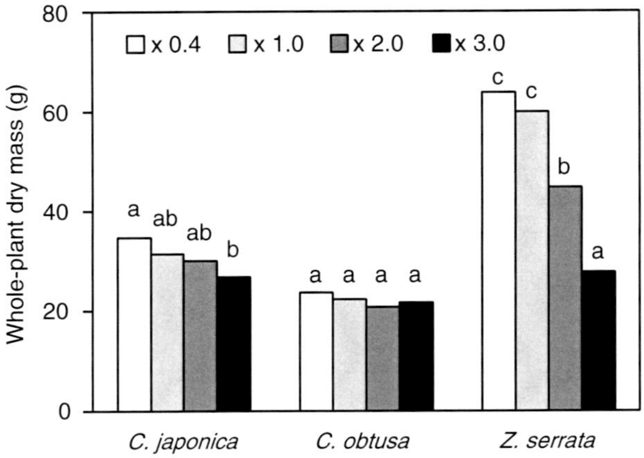

Fig. 5.2 Whole-plant dry mass of *Cryptomeria japonica*, *Chamaecyparis obtusa*, and *Zelkova serrata* seedlings exposed to four levels of ozone for 24 weeks. The seedlings were exposed to four simulated profiles with diurnal fluctuations of ozone: 0.4 (×0.4), 1.0 (×1.0), 2.0 (×2.0), and 3.0 (×3.0) times the ambient ozone concentrations. The average concentrations of ozone during the daytime (12 h) at ×0.4, ×1.0, ×2.0 and ×3.0 treatments were 16, 39, 74, and 114 nmol mol^{-1}, respectively. Values with different letters are significantly different at $P<0.05$ (Data source: Matsumura et al. 1996)

Wang et al. 2015), evergreen broad-leaved (Matsumura and Kohno 2003; Watanabe et al. 2008), and evergreen conifer (Kohno and Matsumura 1999; Matsumura 2001; Matsumura et al. 1996, 1998, 2005; Nakaji and Izuta 2001; Nakaji et al. 2004; Watanabe et al. 2006).

The relationships between O_3 exposure and growth responses of tree species have been studied since the late 1990s. These studies have revealed big differences of sensitivity to O_3 between forest tree species in Japan (Table 5.1). As a general trend of differences between functional types, the O_3 sensitivity of deciduous tree species is higher than that of evergreen tree species. As a general trend between functional types, O_3 sensitivity of deciduous tree species is higher than that of evergreen tree species. The foliage of evergreen tree species is exposed various biotic and abiotic stresses such as strong wind, low temperature, herbivore and diseases as compared to that of deciduous tree species due to their longer leaf longevity. Therefore, the foliage of evergreen trees should have higher physical and chemical tolerance to these stress factors, and this might confer higher tolerance to O_3 exposure. This tendency is also found in trees in other world regions such as Europe and China (Büker et al. 2015; Zhang et al. 2012). However, there is a big variation in O_3 susceptibility within functional types. For example, *Castanopsis sieboldii*, a representative evergreen tree species in warm temperate forests in Japan, has high susceptibility to O_3 exposure, which is comparable to that of *F. crenata* (Watanabe et al. 2008).

O_3, which enters the leaf through stomata, causes damage to plants. Accordingly, the effects of O_3 are described in the following formula,

Table 5.1 Classification of ozone sensitivity for Japanese forest tree species

Sensitivity class	Functional type		Species	AOT40 corresponding to 10% reduction of growth
High	Broad-leaved	Deciduous	*Populus maximowiczii* *Populus nigra* *Fagus crenata* *Zelkova serrata*	8–15 ppm h
		Evergreen	*Castanopsis sieboldii*	
	Coniferous	Deciduous	*Larix kaempferi*[a]	
		Evergreen	*Pinus densiflora*	
Moderate	Broad-leaved	Deciduous	*Quercus serrata* *Betula platyphylla* var. *japonica*	16–30 ppm h
		Evergreen	*Quercus myrsinaefolia* *Cinnamomum camphora*	
	Coniferous	Evergreen	*Abies homolepis*	
Low	Broad-leaved	Deciduous	*Quercus mongolica* var. *crispula*[b]	31 ppm h<
		Evergreen	*Lithocarpus edulis* *Machilus thunbergii*	
	Coniferous	Evergreen	*Pinus thunbergii* *Cryptomeria japonica* *Chamaecyparis obtuse*	

After Kohno et al. (2005), Yamaguchi et al. (2011)
High sensitivity: dry mass growth was significantly reduced by the ambient level of ozone
Moderate sensitivity: dry mass growth was significantly reduced by 1.5 or 2.0 times the ambient level of ozone
Low sensitivity: dry mass growth was significantly reduced by 1.5 or 2.0 times the ambient level of ozone
AOT40: accumulated exposure over a threshold of 40 nmol mol^{-1} O_3 during daylight hours for 6 months (April–September)
[a]*Larix leptolepis*
[b]*Quercus mongolica* var. *grosseserata* or *Q. crispula*

$$O_3 \text{ effect} = \text{stomatal } O_3 \text{ uptake} \times \text{damage caused by a unit of } O_3 \text{ uptake}$$

It should be noted that both factors explaining the O_3 effect are regulated by complex physiological mechanisms. In Europe, analyses of accumulated stomatal O_3 uptake and growth reduction have been carried out since the first decade of the 2000s (Karlsson et al. 2004, 2007; Büker et al. 2015). On the other hand, uptake-based analysis of O_3 effects on Japanese forest tree species is an on-going study (see Sect. 5.2.2.1). The latter factor in the formula may be related to the detoxification of O_3 and reactive oxygen species derived from O_3, and processes of repair and compensation of O_3 damage (Fuhrer and Booker 2003); however, the worldwide knowledge is not yet complete enough to explain this latter factor in the formula.

5.2.1.2 Leaf Turnover

Ozone affects leaf turnover in trees, which can continuously produce new leaves during a growing season. In the experiment of Pell et al. (1994), O_3-induced acceleration of abscission of old leaves, and simultaneous increase of new leaf production, were observed in *Populus tremuloides* seedlings. These phenomena are considered as a kind of compensation response to elevated O_3 stress. A similar tendency was observed in *C. sieboldii* seedlings in the experiment of Watanabe et al. (2008) (Fig. 5.3). In their experiment, *C. sieboldii* seedlings were grown under 12 different experimental conditions, comprised of four gas treatments (charcoal-filtered air and three levels of O_3 at 1.0, 1.5, and 2.0 times the ambient concentration), in open-top chambers during the two growing seasons from April 2004 to November 2005. The exposure to O_3 significantly increased the abscission of the old leaves of *C. sieboldii* seedlings. But new leaf emergence was stimulated with increasing levels of O_3 exposure. In addition, the net photosynthetic rate of second-flush leaves was higher than that of first-flush leaves. Ozone-induced stimulation of new leaf production was also observed in the seedlings of *F. crenata* (Watanabe et al. 2010b), *Betula platyphylla* var. *japonica* (hereafter *B. platyphylla*; Hoshika et al. 2013b), *Quercus serrata,* and *Quercus mongolica* var. *crispula* (hereafter *Q. mongolica*; Kitao et al. 2015). However, the efficiency of the compensation response

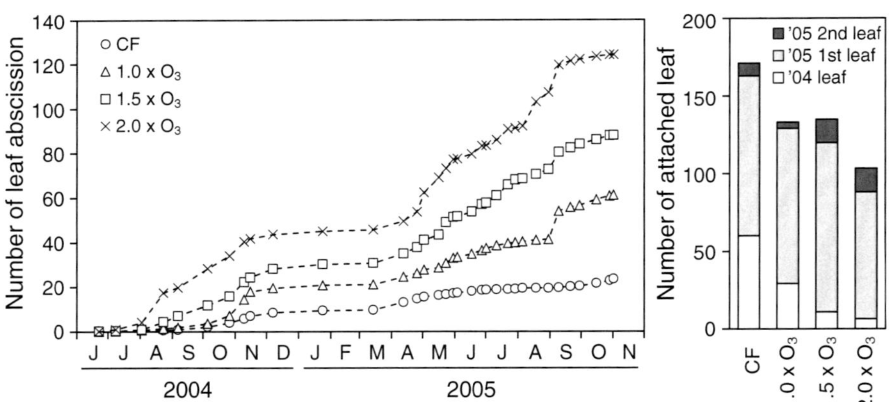

Fig. 5.3 Time course of leaf abscission (*left*) and number of attached leaves in each age class, on 1 November 2005, of *Castanopsis sieboldii* seedlings grown under charcoal-filtered air (CF) or 1.0, 1.5, and 2.0 times ambient concentration of ozone: '04 leaf, emerged in 2004; '05 first leaf, emerged in May 2005; '05 second leaf, emerged from July to October 2005 (Data source for the right figure is Watanabe et al. (2008), while data for the left figure is unpublished data from Watanabe et al. (2008))

to O_3 stress is not clear, because with old leaf falling there is loss of carbon and new leaf production requires another carbon cost. Actually, the total number of *C. sieboldii* leaves was decreased by the exposure to O_3, even with the increase of second-flush leaves (Fig. 5.3, right).

5.2.1.3 Carbon Allocation

Ozone changes the carbon allocation in tree species between above- and below-ground organs. In many tree species, O_3 induces greater inhibition of root growth than above-ground growth, resulting in an increase in the ratio of shoot biomass to root biomass (S/R) under elevated O_3 (Wittig et al. 2009). Matsumura et al. (1996) demonstrated increases of S/R in *Z. serrata* seedlings with increasing accumulated O_3 exposure (concentration × time) in a 24-week O_3 fumigation study. Also, an increase of S/R has been reported in other tree species in Japan, such as *C. japonica* (Miwa et al. 1993; Matsumura et al. 1998; Kohno and Matsumura 1999), *C. obtusa* (Kohno and Matsumura 1999), *Chamaecyparis pisifera* (Kohno and Matsumura 1999), *B. platyphylla* (Matsumura et al. 1998), *Abies homolepis* (Matsumura et al. 1998), *F. crenata* (Yamaguchi et al. 2007), *Q. serrata*, and *Q. mongolica* (Kitao et al. 2015). This change in carbon allocation might be a result of a compensation response to maintain growth under elevated O_3 conditions (Kitao et al. 2015). However, the mechanism of the change in carbon allocation of tree species under elevated O_3 conditions is not understood yet. If the carbon allocation to below-ground organs (i.e., roots) is decreased, carbon input from the root system to soil brought about by root turnover and exudate may be decreased. On the other hand, relative above-ground carbon input from leaf and branch abscissions increased. The decomposition processes, including animal and microbial decomposers, are quite different between carbon from above ground and that from below ground. Therefore, the changes in carbon allocation may be a critical matter from the viewpoint of the carbon cycle in a forest.

5.2.1.4 Phenology

The acceleration of foliar senescence is one of the common responses of trees to elevated O_3 conditions (e.g. Pell et al. 1999), although the phenomena induced by O_3 are somewhat different to those of natural senescence (Matyssek and Sandermann 2003). Although the acceleration of foliar senescence and the above-mentioned old leaf abscission for producing new leaves, as a kind of compensation response under elevated O_3 conditions, might occur via the same mechanism, any difference has not been clarified. Yonekura et al. (2004) examined the effects of O_3 on the phenological characteristics of *F. crenata* seedlings. The seedlings were exposed to 60 nmol mol^{-1} O_3 for 7 h per day during one growing season (May to October) in naturally lit growth chambers. All the seedlings were removed from the growth chambers after the growing season, and grown under field conditions until the next spring. The exposure to O_3 during the growing season induced early leaf fall. In the next spring, bud break of the seedlings was delayed, although these authors did not conducted O_3 exposure during the bud-break observation. This phenomenon is called the "carry-over effect". A similar phenomenon was observed in mature *Fagus sylvatica* in the Kranzberg Forest in Germany (Nunn et al. 2005). The carry-over effect, not only in relation to phenology but also in relation to growth, morphology, and physiology, was reported in *B. pendula* seedlings in northern Europe (Oksanen and Saleen 1999).

From the ecological viewpoint, the impact of O_3 on phenology is highly important for forest dynamics. The light environment of the deciduous forest floor undergoes seasonal changes. And seedlings in the forest floor respond to the changes in light condition (e.g. Kitaoka and Koike 2004). Because the responses to changes in light condition are species-specific, shortening of the leafy period in canopy trees induced by elevated O_3 will affect seedlings in the forest floor in various ways, and may cause a change in species composition, indicating change in forest dynamics. On the other hand, the phenological changes of seedlings in the forest floor under elevated O_3 conditions also have the potential to affect forest dynamics. For example, a phenological gap is important for the seedlings of *F. crenata* (Terazawa and Koyama 2008). The leaf emergence of *F. crenata* is earlier than that of other tree species such as *Q. mongolica* and *Magnolia obovate* (= *M. hypoleuca*). Even though the seedlings of *F. crenata* in the forest floor are covered by upper-storey trees of other species, they can use light resources before the leaf emergence of upper-storey trees. However, if the leaf emergence of *F. crenata* seedlings is delayed under elevated O_3 conditions, the seedlings may not use the phenological gap efficiently and the growth of the seedlings may be inhibited.

5.2.2 O_3 Effects on Physiology

5.2.2.1 Stomatal Functions

Stomata are small pores on leaves whose aperture is actively regulated by the plant in response to multiple abiotic and biotic factors, and stomatal conductance, which is an index of stomatal apertures, is a major determinant of CO_2 uptake (photosynthesis) and water loss (transpiration). Ozone enters leaves through the stomata and causes damage to physiological processes such as photosynthesis (Reich 1987). Stomatal O_3 flux or uptake is considered to be closely related to O_3 impacts on forest trees (Mills et al. 2010). Stomatal conductance is one of the most important factors for the estimation of stomatal O_3 flux. In the 2000s, modeling studies have been developed to estimate stomatal O_3 flux for assessing the impact of O_3 in Japanese forest trees. A multiplicative numerical approach (Jarvis-type model; Jarvis 1976) has been commonly used to estimate stomatal conductance (Mills et al. 2010). This approach assumes that stomatal conductance is a multiplicative function of leaf age and environmental factors (e.g., light intensity, temperature, and humidity deficits and soil-moisture stress). Several efforts to estimate stomatal O_3 flux have thus been made in forest tree species in Japan (e.g. Hoshika et al. 2012c; Azuchi et al. 2014; Kinose et al. 2014). Kinose et al. (2014) reported the importance of a species-specific stomatal response for determining stomatal O_3 flux. In their study, the cumulative stomatal uptake of O_3 (CUO) of *F. crenata*, *Q. serrata*, and *Q. mongolica* seedlings in spring, from April to May, was relatively low, whereas the O_3 concentration was relatively high. On the other hand, the CUO of *B. platyphylla* seedlings was relatively high, mainly because of rapid leaf maturation and a lower optimal temperature for stomatal opening. In addition to the Jarvis-type model, a model based on the empirical relationship between the photosynthetic rate and stomatal conductance (i.e., the Ball-Woodrow-Berry model; Ball et al. 1987) was recently applied to estimate stomatal O_3 flux (Hoshika et al. 2015a). These modeling studies also found that O_3 affected stomatal conductance. Hoshika et al. (2012c) reported that O_3 decreased the maximum stomatal conductance of *F. crenata* leaves. Although simple functions are used for estimating O_3 effects on stomatal conductance (Hoshika et al. 2012c; Azuchi et al. 2014; Kinose et al. 2014), the actual stomatal responses to O_3 are more complex than those used in the current models (Hoshika et al. 2015c). Also, the mechanisms of O_3 effects on stomatal conductance are still under investigation (Kangasjärvi et al. 2005; Mills et al. 2009). Stomatal behavior under elevated O_3 is thus a major concern.

According to the meta-analysis of O_3 effects on tree species done by Wittig et al. (2007), O_3 was found to reduce stomatal conductance. There are several reports on lower stomatal conductance in the leaves of Japanese forest tree species under elevated O_3 (e.g. Hoshika et al. 2012c, 2013b). The reduction of stomatal conductance may limit stomatal O_3 flux into leaves. This raises the question of whether O_3-induced stomatal closure could act as an avoidance response to O_3 stress. Hoshika et al. (2013a) examined this hypothesis for *F. crenata* using an optimal stomatal model, which was based on the optimization theory of stomatal conductance for maximizing carbon gain while minimizing accompanying water loss and O_3 flux. Eleven-year-old saplings of *F. crenata* were exposed to 60 nmol mol^{-1} O_3 by a free-air fumigation technique (see Sect. 5.4) for daylight hours during the growing season. Analysis based on the optimal stomatal model suggested that O_3 flux was efficiently limited by stomatal closure in early summer (June). Their results indicate that O_3-induced stomatal closure in early summer could be considered as an avoidance response to O_3 stress, to allow maximum photosynthetic capacity to be reached. It should be noted, however, that it is not likely that sufficient evolutionary time has elapsed to develop plant adaptation to elevated O_3, which is a "new" stress with a short history as compared with times for plant evolution (Jacob 1999). Therefore, such stomatal closure may be just a reaction to O_3 and fortunately it avoids O_3 stress for trees. In late summer (August) and autumn (early October), on the other hand, the decrease of stomatal conductance by O_3 exposure became small. Previous studies suggest that O_3 also causes less efficient stomatal control, especially a weaker capacity to close stomata, so-called stomatal sluggishness (Paoletti and Grulke 2005). Yamaguchi et al. (2007) found that O_3 increased the stomatal conductance of *F. crenata* seedlings under light-saturated conditions in autumn (Fig. 5.4). Watanabe et al. (2014b) pointed out an inconsistent stomatal response to O_3 (60 nmol mol^{-1} during daylight hours), due to

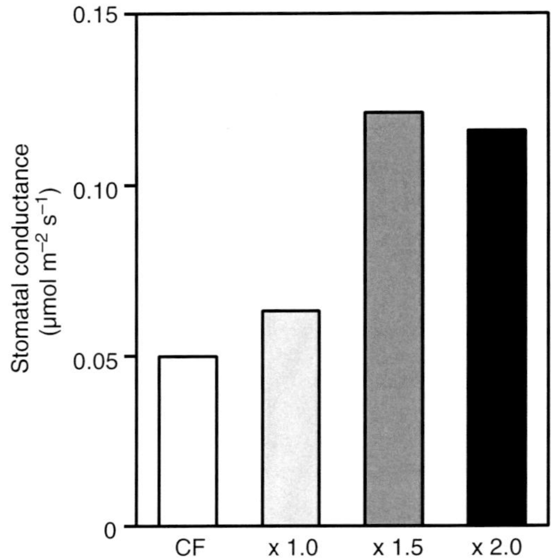

Fig. 5.4 Stomatal conductance of *Fagus crenata* seedlings in September 2005. The seedlings were grown under charcoal-filtered air (CF) or 1.0, 1.5, and 2.0 times ambient concentration of ozone for two growing seasons, from April 2004 to September 2005 (Data source: Yamaguchi et al. 2007)

both stomatal closure and stomatal sluggishness, in *Betula maximowicziana* seedlings. Hoshika et al. (2015b) reported that O_3-induced stomatal closure decreased with increasing leaf age in *F. crenata*. Also, O_3 may increase night-time stomatal conductance, as reportedly occurs in *F. crenata* in October (Hoshika et al. 2013a).

The impairment of stomatal control may lead to a slower stomatal response to environmental stimuli. Hoshika et al. (2012b) found a significantly slower stomatal response to light variation (alternative light exposure of photosynthetic photon flux density between 100 and 1,500 µmol m^{-2} s^{-1}) for *F. crenata* saplings. Although the mechanisms of stomatal sluggishness are still not known, Omasa (1990) suggested that such a stomatal response was caused by a slight increase in the permeability of epidermal cell membranes and alterations of the osmotic pressure modulating a balance in turgor between guard and subsidiary cells. Mills et al. (2009) found that O_3 reduced stomatal sensitivity to abscisic acid (ABA). Ozone-induced ethylene synthesis might also be related to stomatal sluggishness, because ethylene is known to inhibit the action of ABA in stomatal closure (Tanaka et al. 2005; Wilkinson and Davies 2010). Moreover, it was recently demonstrated that gene expression under O_3 impact inhibited the CO_2 signaling involved in stomatal closure (Dumont et al. 2014). The effect of elevated O_3 on stomatal reaction to changes in environmental conditions should be considered in dynamic models for the estimation of stomatal O_3 flux.

In this section, we have explained two kinds of stomatal sluggishness, one where stomata do not close to a sufficient level (steady-state evidence) and the other in which there is a slower stomatal response (kinetic evidence). Whether or not these two types of stomatal sluggishness are supported by the same mechanism is a future topic for study.

5.2.2.2 Photosynthesis

Approximately 40 % of a plant's dry mass consists of carbon, fixed during photosynthesis. This process of CO_2 assimilation with water and light energy is vital for the growth and survival of substantially all plants (Lambers et al. 2008). Many studies have reported decreased leaf photosynthetic rates (usually in light-saturated conditions) under elevated O_3. A decreased photosynthetic rate is considered to be one of the most important factors that induces growth reduction in trees. According to Watanabe et al. (2006), reductions in net photosynthetic rates by exposure to O_3 were observed in *Larix kaempferi* and *P. densiflora* seedlings, which showed relatively large growth reduction. On the other hand, in *C. japonica* seedlings, the extent of growth reduction was relatively small and there was no significant reduction in the net photosynthetic rate under elevated O_3. Similar tendencies were reported in *F. crenata* (sensitive to O_3) and *Q. mongolica* (tolerant to O_3) saplings (Watanabe et al. 2013).

The rate of photosynthetic carbon assimilation is determined by both the supply and the demand for CO_2. The supply of CO_2 to the chloroplasts from ambient air is governed by diffusion in the gas and liquid phases. Stomatal conductance is one of the main parameters controlling CO_2 diffusion in the gas phase and therefore this process controls the supply of CO_2 to the chloroplasts. The demand for CO_2 is determined by the rate of CO_2 assimilation in the chloroplast, which is governed by the component structures of the chloroplast and by biochemical reactions, such as carboxylation in the stroma and electron transport and adenosine triphosphate (ATP) synthesis in thylakoid membranes. The question is: which factor (i.e., supply or demand) mainly limits the rate of photosynthesis under elevated O_3?

As mentioned above, O_3 generally reduces stomatal conductance (Wittig et al. 2007). On the other hand, there is a high correlation between the net photosynthetic rate and stomatal conductance (Ball et al. 1987) under normal ambient conditions (i.e., no O_3 effect). The ratio of the intercellular CO_2 concentration (C_i) to the ambient CO_2 concentration (C_a), i.e., C_i/C_a, at a given vapor pressure deficit is constant under a certain range of atmospheric CO_2 concentrations (e.g. Sage 1994). If the biochemical assimilation capacity decreases, stomatal conductance also decreases for constant C_i/C_a. Therefore, it should be noted that the co-reduction of the net photosynthetic rate and stomatal conductance may not indicate stomatal limitation of photosynthesis.

The analysis of the C_i-response curve of the net photosynthetic rate (A/C_i curve) is one of the strongest methods with which to quantify the supply and demand for CO_2 in photosynthesis (Fig. 5.5; Farquhar et al. 1980). The maximum rate of carboxylation (V_{cmax}) and the maximum rate of electron transport (J_{max}) are representative parameters of CO_2 demand, and stomatal limitation is the representative parameter of CO_2 supply. Detailed information on photosynthesis and the general environmental responses (light, humidity and CO_2 responses) of forest tree species are described in specialized books (e.g. Lambers et al. 2008). Limitation by the rate of triose phosphate utilization (TPU) is also considered as a factor that regulates the photosynthetic rate (Sharkey 1985). But there is no report on the effect of O_3 on the rate of TPU utilization for Japanese forest tree species.

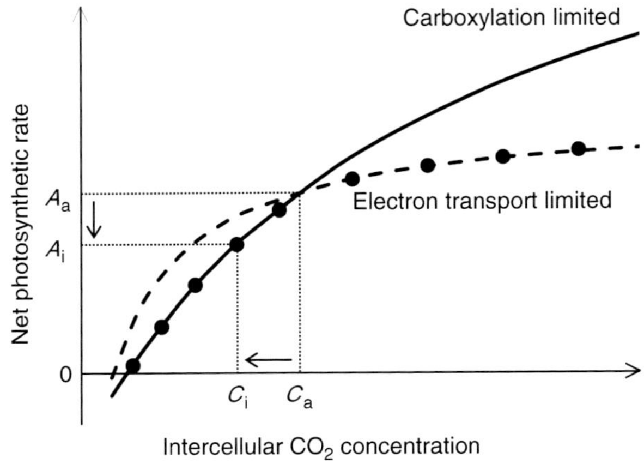

Fig. 5.5 Idealized intercellular CO_2 concentration-response curve of the net photosynthetic rate (A/C_i curve). C_a ambient CO_2 concentration, C_i intercellular CO_2 concentration, A_a theoretical net photosynthetic rate at $C_i = C_a$ (absence of limitation in diffusion of CO_2 from ambient air to intercellular air space), A_i net photosynthetic rate at C_i. The net photosynthetic rate at given C_i is plotted on the *lower line* of carboxylation-limited or electron transport-limited photosynthesis as *filled circles*. Stomatal limitation (L_s, %) is calculated as $L_s = (A_a - A_i) / A_a \times 100$

Although many simulations of the carbon cycle in vegetation use Farquhar's biochemical model for the calculation of photosynthesis, to date only a limited number of studies have determined the parameters of this model. Watanabe et al. (2013) analyzed the A/C_i curve in the leaves of *F. crenata* saplings exposed to free air O_3 enrichment (see Sect. 5.4) (Fig. 5.6). Elevated O_3 (60 nmol mol^{-1} during daytime) significantly reduced the light-saturated net photosynthetic rate (A_{sat}), V_{cmax}, and J_{max}, while there was no significant effect of O_3 on stomatal conductance and stomatal limitation. These results indicate the O_3-induced reduction of photosynthetic activity in *F. crenata* was due not to stomatal closure, but to biochemical limitations. In addition, the C_i/C_a was increased under elevated O_3, indicating that the stomata did not close enough to keep the C_i/C_a constant (i.e., there was stomatal sluggishness, see subsection "*Stomatal functions*" above). Ozone-induced decreases of V_{cmax} or carboxylation efficiency, which is calculated as the initial slope of the A/C_i curve and is closely correlated with V_{cmax}, have also been reported in the leaves of other Japanese forest tree species: *F. crenata* (Izuta et al. 1996; Yonekura et al. 2001b; Yamaguchi et al. 2007), *B. platyphylla* (Matsumura et al. 1998), *B. maximowicziana* (Watanabe et al. 2014b), *Z. serrata* (Matsumura et al. 1998), and *P. densiflora* (Nakaji and Izuta 2001).

Nitrogen is a nutrient that is strongly related to biochemical assimilation capacity (e.g. Lambers et al. 2008). A large fraction of the nitrogen in leaves is incorporated into proteins associated with the photosynthetic process (Evans 1989; Evans and Seemann 1989), and there is generally a positive correlation between the photosynthetic rate and the nitrogen content of leaves (e.g. Kitaoka and Koike 2004). The net photosynthetic rate (A) is described as follows:

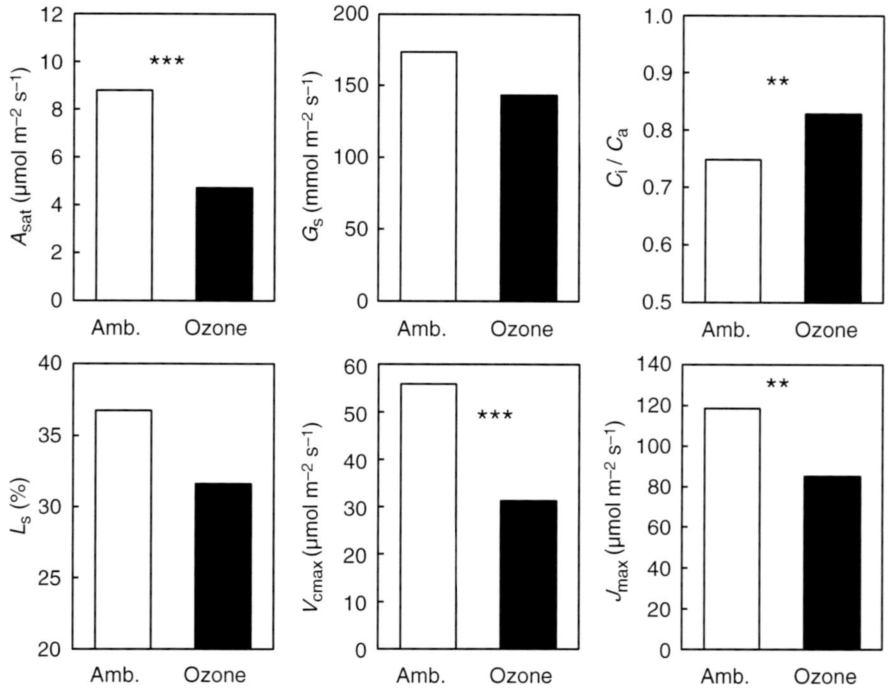

Fig. 5.6 Leaf photosynthetic traits of *F. crenata* saplings grown under ambient and elevated concentrations of O_3. A_{sat} light-saturated net photosynthetic rate at 380 μmol mol^{-1} CO_2, G_s stomatal conductance to water vapor, C_i/C_a ratio of intercellular CO_2 concentration to ambient CO_2 concentration, L_s stomatal limitation of photosynthesis, V_{cmax} maximum rate of carboxylation, J_{max} maximum rate of electron transport. **, $P<0.01$, ***, $P<0.001$ (Data source: Watanabe et al. 2013)

$$A\left(\mu\text{mol m}^{-2}\text{s}^{-1}\right) = \text{PNUE}\left(\mu\text{mol mol}^{-1}\text{s}^{-1}\right) \times \text{leaf nitrogen content}\left(\text{mol m}^{-2}\right),$$

where PNUE is photosynthetic nitrogen use efficiency (net photosynthetic rate per unit leaf nitrogen content). Generally, exposure to O_3 has little effect on leaf nitrogen content, except in autumn. In autumn, nitrogen resorption, and therefore decreases in leaf nitrogen content, may start earlier in leaves exposed to O_3 due to the O_3-induced acceleration of foliar senescence (see Sect. 5.2.1.4). On the other hand, several studies have reported decreased PNUE in woody species native to Japan under elevated O_3 (Watanabe et al. 2007, 2015; Yamaguchi et al. 2007; Hoshika et al. 2013b). These findings indicate that O_3 reduced leaf nitrogen allocation to the photosynthetic apparatus. As shown in Fig. 5.7, a decrease of the nitrogen fraction in the photosynthetic apparatus under elevated O_3 was demonstrated in the leaves of *F. crenata* saplings (Watanabe et al. 2013).

Yonekura et al. (2001b) monitored the time course of gas exchange traits in the leaves of *F. crenata* seedlings (A/C_i curve and light response curve of A) under elevated O_3. The negative effect of O_3 was first observed in regard to carboxylation

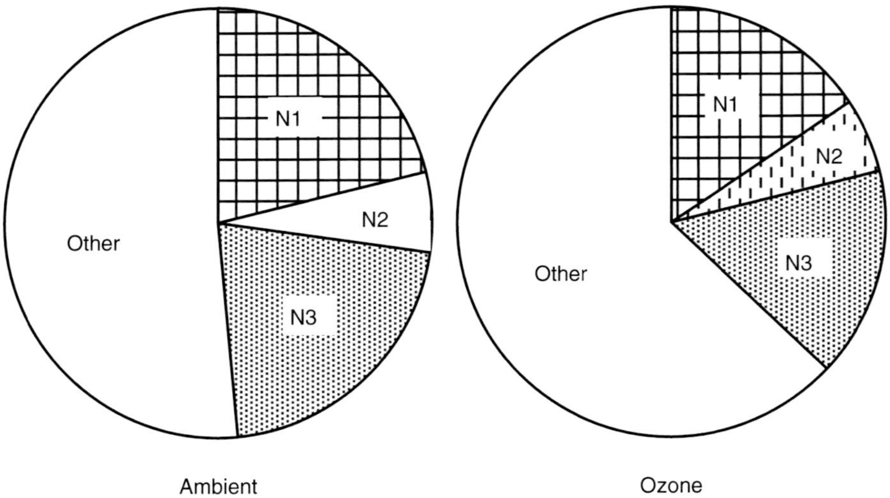

Fig. 5.7 Nitrogen fraction in leaves of *Fagus crenata* saplings. N1: Rubisco, N2: bioenergetics (electron carriers except for photosystems, coupling factor, and Calvin cycle enzymes except for Rubisco), N3: light-harvesting complex and photosystems (After Watanabe et al. 2013)

efficiency (initial slope of A/C_i curve) from mid-June, and in A_{sat} and light and CO_2 saturated A (which has a high correlation with J_{max}), which was decreased under elevated O_3 from mid-July. These results suggest that O_3 first causes a reduction of carboxylation capacity (i.e., a reduction in ribulose-1,5-bisphosphate carboxylase/oxygenase [Rubisco] content and/or its specific activity), and then a reduction in the capacity of other photosynthetic functions is induced as a result of the feedback regulation of the photosynthetic rate to match photosynthetic capacities with carboxylation capacity. According to Yamaguchi et al. (2007, 2010a), O_3 decreased the ratio of soluble protein content (mainly Rubisco) to nitrogen content in the leaves of *F. crenata* seedlings. Furthermore, they reported that the decrease in Rubisco in the leaves of *F. crenata* seedlings under elevated O_3 was accompanied by an increase in the acidic amino acid content, indicating the degradation of Rubisco protein. Brendley and Pell (1998) also found O_3-induced acceleration of proteolysis in a hybrid poplar (*Populus maximowiczii* A. Henry × *trichocarpa* Torr. & A. Gray). Reduction in Rubisco content was also observed in other Japanese forest tree species (*Q. serrata*; Watanabe et al. 2007, *P. densiflora*; Nakaji and Izuta 2001, *C. sieboldii*; Watanabe et al. unpublished data).

5.2.2.3 Respiration

The response of the dark respiration rate (R_d) in the leaves of Japanese forest tree species to elevated O_3 varied between experiments. Izuta et al. (1996) reported over 40 % reduction in R_d of *F. crenata* seedlings exposed to 75 and 150 nmol mol^{-1} O_3

(6 h per day for 18 weeks), although the reduction was not statistically significant. According to Wittig et al. (2007), R_d generally decreases under elevated O_3. In contrast, however, increased R_d was observed in *F. crenata* and *Q. mongolica* saplings under elevated O_3 (Watanabe et al. 2013). No significant effects of O_3 on R_d were reported in *C. japonica*, *C. obtusa*, and *Z. serrata* seedlings (Matsumura et al. 1996) and *F. crenata* seedlings (Yonekura et al. 2001b). Although a conclusive mechanism of these differences has not been clarified, the extent of damage caused by O_3 may be one of the factors responsible for this discordance. If the O_3 damage is severe, all metabolic processes, including R_d, would be decreased. On the other hand, minor damage by O_3 may induce an increase in R_d to enhance the detoxification capacity of O_3-related reactive oxygen species, and/or the repair of damaged tissues (Landolt et al. 1997; Matyssek and Sandermann 2003). There is no study about the effects of O_3 on photorespiration in the leaves of Japanese forest tree species.

5.3 Risk Assessment of Ozone Impact

Risk assessments of the impact of O_3 on tree growth in Japan have been conducted, based on experimental studies, since 2003 (Takagi and Ohara 2003). Kohno et al. (2005) estimated the critical level of O_3 for forest trees, based on experimental studies of 18 tree species, and they mapped the area in which the level of O_3 corresponds to at least a 10 % growth reduction of O_3-sensitive tree species throughout Japan.

Based on an experimental study on the combined effects of O_3 and soil nitrogen loading (see Sect. 6.3 and Yamaguchi et al. 2007 for details), the risk of a negative O_3 impact on the growth of *F. crenata,* with consideration of changes in O_3 sensitivity caused by nitrogen deposition, was determined (Watanabe et al. 2012). The values for the average and maximum O_3-induced relative growth reduction in *F. crenata* across Japan were estimated to be 3.2 % and 9.7 %, respectively. On the other hand, when the value for nitrogen deposition was assumed to be zero, the estimated values for average and maximum relative growth reduction were 2.3 % and 5.7 %, respectively. The inclusion of atmospheric nitrogen deposition data thus increased the estimated values for average and maximum relative growth reduction (by 38 % and 71 %, respectively). These authors emphasize that a change in the sensitivity to O_3 associated with atmospheric nitrogen deposition is an important consideration in the risk assessment of the impact of O_3 on the growth of *F. crenata* in Japan.

Watanabe et al. (2010a) indicated that not only the accumulated O_3 exposure but also the variety of tree habitat, differences in sensitivity to O_3, and differences in annual carbon absorption among the tree species must be taken into account to assess the risk of O_3 impact on annual carbon absorption in Japanese conifers. They integrated a forest growth model from a national forest monitoring data set and the results of an experimental study, and estimated the O_3 impact on annual forest carbon absorption in each prefecture throughout Japan for three representative Japanese conifers: *L. kaempferi*, *P. densiflora*, and *C. japonica*. The areas with high O_3-induced reduction in annual carbon absorption did not necessarily correspond to the areas with relatively high O_3 exposure. Widespread distribution of O_3-sensitive tree species, such as *P. densiflora* and *L. kaempferi*, and high annual carbon absorption were important factors that induced a high risk of O_3 impact on annual carbon absorption. The O_3-induced reduction in the total annual forest carbon absorption of the above three representative tree species in Japan was estimated to be 0.8 %.

Risk assessments of the effects of O_3, based on stomatal O_3 flux, which could account for the influence of climatic and ontogenetic factors, are now being developed in Japan (Izuta 2012). Hoshika et al. (2012a) compared the maps of accumulated O_3 exposure (e.g. the accumulated exposure over a threshold of 40 nmol mol^{-1}, AOT40) and cumulative O_3 uptake for *Z. serrata* in Japan and suggested that stomatal closure induced by a high vapor pressure deficit caused a decoupling of stomatal O_3 uptake from high O_3 exposure. This indicates that a stomatal flux-based approach may provide higher precision in the assessment of O_3 impacts on a regional scale (Watanabe and Yamaguchi 2011). Hoshika et al. (2015a) focused on O_3-induced steady-state stomatal sluggishness (see Sect. 5.2.2.1), and examined the effects of O_3-induced stomatal sluggishness on carbon assimilation and transpiration in temperate deciduous forests in the Northern Hemisphere. They conducted a simulation by combining a detailed multi-layer land surface model and a global atmospheric chemistry model. They reported that O_3-induced stomatal sluggishness may change the balance of the carbon and water cycle in temperate deciduous forests. According to their simulation, water use efficiency, i.e., the ratio of net CO_2 assimilation to transpiration, in temperate deciduous forests would decrease by up to 20 % when considering stomatal sluggishness data (i.e., the relationship between minimum conductance and CUO) obtained from the experimental study on *F. crenata* at the Sapporo experimental forest (see Sect. 5.4).

The risk assessment of O_3 impact requires relatively extensive studies. Data from experimental studies, field monitoring, and vegetation surveys are combined. Therefore, there are uncertainties at each level. Here, we discuss some uncertainties in the risk assessment of O_3 impact on Japanese forest tree species. According to previous studies, the extent of responses in plant traits (e.g., reduction of biomass, and leaf-level photosynthesis) to O_3 is species-specific (e.g. Yamaguchi et al. 2011), and even genotype-specific (Paludan-Müller et al. 1999). There are also differences in sensitivities between seedlings and mature trees (Karnosky et al. 2007; Matyssek et al. 2007). Pretzsch et al. (2010) reported comparable growth sensitivity to O_3 in mature *F. sylvatica* trees and juvenile seedlings, although the mechanism of the O_3-induced growth reduction may differ. Other environmental factors may affect the O_3 sensitivity of trees (see Chap. 6). To overcome these uncertainties, we need to develop a mechanism for understanding the effects of O_3 on trees in terms of whole-tree physiology and vegetation production; better modeling of the processes of plant growth and physiology is also needed.

Monitoring stations for O_3 in Japan have been mainly located in urban areas, because the aim of monitoring is the protection of human health. There are limited numbers of monitoring stations in the mountains and rural areas. However, several phenomena indicate that the O_3 concentration in mountainous and rural areas is higher than that in urban regions (e.g. Yamaguchi et al. 2010b). Furthermore, the atmospheric concentration of O_3 in mountainous areas sometimes shows diurnal variation that is different from that in urban areas (mainly flatland); in particular, a typical phenomenon in mountainous areas is that there is little change in O_3 concentration under the inversion layer. Reconsideration of the location of monitoring stations for O_3 is needed for the accurate assessment of O_3 impact on forest trees in Japan.

5.4 Free-Air Ozone Fumigation Experiment

Previous studies have applied chamber experiments to study the effects of O_3. Although chamber experiments offer an advantage as a study mechanism owing to their controllability of O_3 concentration, artifacts may arise in the environmental conditions as a result of differences in micro-meteorological conditions and the absence of biotic stresses such as herbivores and diseases. Most studies in Japan, as well as those around the world, have used pot-grown seedlings due to the limitations of the study facility. Accurate simulation of forest conditions is therefore questionable (Kolb and Matyssek 2001).

Free-air O_3 fumigation with field-grown trees is a novel technique for assessing the effects of O_3 in field conditions without volume restrictions on root growth. Studies employing this technology have been conducted in Europe and the United States since the 1990s (e.g. Karnosky et al. 2007; Matyssek et al. 2007; Oksanen et al. 2007; Díaz-de-Quijano et al. 2012). Although the concerns regarding O_3 in the Asian region are acute and important, no free-air fumigation study of forest tree species had been reported in this region until recent years. Three free-air O_3 fumigation experiments are running at present in Japan; two systems are located in the Sapporo experimental forest, Hokkaido University, Sapporo (43°04′ N, 141°20′ E, 15 m above sea level) (e.g. Watanabe et al. 2013; Koike et al. 2013) and the other one is located in the nursery of the Forestry and Forest Products Research Institute in Tsukuba (36°00′ N, 140°80′ E, 20 m a.s.l.) (Kitao et al. 2015). Some of the results of the free-air O_3 fumigation experiments have already been described (e.g. Hoshika et al. 2012b, c; Watanabe et al. 2013). Here, we introduce the details of the fumigation system and the results of canopy-scale study in the Sapporo experimental forest (Watanabe et al. 2014a, 2015). The results from the Tsukuba experiment are shown in Chapter 6.

The size of the system in the Sapporo experimental forest is a 5.5×7.2 m rectangle, and the height is 5.5 m (Figs. 5.8 and 5.9). In this system, 10-year-old saplings (as of 2011) of *F. crenata* and *Q. mongolica*, and 3-year seedlings of *B. platyphylla*, which are representative species in northern Japan, were grown. *F. crenata* and *Q. mongolica* have similar distribution and growth traits, which are late successional with shade tolerance, although *Q. mongolica* prefers a relatively higher light condition than *F. crenata* (Hokkaido Forest Tree Breeding Association 2008).

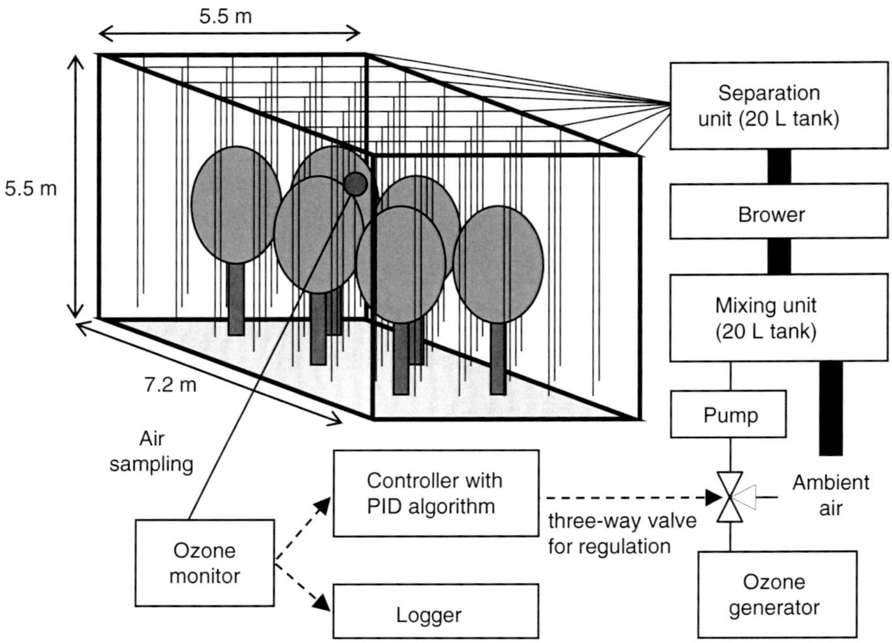

Fig. 5.8 Diagram of free-air ozone fumigation system located in the Sapporo experimental forest, northern Japan (After Watanabe et al. 2013)

Fig. 5.9 Overview of free-air ozone fumigation system at Sapporo experimental forest (*left*), inside of the system (*upper right*), and gas exchange measurement in leaves of *Fagus crenata* saplings (*lower right*)

On the other hand, *B. platyphylla* is a typical pioneer species in Northern Japan. The target O_3 concentration was 60 nmol mol^{-1} during the daylight hours of the growing season (i.e., the leafy period). The O_3 generated from an O_3 generator was mixed with ambient air, using a three-way valve to control the concentration. The air containing O_3 was then diluted with ambient air in a mixing unit and passed into the canopies through 48 fluorine resin tubes hanging from a fixed grid above the trees down to a height of 50 cm above the ground. Each tube has ten holes (2-mm diameter) at 50-cm intervals. The signal of O_3 monitoring at canopy height was used as feedback to the three-way valve so as to regulate the O_3 concentration, using the proportional-integral-differential (PID) algorithm.

When we consider the effects of O_3 on whole-canopy carbon uptake, it is important to understand the differences in O_3 susceptibility between canopy positions. There are considerable variations of leaf traits within a developed canopy. One of the most important factors that determine leaf traits is light intensity. For example, thick leaves with a higher leaf mass per area (LMA) and higher area-based nitrogen content (N_{area}) are produced under high light conditions (e.g. Evans and Poorter 2001; Kitaoka et al. 2009a, b; Poorter et al. 2009; Niinemets et al. 2015). As a result, photosynthetic parameters, such as V_{cmax} and J_{max}, are generally higher in leaves exposed to higher irradiance (Iio et al. 2005; Rodríguez-Calcerrada et al. 2008; Niinemets et al. 2015). Watanabe et al. (2014a, 2015) investigated the response of A_{sat} of upper and lower canopy leaves of *F. crenata* and *Q. mongolica* saplings in a free-air O_3 exposure experiment (Fig. 5.10). There were considerable differences between species. Ozone-induced significant reduction in A_{sat} of *F. crenata* was

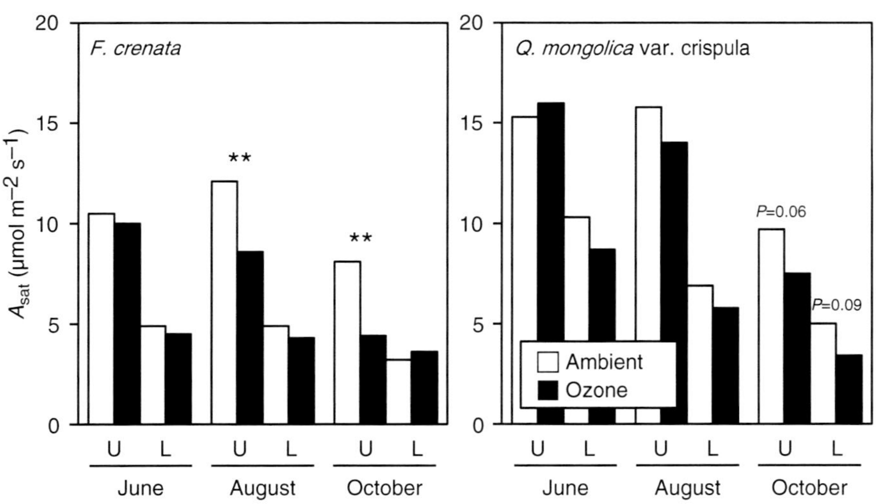

Fig. 5.10 Light-saturated net photosynthetic rate (A_{sat}) in upper (U) and lower (L) canopy leaves of *Fagus crenata* and *Quercus mongolica* var. *crispula* saplings grown under ambient and elevated concentrations of ozone (60 nmol mol^{-1} during daylight hours). Significant reduction in A_{sat} of *F. crenata* under elevated O_3 was observed only in upper canopy leaves (**, $P<0.01$), while A_{sat} in both the upper and lower canopy leaves of *Q. mongolica* showed similar marginal reduction by exposure to O_3 in October (Data source: Watanabe et al. 2014a, 2015)

observed only in upper canopy leaves in August and October, indicating higher O_3 susceptibility in upper canopy leaves. On the other hand, a marginal decrease of A_{sat} under elevated O_3 was found in both the upper and lower canopy leaves of *Q. mongolica* in October. Species-specific differences in O_3 susceptibility between canopy positions have already been indicated in studies in Europe and the United States (Tjoelker et al. 1993, 1995; Kitao et al. 2009). However, there were big variations in environmental conditions among the previous studies that evaluated O_3 sensitivity, and therefore uncertainties were raised regarding species specific comparisons. On the other hand, the findings of Watanabe et al. (2014a, 2015) were obtained from the same experimental site during the same growing season. In addition, relative light intensities in the lower canopy were similar in the two species examined, at around 10–15 %. Therefore, the results strongly confirm the species specificity in the difference of O_3 sensitivity between upper and lower canopy leaves even in the same environmental conditions.

Watanabe et al. (2014a) estimated the canopy carbon budget for one growing season of *F. crenata* saplings under elevated O_3 (Fig. 5.11). They determined the light-response curve of the photosynthetic rate and respiration rate of leaves in various light conditions. Then the canopy-level photosynthetic carbon gain (PCG) and respiratory carbon loss (RCL) were calculated using a canopy photosynthesis model (Monsi and Saeki 1953; Hirose 2005). The canopy net carbon gain (NCG) was reduced by 12.4 % by the exposure to O_3 (60 nmol mol^{-1} during daytime) during one growing season. The main factor in the O_3-induced reduction in NCG in late summer was the increased RCL. However, the contributions of the reduced PCG and the increased RCL to the NCG were almost the same in autumn. Watanabe et al. (2014a) concluded that contributions of changes in PCG and RCL to NCG under elevated O_3 were different between seasons.

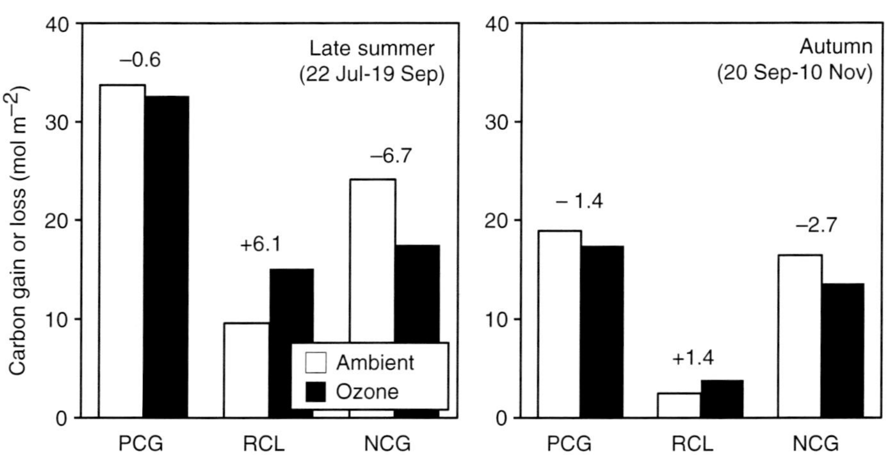

Fig. 5.11 Carbon budget per unit ground area (mol m^{-2}) due to photosynthesis and respiration in canopy of *Fagus crenata* grown under ambient and elevated concentrations of ozone. *PCG* photosynthetic carbon gain, *RCL* respiratory carbon loss, *NCG* net carbon gain. The value on the bar is the difference between ambient and elevated ozone (Data source: Watanabe et al. 2014a)

5.5 Biotic Relations

5.5.1 Defense Capacity Against Biotic Stresses

The defense capacity of plants against disease and insect attack is modified by environmental conditions (e.g. Percy et al. 2002; Oßwald et al. 2012; Watanabe et al. 2014c). There are several lines of evidence showing changes in the concentrations of defense chemicals in leaves under elevated O_3 (Percy et al. 2002; Peltonen et al. 2005). Sakikawa et al. (2016) reported that damage of leaves of *B. platyphylla* saplings by an insect herbivore (mainly by the leaf beetle, *Agelastica coerulea*) was decreased under free-air O_3 exposure. Similar trends were observed in seedlings of the same species by Vanderstock et al. (2016). At the present time, however, information on the effects of O_3 on plant-insect/disease interaction in Japanese forest tree species is very limited. Further research is needed.

5.5.2 Symbiosis with Ectomycorrhizal Fungi

The majority of below-ground root systems in woody plants are symbiotically colonized by ectomycorrhizal (ECM) fungi (Marschner 2012). Up to 30% of the total carbohydrate assimilated through the photosynthetic process can be used in the growth and maintenance of ECM (Hampp and Nehls 2001). In turn, ECM fungi usually act as efficient patronage for the root system of the host plant by absorbing water and essential nutrients, such as phosphorus and sometimes nitrogen (e.g., Quoreshi et al. 2003; Cairney 2011). Ozone usually has negative effects on photosynthetic assimilation. As a result, the amount of carbon allocated to below-ground systems is reduced (Grantz and Farrar 2000; King et al. 2005). These effects are expected to decrease ECM colonization and to affect species-host compatibility. In fact, there is evidence of a reduction in ECM colonization caused by exposure to O_3. In the O_3 fumigation study reported by Yamaguchi et al. (2007, 2010a), the ECM colonization rates (the ratio of ECM fine root tip to all fine root tip) in *F. crenata* seedlings were also determined, and a trend of significant decrease in the ECM colonization rate was found (Fig. 5.12). Wang et al. (2015) also reported a reduction in ECM colonization under elevated O_3 (60 nmol mol^{-1} during daylight hours) in hybrid larch F_1 (*Larix gmelinii* var. *japonica* × *L. kaempferi*) seedlings. Furthermore, they revealed changes in the composition of the ECM community under elevated O_3 – O_3 decreased the fraction of *Tomentella* sp., while the fraction of *Cadophora finlandia* was increased.

5.6 Future Perspectives

A relatively large data set is available for the effects of O_3 on forest tree species in Japan, compared with data for Europe and North America. However, a large part of the data was obtained from juvenile seedlings under controlled chamber conditions

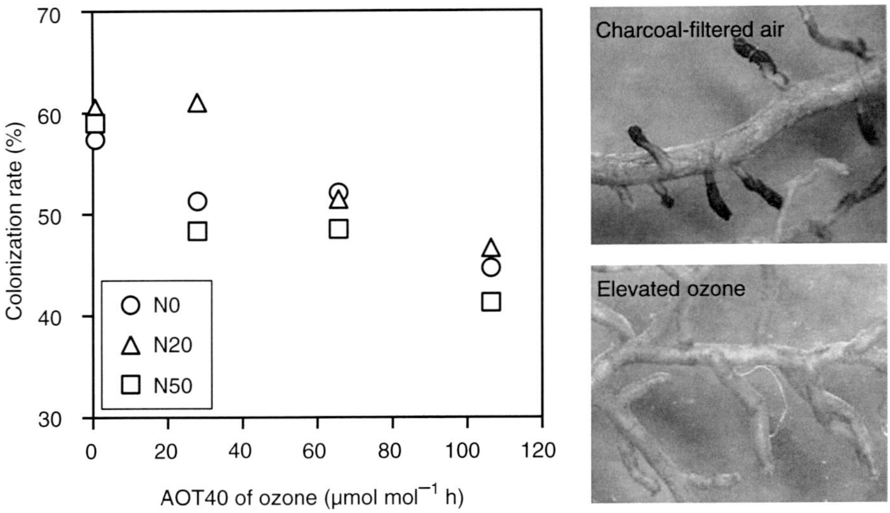

Fig. 5.12 The relationship between accumulated exposure over a threshold of 40 nmol mol^{-1} (AOT40) of ozone and the rate of ectomycorrhizal colonization in the fine root tips of *Fagus crenata* seedlings (*left*). Seedlings were exposed to four levels of ozone in combination with three levels of nitrogen loading: 0, 20, and 50 kg ha^{-1} year^{-1} (N0, N20, and N50, respectively). Fine root tips of *F. crenata* seedlings grown under charcoal-filtered air (*upper* photo) and elevated ozone (*lower* photo). Little ectomycorrhizal colonization was observed in the root tips of *F. crenata* seedlings grown under elevated ozone (Yamaguchi et al. (unpublished data))

with short-term exposure (1–3 years). It is clear that long-term (>10 years) free-air O_3 fumigation study with mature trees at the vegetation scale will be the direction of study in the near future. On the other hand, an epidemiological approach would also be useful for understanding the impact of O_3 on large trees in a forest (e.g. Karlsson et al. 2006). The free-air fumigation method requires expensive experimental facilities for realistic O_3 simulation and only a few individual trees can usually be investigated. Studies for understanding the physiological mechanisms of O_3 effects in an environmental control chamber are, of course, still important.

References

Azuchi F, Kinose Y, Matsumura T, Kanomata T, Uehara Y, Kobayashi A, Yamaguchi M, Izuta T (2014) Modeling stomatal conductance and ozone uptake of *Fagus crenata* grown under different nitrogen loads. Environ Pollut 184:481–487

Ball JT, Woodrow IE, Berry JA (1987) A model predicting stomatal conductance and its contribution to the control of photosynthesis under different environmental conditions. In: Biggens J (ed) Progress in photosynthesis research. Martinus-Nijhoff Publishers, Dordrecht, pp 221–224

Brendley BW, Pell EJ (1998) Ozone-induced changes in biosynthesis of Rubisco and associated compensation to stress in foliage of hybrid poplar. Tree Physiol 18:81–90

Büker P, Feng Z, Uddling J, Briolat A, Alonso R, Braun S, Elvira S, Gerosa G, Karlsson PE, Le Thiec D, Marzuoli R, Mills G, Oksanen E, Wieser G, Wilkinson M, Emberson LD (2015) New flux based dose–response relationships for ozone for European forest tree species. Environ Pollut 206:163–174

Cairney JWG (2011) Ectomycorrhizal fungi: the symbiotic route to the root for phosphorus in forest soils. Plant and Soil 344:51–71

Díaz-de-Quijano M, Schaub M, Bassin S, Volk M, Peñuelas J (2012) Ozone visible symptoms and reduced root biomass in the subalpine species *Pinus uncinata* after two years of free-air ozone fumigation. Environ Pollut 169:250–257

Dumont J, Cohen D, Gérard J, Jolivet Y, Dizengremel P, Le Thiec D (2014) Distinct responses to ozone of abaxial and adaxial stomata in three Euramerican poplar genotypes. Plant Cell Environ 37:2064–2076

Evans JR (1989) Photosynthesis and nitrogen relationships in leaves of C3 plants. Oecologia 78:9–19

Evans JR, Poorter H (2001) Photosynthetic acclimation of plants to growth irradiance: the relative importance of specific leaf area and nitrogen partitioning in maximizing carbon gain. Plant Cell Environ 24:755–767

Evans JR, Seemann J (1989) The allocation of protein nitrogen in the photosynthetic apparatus: cost, consequences and control. In: Briggs WR (ed) Photosynthesis. Alan R. Liss, New York, pp 183–205

Farquhar GD, von Caemmerer S, Berry JA (1980) A biochemical model of photosynthetic CO_2 assimilation in leaves of C3 species. Planta 149:78–90

Fine PVA, Ree RH (2006) Evidence for a time-integrated species-area effect on the latitudinal gradient in tree diversity. Am Nat 168:796–804

Fuhrer J, Booker F (2003) Ecological issues related to ozone: agricultural issues. Environ Int 29:141–154

Fujinuma Y, Furukawa A, Totsuka T, Tazaki T (1987) Uptake of ozone by various street trees. Environ Cont Biol 25:31–40

Furukawa A (1991) Inhibition of photosynthesis of *Populus euramericana* and *Helianthus annuus* by SO_2, NO_2 and O_3. Ecol Res 6:79–86

Furukawa A, Totsuka T, Katase M, Ushijima T (1983) Inhibition of photosynthesis of poplar species by ozone. J Jpn For Soc 65:321–326

Grantz DA, Farrar JF (2000) Ozone inhibits phloem loading from a transport pool: compartmental efflux analysis in Pima cotton. Aust J Plant Physiol 27:859–868

Hampp R, Nehls U (2001) Physiology of tree root/fungus symbiosis. In: Huttunen S, Heikkilä H, Bucher J, Sundberg B, Jarvis P, Matyssek R (eds) Trends in European forest tree physiology research. Kluwer Academic Publishers, Dordrecht, pp 53–62

Hirose T (2005) Development of the Monsi-Saeki theory on canopy structure and function. Ann Bot 95:483–494

Hokkaido Forest Tree Breeding Association (2008) Forest tree breeding and forest genetic resources in Hokkaido. Hokkaido Forest Tree Breeding Association, Ebetsu (in Japanese)

Hoshika Y, Paoletti E, Omasa K (2012a) Parameterization of *Zelkova serrata* stomatal conductance model to estimate stomatal ozone uptake in Japan. Atmos Environ 55:271–278

Hoshika Y, Wanatabe M, Inada N, Koike T (2012b) Ozone-induced stomatal sluggishness develops progressively in Siebold's beech (*Fagus crenata*). Environ Pollut 166:152–156

Hoshika Y, Watanabe M, Inada N, Koike T (2012c) Modeling of stomatal conductance for estimating ozone uptake of (*Fagus crenata*) under experimentally enhanced free-air ozone exposure. Water Air Soil Pollut 223:3893–3901

Hoshika Y, Watanabe M, Inada N, Koike T (2013a) Model-based analysis of avoidance of ozone stress by stomatal closure in Siebold's beech (*Fagus crenata*). Ann Bot 112:1149–1158

Hoshika Y, Watanabe M, Inada N, Mao Q, Koike T (2013b) Photosynthetic response of early and late leaves of white birch (*Betula platyphylla* var. *japonica*) grown under free-air ozone exposure. Environ Pollut 182:242–247

Hoshika Y, Katata G, Deushi M, Watanabe M, Koike T, Paoletti E (2015a) Ozone-induced stomatal sluggishness changes carbon and water balance of temperate deciduous forests. Sci Rep 5:9871
Hoshika Y, Watanabe M, Inada N, Koike T (2015b) Effects of ozone-induced stomatal closure on ozone uptake and its changes due to leaf age in sun and shade leaves of Siebold's beech. J Agric Meteorol 71:218–226
Hoshika Y, Watanabe M, Kitao M, Haberle K-H, Grams TEE, Koike T, Matyssek R (2015c) Ozone induces stomatal narrowing in European and Siebold's beeches: a comparison between two experiments of free-air ozone exposure. Environ Pollut 196:527–533
Iio A, Fukasawa H, Nose Y, Kato S, Kakubari Y (2005) Vertical, horizontal and azimuthal variations in leaf photosynthetic characteristics within a *Fagus crenata* crown in relation to light acclimation. Tree Physiol 25:533–544
Izuta T (1998) Ecophysiological responses of Japanese forest tree species to ozone, simulated acid rain and soil acidification. J Plant Res 111:471–480
Izuta T (2003) Air pollution impacts on vegetation in Japan. In: Emberson L, Ashmore M, Murray F (eds) Air pollution impacts on crops and forests: a global assessment. Imperial College Press, London, pp 89–101
Izuta T (2006) Plants and environmental stresses. Corona Publishing, Tokyo (in Japanese)
Izuta T (2012) Evaluation of ozone effects on plant based on leaf ozone uptake. J Jpn Soc Atmos Environ 47:A12–A15 (in Japanese)
Izuta T, Umemoto M, Horie K, Aoki M, Totsuka T (1996) Effects of ambient levels of ozone on growth, gas exchange rates and chlorophyll contents of *Fagus crenata* seedlings. J Jpn Soc Atmos Environ 31:95–105
Jacob DJ (1999) Introduction to atmospheric chemistry. Harvard University Press, Cambridge, MA
Jarvis PG (1976) Interpretation of variations in leaf water potential and stomatal conductance found in canopies in field. Philos Trans Royal Soc London B 273:593–610
Kangasjärvi J, Jaspers P, Kollist H (2005) Signalling and cell death in ozone-exposed plants. Plant Cell Environ 28:1021–1036
Karlsson PE, Uddling J, Braun S, Broadmeadow M, Elvira S, Gimeno BS, Le Thiec D, Oksanen E, Vandermeiren K, Wilkinson M, Emberson L (2004) New critical levels for ozone effects on young trees based on AOT40 and stimulated cumulative leaf uptake of ozone. Atmos Environ 38:2283–2294
Karlsson PE, Örlander G, Langvall O, Uddling J, Hjorth U, Wiklander K, Areskoug B, Grenfelt P (2006) Negative impact of ozone on the stem basal area increment of mature Norway spruce in south Sweden. For Ecol Manage 232:146–151
Karlsson PE, Braun S, Broadmeadow M, Elvira S, Emberson L, Gimeno BS, Le Thiec D, Novak K, Oksanen E, Schaub M, Uddling J, Wilkinson M (2007) Risk assessments for forest trees: the performance of the ozone flux versus the AOT concepts. Environ Pollut 146:608–616
Karnosky D-F, Werner H, Holopainen T, Percy K, Oksanen T, Oksanen E, Heerdt C, Fabian P, Nagy J, Heilman W, Cox R, Nelson N, Matyssek R (2007) Free-air exposure systems to scale up ozone research to mature trees. Plant Biol 9:181–190
King JS, Kubiske ME, Pregitzer KS, Hendrey GR, McDonald EP, Giardina CP, Quinn VS, Karnosky DF (2005) Tropospheric O_3 compromises net primary production in young stands of trembling aspen, paper birch and sugar maple in response to elevated atmospheric CO_2. New Phytol 168:623–635
Kinose Y, Azuchi F, Uehara Y, Kanomata T, Kobayashi A, Yamaguchi M, Izuta T (2014) Modeling of stomatal conductance to estimate stomatal ozone uptake by *Fagus crenata*, *Quercus serrata*, *Quercus mongolica* var. *crispula* and *Betula platyphylla*. Environ Pollut 194:235–245
Kitao M, Löw M, Heerdt C, Grams TEE, Häberle K-H, Matyssek R (2009) Effects of chronic elevated ozone exposure on gas exchange responses of adult beech trees (*Fagus sylvatica*) as related to the within-canopy light gradient. Environ Pollut 157:537–544
Kitao M, Komatsu M, Yazaki K, Kitaoka S, Tobita H (2015) Growth overcompensation against O_3 exposure in two Japanese oak species, *Quercus mongolica* var. *crispula* and *Quercus serrata*, grown under elevated CO_2. Environ Pollut 206:133–141

Kitaoka S, Koike T (2004) Invasion of broad-leaf tree species into a larch plantation: seasonal light environment, photosynthesis and nitrogen allocation. Physiol Plant 121:604–611
Kitaoka S, Watanabe M, Watanabe Y, Kayama M, Nomura M, Sasa K (2009a) Growth of regenerated tree seedlings associated with microclimatic change in a mature larch plantation after harvesting. Landsc Ecol Eng 5:137–145
Kitaoka S, Watanabe Y, Koike T (2009b) The effects of cleared larch canopy and nitrogen supply on gas exchange and leaf traits in deciduous broad-leaved tree seedlings. Tree Physiol 29:1503–1511
Kohno Y, Matsumura H (1999) Combined effect of simulated acid rain and ozone on the growth of Japanese conifer seedlings. J Jpn Soc Atmos Environ 34:74–85 (in Japanese)
Kohno Y, Matsumura H, Ishii T, Izuta T (2005) Establishing critical levels of air pollutants for protecting East Asian vegetation, a challenge. In: Omasa K, Nouchi I, De Kok LJ (eds) Plant responses to air pollution and global change. Springer, Tokyo, pp 243–250
Koike T, Mao Q, Inada N, Kawaguchi K, Hoshika Y, Kita K, Watanabe M (2012) Growth and photosynthetic responses of cuttings of a hybrid larch (*Larix gmelinii* var. *japonica* × *L. kaempferi*) to elevated ozone and/or carbon dioxide. Asian J Atmos Environ 6:104–110
Koike T, Watanabe M, Hoshika Y, Kitao M, Matsumura H, Funada R, Izuta T (2013) Effects of ozone on forest ecosystems in East and Southeast Asia. In: Matyssek R, Clarke N, Cudlin P, Mikkelsen TN, Tuovinen J-P, Wieser G, Paoletti E (eds) Climate change, air pollution and global challenges: understanding and perspectives from forest research. Elsevier, Oxford, pp 371–390
Kolb TE, Matyssek R (2001) Limitations and perspectives about scaling ozone impacts in trees. Environ Pollut 115:373–392
Lambers H III, Chapin FS, Pons TL (2008) Plant physiological ecology, 2nd edn. Springer, New York
Landolt W, Günthardt-Goerg MS, Pfenninger I, Einig W, Hampp R, Maurer S, Matyssek R (1997) Effect of fertilization on ozone-induced change in the metabolism of birch (*Betula pendula*) leaves. New Phytol 137:389–397
Marschner P (2012) Marschner's mineral nutrition of higher plants, 3rd ed. Elsevier, London
Matsumura H (2001) Impacts of ambient ozone and/or acid mist on the growth of 14 tree species: an open-top chamber study conducted in Japan. Water Air Soil Pollut 130:959–964
Matsumura H, Kohno Y (2003) Effects of sulfur dioxide and/or ozone on Japanese evergreen broad-leaved tree species. Research report of Central Research Institute of Electric Power Industry U02021. (in Japanese with English summary)
Matsumura H, Aoki H, Kohno Y, Izuta T, Totsuka T (1996) Effects of ozone on dry weight growth and gas exchange rate of Japanese cedar, Japanese cypress and Japanese zelkova seedlings. J Jpn Soc Atmos Environ 31:247–261 (in Japanese)
Matsumura H, Kobayashi T, Kohno Y (1998) Effects of ozone and/or simulated acid rain on dry weight and gas exchange rates of Japanese cedar, Nikko fir, Japanese white birch and Japanese zelkova seedlings. J Jpn Soc Atmos Environ 33:16–35 (in Japanese)
Matsumura H, Mikami C, Sakai Y, Murayama K, Izuta T, Yonekura T, Miwa M, Kohno Y (2005) Impacts of elevated O_3 and/or CO_2 on growth of *Betula platyphylla*, *Betula ermanii*, *Fagus crenata*, *Pinus densiflora* and *Cryptomeria japonica* seedlings. J Agric Meteorol 60: 1121–1124
Matyssek R, Sandermann H (2003) Impact of ozone on trees: an ecophysiological perspective. Prog Bot 64:349–404
Matyssek R, Bytnerowicz A, Karlsson PE, Paoletti E, Sanz M, Schaub M, Wieser G (2007) Promoting the O_3 flux concept for European forest trees. Environ Pollut 146:587–607
Mills G, Hayes F, Wilkinson S, Davies WJ (2009) Chronic exposure to increasing background ozone impairs stomatal functioning in grassland species. Glob Chang Biol 15:1522–1533
Mills G, Pleijel H, Büker P, Braun S, Emberson LD, Harmens H, Hayes F, Simpson D, Grünhage L, Karlsson PE, Danielsson H, Bermejo V, Gonzalez Fernandez I (2010) Mapping Critical Levels for Vegetation. Revision undertaken in Summer 2010 to include new flux-based critical

levels and response functions for ozone, in: Mapping Manual 2004. International Cooperative Programme on Effects of Air Pollution on Natural Vegetation and Crops
Miwa M, Izuta T, Totsuka T (1993) Effects of simulated acid rain and/or ozone on the growth of Japanese cedar seedlings. J Jpn Soc Atmos Environ 28:279–287 (in Japanese)
Monsi M, Saeki T (1953) Über den Lichtfaktor in den Pflanzengesellschaften und seine Bedeutung für die Stoffproduktion. Jpn J Bot 14:22–52
Nakaji T, Izuta T (2001) Effects of ozone and/or excess soil nitrogen on growth, needle gas exchange rates and Rubisco contents of *Pinus densiflora* seedlings. Water Air Soil Pollut 130:971–976
Nakaji T, Kobayashi T, Kuroha M, Omori K, Matsumoto Y, Yonekura T, Watanabe K, Utriainen J, Izuta T (2004) Growth and nitrogen availability of red pine seedlings under high nitrogen load and elevated ozone. Water, Air, Soil Pollut, Focus 4:277–287
Niinemets U, Keenan TF, Hallik L (2015) A worldwide analysis of within-canopy variations in leaf structural, chemical and physiological traits across plant functional types. New Phytol 205:973–993
Nouchi I (2001) Plant responses to changing environment. Yokendo co. ltd., Tokyo (in Japanese)
Nouchi I, Odaira T, Sawada T, Oguchi K, Komeiji T (1973) Plant ozone injury symptoms. J Jpn Soc Air Pollut 8:113–119 (In Japanese with English summary)
Nunn AJ, Reiter UM, Häberle K-H, Langebartels C, Bahnweg G, Pretzsch H, Sandermann H, Matyssek R (2005) Response patterns in adult forest trees to chronic ozone stress: identification of variations and consistencies. Environ Pollut 136:365–369
Ohwi J (1983) New atlas of Japanese plants. Shibundo, Tokyo (In Japanese)
Oksanen E, Saleen A (1999) Ozone exposure results in various carry-over effects and prolonged reduction in biomass in Birch (*Betula pendula* Roth). Plant Cell Environ 22:1401–1411
Oksanen E, Kontunen-Soppela S, Riikonen J, Peltonen P, Uddling J, Vapaavuori E (2007) Northern environment predisposes birches to ozone damage. Plant Biol 9:191–196
Omasa K (1990) Study on changes in stomata and their surrounding cell using a non-destructive light microscope system: responses to air pollutants. J Agric Meteorol 45:251–257
Oßwald W, Fleischmann F, Treutter D (2012) Host-parasite interaction and trade-offs between growth – and defence-related metabolism under changing environments. In: Matyssek R, Schnyer H, Oßwald W, Ernst D, Munch JC, Pretzsch H (eds) Growth and defence in plants: resource allocation at multiple scales. Springer, Berlin, pp 53–83
Paludan-Müller G, Saxe H, Leverenz JW (1999) Responses to ozone in 12 provenances of European beech (*Fagus sylvatica*): genotypic variation and chamber effects on photosynthesis and dry-matter partitioning. New Phytol 144:261–273
Paoletti E, Grulke NE (2005) Does living in elevated CO_2 ameliorate tree response to ozone? A review on stomatal responses. Environ Pollut 137:483–493
Pell EJ, Temple PJ, Friend AL, Mooney HA, Winner WE (1994) Compensation as a plant response to ozone and associated stresses: an analysis of ROPIS experiments. J Environ Qual 23:429–436
Pell EJ, Sinn JP, Brendley BW, Samuelson L, Vinten-Johansen C, Tien M, Skillman J (1999) Differential response of four tree species to ozone-induced acceleration of foliar senescence. Plant Cell Environ 22:779–790
Peltonen PA, Vapaavuori E, Julkunen-Tiitto R (2005) Accumulation of phenolic compounds in birch leaves is changed by elevated carbon dioxide and ozone. Glob Chang Biol 11:1305–1324
Percy KE, Awmack CS, Lindroth RL, Kubiske ME, Kopper BJ, Isebrands JG, Pregitzer KS, Hendrey GR, Dickson RE, Zak DR, Oksanen E, Sober J, Harrington R, Karnosky DF (2002) Altered performance of forest pests under CO_2 – and O_3-enriched atmospheres. Nature 420:403–407
Poorter H, Niinemets Ü, Poorter L, Wright IJ, Villar R (2009) Causes and consequences of variation in leaf mass per area (LMA): a meta-analysis. New Phytol 182:565–588
Pretzsch H, Dieler J, Matyssek R, Wipfler P (2010) Tree and stand growth of mature Norway spruce and European beech under long-term ozone fumigation. Environ Pollut 158: 1061–1070

Quoreshi AM, Maruyama Y, Koike T (2003) The role of mycorrhiza in forest ecosystems under CO_2-enriched atmosphere. Eurasian J For Res 6:171–176
Reich PB (1987) Quantifying plant response to ozone: a unifying theory. Tree Physiol 3:63–91
Rodríguez-Calcerrada J, Reich PB, Rosenqvist E, Pardos JA, Cano FJ, Aranda I (2008) Leaf physiological versus morphological acclimation to high-light exposure at different stages of foliar development in oak. Tree Physiol 28:761–771
Sage RF (1994) Acclimation of photosynthesis to increasing atmospheric CO_2: the gas exchange perspective. Photosynth Res 39:351–368
Sakikawa T, Shi C, Nakamura M, Watanabe M, Oikawa M, Satoh F, Koike T (2016) Leaf phenology and insect grazing of Japanese white birch saplings grown under free-air ozone exposure. J Agric Meteorol 72:80–84
Shan Y, Izuta T, Aoki M, Totsuka T (1997) Effects of O_3 and soil acidification, alone and in combination, on growth, gas exchange rate and chlorophyll content of red pine seedlings. Water Air Soil Pollut 97:355–366
Sharkey TD (1985) Photosynthesis in intact leaves of C3 plants: physics, physiology and rate limitations. Bot Rev 51:53–105
Takagi K, Ohara T (2003) Estimation of ozone impact on plants by damage functions in the Kanto area. J Jpn Soc Atmos Environ 38:205–216 (in Japanese with English summary)
Tanaka Y, Sano T, Tamaoki M, Nakajima N, Kondo N, Hasezawa S (2005) Ethylene inhibits abscisic acid-induced stomatal closure in *Arabidopsis*. Plant Physiol 138:2337–2343
Terazawa K, Koyama H (2008) Applied ecology for restoration of beech forests. Bun-ichi Sogo Shuppan, Tokyo (in Japanese)
Tjoelker MG, Volin JC, Oleksyn J, Reich PB (1993) Light environment alters response to ozone stress in seedlings of *Acer saccharum* Marsh. and hybrid *Populus* L. I. In situ net photosynthesis, dark respiration and growth. New Phytol 124:627–636
Tjoelker MG, Volin JC, Oleksyn J, Reich PB (1995) Interaction of ozone pollution and light effects on photosynthesis in a forest canopy experiment. Plant Cell Environ 18:895–905
Vanderstock A, Agathokleous E, Inoue W, Eguchi N, Nakamura M, Satoh F, Kanie S, Koike T (2016) Preliminary survey on insect grazing in white birch stands under free-air O_3 fumigation. Boreal For Res 64:27–29
Wang X, Qu L, Mao Q, Watanabe M, Hoshika Y, Koyama A, Kawaguchi K, Tamai Y, Koike T (2015) Ectomycorrhizal colonization and growth of the hybrid larch F_1 under elevated O_3 and CO_2. Environ Pollut 197:116–126
Watanabe M, Yamaguchi M (2011) Risk assessment of ozone impact on 6 Japanese forest tree species with consideration of nitrogen deposition. Jpn J Ecol 61:89–96 (in Japanese)
Watanabe M, Yamaguchi M, Iwasaki M, Matsuo N, Naba J, Tabe C, Matsumura H, Kohno Y, Izuta T (2006) Effects of ozone and/or nitrogen load on the growth of *Larix kaempferi*, *Pinus densiflora* and *Cryptomeria japonica* seedlings. J Jpn Soc Atmos Environ 41:320–334
Watanabe M, Yamaguchi M, Tabe C, Iwasaki M, Yamashita R, Funada R, Fukami M, Matsumura H, Kohno Y, Izuta T (2007) Influences of nitrogen load on the growth and photosynthetic responses of *Quercus serrata* seedlings to O_3. Trees 21:421–432
Watanabe M, Yamaguchi M, Matsumura H, Kohno Y, Izuta T (2008) Effects of ozone on growth and photosynthesis of *Castanopsis sieboldii* seedlings grown under different nitrogen loads. J Agric Meteorol 24:143–155
Watanabe M, Matsuo N, Yamaguchi M, Matsumura H, Kohno Y, Izuta T (2010a) Risk assessment of ozone impact on the carbon absorption of Japanese representative conifers. European J For Res 129:421–430
Watanabe M, Umemoto-Yamaguchi M, Koike T, Izuta T (2010b) Growth and photosynthetic response of *Fagus crenata* seedlings to ozone and/or elevated carbon dioxide. Landsc Ecol Eng 6:181–190
Watanabe M, Yamaguchi M, Matsumura H, Kohno Y, Izuta T (2012) Risk assessment of ozone impact on *Fagus crenata* in Japan: consideration of atmospheric nitrogen deposition. European J For Res 131:475–484

Watanabe M, Hoshika Y, Inada N, Wang X, Mao Q, Koike T (2013) Photosynthetic traits of Siebold's beech and oak saplings grown under free air ozone exposure in northern Japan. Environ Pollut 174:50–56
Watanabe M, Hoshika Y, Inada N, Koike T (2014a) Canopy carbon budget of Siebold's beech (*Fagus crenata*) sapling under free air ozone exposure. Environ Pollut 184:682–689
Watanabe M, Hoshika Y, Koike T (2014b) Photosynthetic responses of Monarch birch seedlings to differing timing of free air ozone fumigation. J Plant Res 127:339–345
Watanabe M, Kitaoka S, Eguchi N, Watanabe Y, Satomura T, Takagi K, Satoh F, Koike T (2014c) Photosynthetic traits and growth of *Quercus mongolica* var. *crispula* sprouts attacked by powdery mildew under free air CO_2 enrichment. European J For Res 133:725–733
Watanabe M, Hoshika Y, Inada N, Koike T (2015) Difference in photosynthetic responses to free air ozone fumigation between upper and lower canopy leaves of Japanese oak (*Quercus mongolica* var. *crispula*) saplings. J Agric Meteorol 71:227–231
Wilkinson S, Davies W (2010) Drought, ozone, ABA and ethylene: new insights from cell to plant community. Plant Cell Environ 33:510–525
Wittig VE, Ainsworth EA, Long SP (2007) To what extent do current and projected increases in surface ozone affect photosynthesis and stomatal conductance of trees? A meta-analytic review of the last 3 decades of experiments. Plant Cell Environ 30:1150–1162
Wittig VE, Ainsworth EA, Naidu SL, Karnosky DF, Long SP (2009) Quantifying the impact of current and future tropospheric ozone on tree biomass, growth, physiology and biochemistry: a quantitative meta-analysis. Glob Chang Biol 15:396–424
Yamaguchi M, Watanabe M, Iwasaki M, Tabe C, Matsumura H, Kohno Y, Izuta T (2007) Growth and photosynthetic responses of *Fagus crenata* seedlings to O_3 under different nitrogen loads. Trees 21:707–718
Yamaguchi M, Watanabe M, Matsumura H, Kohno Y, Izuta T (2010a) Effects of ozone on nitrogen metabolism in the leaves of *Fagus crenata* seedlings under different soil nitrogen loads. Trees 24:175–184
Yamaguchi T, Noguchi I, Eguchi M (2010b) Ambient ozone concentration around Lake Mashu, Hokkaido, Japan. Trans Meet Hokkaido Branch Jpn For Soc 58:123–124 (In Japanese)
Yamaguchi M, Watanabe M, Matsumura H, Kohno Y, Izuta T (2011) Experimental studies on the effects of ozone on growth and photosynthetic activity of Japanese forest tree species. Asian J Atmos Environ 5:65–78
Yonekura T, Dokiya Y, Fukami M, Izuta T (2001a) Effects of ozone and/or soil water stress on growth and photosynthesis of *Fagus crenata* seedlings. Water Air Soil Pollut 130:965–970
Yonekura T, Honda Y, Oksanen E, Yoshidome M, Watanabe M, Funada R, Koike T, Izuta T (2001b) The influences of ozone and soil water stress, singly and in combination, on leaf gas exchange rates, leaf ultrastructural characteristics and annual ring width of *Fagus crenata* seedlings. J Jpn Soc Atmos Environ 36:333–351
Yonekura T, Yoshidome M, Watanabe M, Honda Y, Ogiwara I, Izuta T (2004) Carry-over effects of ozone and water stress on leaf phenological characteristics and bud frost hardiness of *Fagus crenata* seedlings. Trees 18:581–588
Zhang W, Feng Z, Wang X, Niu J (2012) Responses of native broadleaved woody species to elevated ozone in subtropical China. Environ Pollut 163:149–157

Chapter 6
Combined Effects of Ozone and Other Environmental Factors on Japanese Trees

Makoto Watanabe, Yasutomo Hoshika, Takayoshi Koike, and Takeshi Izuta

Abstract Plant responses to ozone (O_3) are highly dependent on other environmental factors. Elevated atmospheric CO_2 concentrations may reduce stomatal O_3 uptake in some tree species in certain environmental conditions. In addition, changes in the carbon availability and allocation pattern within a plant body under elevated CO_2 may confer compensative capacity against O_3 stress. The influence of the soil nitrogen load on growth and photosynthetic response to O_3 is highly species-specific. The soil nitrogen load enhanced the negative impact of O_3 on *Fagus crenata*, whereas lower O_3-induced growth reduction of *Larix kaempferi* was observed under a higher nitrogen load. It is predicted that soil-water stress would induce lower stomatal O_3 uptake due to stomatal closure, and therefore the negative impacts of O_3 on trees would decrease under water-stress conditions. In fact, some antagonistic effects of O_3 and soil water stress were observed in *F. crenata*, although these effects may depend on the severity of the water stress. The number of studies on the combined effects of O_3 and other environmental factors on Japanese forest tree species is limited and further research is crucial to accumulate the required datasets.

Keywords Ozone • Japanese forest tree species • Elevated CO_2 • Soil nitrogen load • Soil water stress

---

M. Watanabe (✉) • T. Izuta
Institute of Agriculture, Tokyo University of Agriculture and Technology, Fuchu, Tokyo 183-8509, Japan
e-mail: nab0602@cc.tuat.ac.jp; izuta@cc.tuat.ac.jp

Y. Hoshika
Institute for Sustainable Plant Protection, National Research Council of Italy, Via Madonna del Piano, Sesto Fiorentino 50019, Italy
e-mail: hoshika0803@gmail.com

T. Koike
Silviculture and Forest Ecological Studies, Hokkaido University, Sapporo, Hokkaido 060-8589, Japan
e-mail: tkoike@for.agr.hokudai.ac.jp

T. Izuta (ed.), *Air Pollution Impacts on Plants in East Asia*,
DOI 10.1007/978-4-431-56438-6_6

6.1 Introduction

Forest ecosystems are suffering not only from a single stress factor but also from multiple stresses simultaneously. Since the industrial revolution, human activity has drastically changed our atmospheric environment. In particular, elevated atmospheric CO_2 concentrations, increasing nitrogen depositions in forests, and changes in water availability as a result of climate change are critical factors that affect forest health (Paoletti et al. 2010). Plant responses to O_3 are highly dependent on other environmental factors. The effects of O_3 on the eco-physiological traits of forest trees might be modified by these other factors. Here, we introduce experimental studies that deal with the combined effects of O_3 and elevated atmospheric CO_2, soil nitrogen load, and soil water stress on Japanese forest tree species.

6.2 Combined Effects of O_3 and Elevated Atmospheric CO_2

Atmospheric CO_2 levels have increased dramatically since the industrial revolution and have now reached 400 μmol mol^{-1} (Monastersky 2013). This increase is predicted to continue throughout this century (Stocker et al. 2013). In contrast to the effects of O_3, elevated CO_2 may enhance tree growth, if other factors such as nutrient and water availability do not constrain growth (e.g. Eguchi et al. 2004; Norby et al. 2010; Kitao et al. 2005). Whether recent increasing CO_2 concentrations ameliorate the negative impact of O_3 is a subject of continuous debate (Matyssek and Sandermann 2003). Generally, a higher atmospheric CO_2 concentration induces stomatal closure and therefore decreases O_3 uptake through stomata (Ainsworth and Rogers 2007). Greater amounts of carbohydrate due to a higher photosynthetic rate under elevated CO_2 may confer better detoxification and repair capacities against O_3 stress (Riikonen et al. 2004). Several studies of the combined effects of O_3 and elevated CO_2 on Japanese forest tree species have been carried out based on these hypotheses, mainly in this century.

Matsumura et al. (2005) conducted a study on the combined effects of elevated O_3 and CO_2 on the growth of three broad-leaved deciduous trees [*Betula platyphylla* var *japonica* (hereafter *B. platyphylla*), *Betula ermanii*, and *Fagus crenata*] and two evergreen conifers (*Pinus densiflora* and *Cryptomeria japonica*). The seedlings were exposed to ambient or elevated O_3 and CO_2 (1.5 times ambient concentrations of both gases) during two growing seasons, in open-top chambers. Significant interaction between elevated O_3 and CO_2 treatments was found for the whole-plant dry mass of *B. platyphylla* at the end of the experiment. While the exposure to O_3 significantly decreased the dry-matter growth of this species under ambient conditions, the negative effect of O_3 was not observed under elevated CO_2. However, there were no significant interactions of elevated O_3 and CO_2 treatments for the other four tree species.

The amelioration of O_3 effects under elevated CO_2 was also observed in the study reported by Koike et al. (2012). They studied the effects of elevated O_3 (60 nmol mol^{-1} for 7 h per day) and CO_2 concentrations (600 μmol mol^{-1} for daylight hours)

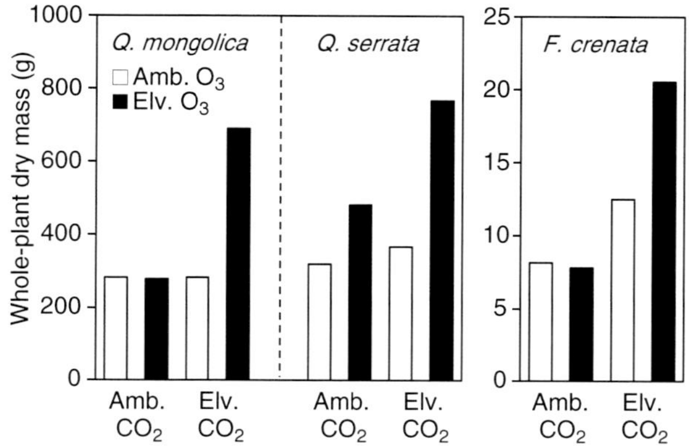

Fig. 6.1 Whole-plant dry mass of *Quercus mongolica* var. *crispula* and *Quercus serrata* and *Fagus crenata* seedlings grown under elevated ozone (O_3) and CO_2 concentrations. See detail for the setup of the experiments in the main text (Data sources: Kitao et al. 2015 for *Q. mongolica* and *Q. serrata* and Watanabe et al. 2010 for *F. crenata*)

on the growth and photosynthesis of the hybrid larch F_1 (*Larix gmelinii* var. *japonica* × *Larix kaempferi*) and its parental (i.e., *L. gmelinii* var. *japonica* and *L. kaempferi*) seedlings, using open-top chambers located at Sapporo, northern Japan. Ozone reduced the height and diameter increments of the main stem, and the needle dry mass of the hybrid larch F_1, while there was almost no effect of O_3 on *L. gmelinii* var. *japonica* and *L. kaempferi*. The decrease in diameter increment of the hybrid larch F_1 was only observed under ambient CO_2 conditions. Under elevated CO_2 conditions, no difference in diameter increments between ambient and elevated O_3 treatments was observed. Reports from another experiment at the same facility in Sapporo indicated no elevated CO_2-induced amelioration of O_3 impact in *B. ermanii*, *B. platyphylla*, *Betula maximowicziana*, and hybrid larch F_1 seedlings (Hoshika et al. 2012; Wang et al. 2015).

In the study of Kitao et al. (2015), two mid-successional oak species, *Quercus mongolica* var. *crispula* (hereafter *Q. mongolica*) and *Quercus serrata* seedlings were grown and fumigated in conditions of elevated O_3 (twice-ambient concentration) in combination with elevated CO_2 (550 μmol mol^{-1}) in an octagonal free-air fumigation system 2 m in height for two growing seasons. The seedlings were directly planted on the ground to avoid any limitation of root growth. Both the oak species showed a significant enhancement of whole-plant dry mass under the combination of elevated O_3 and CO_2 treatments (Fig. 6.1), which was considered to be a result of a preferable biomass partitioning into leaves induced by O_3 (see Sect. 5.2.1.3) and a predominant enhancement of photosynthesis under elevated CO_2. This over-compensative response in the two Japanese oak species resulted in greater plant growth under the combination of elevated O_3 and CO_2 than under elevated CO_2 alone. A similar tendency was observed in *F. crenata* seedlings (Fig. 6.1; Watanabe et al. 2010). In this experiment, 2-year-old seedlings of *F. crenata* were grown in four experimental treatments, comprising two O_3 treatments (charcoal-filtered air and 100 nmol mol^{-1} O_3; 6 h/day, 3 days/week) in combination with two CO_2 treatments (350 and 700 μmol mol^{-1}) for 18 weeks in environmental controlled-growth chambers with an artificial light system. The dry-matter growth of the seedlings was greater under the combination of elevated O_3 and CO_2 than under the elevated CO_2 alone. A marked increase of second-flush leaves was observed under the combination of elevated O_3 and CO_2 concentrations. Watanabe et al. (2010) considered that the greater investment of carbohydrate due to the higher photosynthetic rate in first-flush leaves contributed to the marked increase in second-flush leaf emergence under the combination of elevated O_3 and CO_2.

The results of the above-mentioned studies did not completely support the hypothesis: "elevated CO_2 ameliorates the negative impacts of O_3". Interestingly, the amelioration phenomenon was not consistent even in the same species (e.g. Matsumura et al. 2005 vs Hoshika et al. 2012 for *B. platyphylla*, Koike et al. 2012 vs Wang et al. 2015 for hybrid larch F_1). On the other hand, when the amelioration or compensation of the O_3 impact under elevated CO_2 was observed, an elevated CO_2-induced decrease in stomatal conductance was also found (Matsumura et al. 2005; Watanabe et al. 2010; Kitao et al. 2015), except for the hybrid larch F_1 (Koike et al. 2012), and vice versa. Therefore, modification of stomatal O_3 uptake may be one of the critical functions that determines the amelioration effect produced by elevated CO_2. However, it is hard to explain the over-compensation phenomenon (Watanabe et al. 2010; Kitao et al. 2015) in terms of lower stomatal O_3 uptake. Changes in the availability of carbohydrate within a plant body and changes in the carbon allocation pattern between plant organs under elevated CO_2 may be key functions for understanding the combined effects of O_3 and elevated CO_2.

Although both O_3 and elevated CO_2 may affect secondary metabolism, and therefore may change plant defense capacity against biotic stresses such as pathogens and herbivores (Peltonen et al. 2005), there has been only one report on the combined effects of O_3 and elevated CO_2 on secondary metabolite concentrations in the leaves of Japanese tree species. Karonen et al. (2006) investigated the quantitative response of proanthocyanidins (PAs) to elevated CO_2 and O_3 in the seedlings of *B. platyphylla*, *B. ermanii*, and *F. crenata*. PAs are a major group of phenolic compounds, which are some of the main secondary metabolites in the leaves of many woody plants. The seedlings in the study were the same as those investigated by Matsumura et al. (2005). Total PA concentrations in the leaves of all species examined were similarly affected by the different treatments. PA concentrations were significantly higher in seedlings treated with the combination of O_3 and elevated CO_2 as compared with the other treatments. Significant species × treatment interaction was observed in results for polymeric PA concentrations in *B. platyphylla* and *B. ermanii*. In *B. platyphylla*, leaves treated with the combination of O_3 and elevated CO_2 differed significantly from leaves subjected to all other treatments. The authors concluded that the strongest effect of the O_3 and elevated CO_2 combination on leaf PA content resulted from the additive effect of these environmental factors on phenolic biosynthesis.

6.3 Combined Effects of O_3 and Soil Nitrogen Load

Nitrogen is the element required in the largest amounts by plants after carbon. The availability of nitrogen is therefore a decisive factor for forest production (Magnani et al. 2007), although excess nitrogen availability may have negative impacts on plants (See Chap. 18). Nitrogen deposition from the atmosphere to forest ecosystems has been increasing since the industrial revolution. Generally, there is a spatial correlation between high exposure to O_3 and high nitrogen deposition (Watanabe et al. 2012). In addition, changes in nitrogen availability as a result of increasing nitrogen deposition may affect plant sensitivity to O_3, because nitrogen is a component of enzymes that catalyze many metabolic processes, including the defense against O_3 (Dizengremel et al. 2013). Therefore, it is important to clarify the interaction between elevated O_3 and nitrogen availability for Japanese forest tree species.

Nakaji and Izuta (2001) presented the first report on the combined effects of O_3 and nitrogen load on forest tree species in Japan. They observed additive effects (i.e., no interaction) of O_3 (60 nmol mol^{-1} for 7 h) and excess soil nitrogen load (28–340 kg N ha^{-1} $year^{-1}$) on growth and photosynthetic traits in *Pinus densiflora* seedlings. Similar trends were observed in the same species by Nakaji et al. (2004). The influence of soil nitrogen load on the growth and physiological traits (such as photosynthesis) of six representative forest tree species in Japan (*Q. serrata*, *F. crenata*, *Castanopsis sieboldii*, *L. kaempferi*, *P. densiflora*, and *C. japonica*) were studied (Watanabe et al. 2006, 2007, 2008; Yamaguchi et al. 2007a, b, 2010). The seedlings were assigned to 12 experimental treatments, comprised of the combination of four levels of gas exposure (charcoal-filtered air and three levels of O_3 at 1.0, 1.5, and 2.0 times ambient concentration) and three levels of nitrogen load (0, 20, and 50 kg ha^{-1} $year^{-1}$; N0, N20, and N50, respectively), and grown for two growing seasons. At the end of the experiment, it was found that the nitrogen load did not change the sensitivities of growth to elevated O_3 in *Q. serrata*, *C. sieboldii*, *P. densiflora*, and *C. japonica* seedlings. On the other hand, significant interactions for

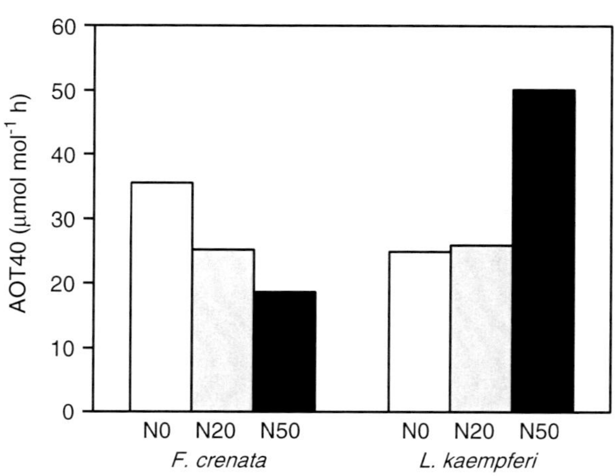

Fig. 6.2 Accumulated exposure over a threshold of 40 nmol mol^{-1} (AOT40) of ozone that induced 5 % reduction in the whole-plant dry mass increment of *Fagus crenata* and *Larix kaempferi* seedlings. The seedlings were grown under three levels of soil nitrogen load (0, 20, and 50 kg ha^{-1} $year^{-1}$; N0, N20, and N50, respectively) (Data source: Watanabe et al. 2008)

dry-matter growth between O_3 and nitrogen load were found in *F. crenata* and *L. kaempferi* seedlings (Fig. 6.2, Watanabe et al. 2006; Yamaguchi et al. 2007a).

There was an increase in the sensitivity of the growth of *F. crenata* to O_3 under a relatively high nitrogen load (Yamaguchi et al. 2007a). To clarify the reason for the enhancement of *F. crenata* O_3 sensitivity with increasing nitrogen load, Yamaguchi et al. (2007a, 2010) studied photosynthetic traits and nitrogen metabolism in the leaves during the second growing season. An ozone-induced reduction in the light-saturated net photosynthetic rate (A_{sat}) was observed with all nitrogen treatments in September. On the other hand, there was a significant interactive effect of O_3 and nitrogen load on A_{sat} in July. The exposure to O_3 significantly reduced the A_{sat} in seedlings with N20 and N50 treatments, but not in those with the N0 treatment. A similar tendency was observed in the concentrations of total soluble protein and ribulose-1,5-bisphosphate carboxylase/oxygenase (Rubisco), while there was no significant interaction of O_3 and nitrogen load for stomatal conductance and area-based nitrogen content. These results indicate that the higher O_3 sensitivity of photosynthesis to elevated O_3 under a relatively high nitrogen load in *F. crenata* seedlings was not due to stomatal closure, but was due to the greater reduction of nitrogen allocation to soluble proteins, including Rubisco, and therefore indicated reduced biochemical assimilation capacity in chloroplasts.

In contrast to findings in *F. crenata*, the O_3-induced reduction in the growth of *L. kaempferi* seedlings was ameliorated by the nitrogen load (Fig. 6.2; Watanabe et al. 2006). Pell et al. (1995) reported that the largest reduction in growth of *Populus tremuloides* seedlings by exposure to O_3 was observed under the condition of nitrogen supply that induced the highest tree growth rate. In the case of *L. kaempferi* seedlings, there was no growth enhancement by the nitrogen loading. There is a possibility that the excessive nitrogen was used for detoxification and repair capacities against O_3 stress. Kam et al. (2015) also studied the combined effects of O_3 (60 nmol mol^{-1} during daytime) and nitrogen load, at 50 kg ha^{-1}, for one growing season on hybrid larch F_1 (*L. gmelinii* var. *japonica* × *L. kaempferi*) and *L. kaempferi* seedlings using a free-air O_3 fumigation system (See Chap. 5). They reported an additive effect of O_3 and nitrogen load on growth in both species, indicating that other environmental conditions affect the influence of nitrogen load on O_3 sensitivity in Japanese forest tree species.

6.4 Combined Effects of O_3 and Soil Water Stress

Water is the most abundant molecule on the Earth's surface. The availability of water is the factor that most strongly restricts plant production in terrestrial ecosystems on a global scale. Low water availability limits the productivity of many natural ecosystems. The progression of climate change at global, regional, and landscape levels modifies the water status of plants. Because periods with relatively high concentrations of atmospheric O_3 often occur in combination with low soil moisture in dry summers, there is the possibility that many tree species are simultaneously affected by O_3 and water stress.

The combined effects of O_3 and soil water stress on the growth and leaf characteristics of *F. crenata* seedlings were investigated for two growing seasons (Yonekura et al. 2001a, b, 2004; Watanabe et al. 2005). Three-year-old seedlings of *F. crenata* were exposed to charcoal-filtered air or 60 nmol mol^{-1} O_3, for 7 h a day, from May to October 1999 in naturally lit growth chambers. Half of the seedlings in each gas treatment group received 250 ml of water at 3-day intervals (well-watered treatment), while the rest received 175 ml of water at the same

Table 6.1 Effects of ozone (O_3) and soil water stress (WS) on growth and leaf physiological traits of *Fagus crenata* seedlings in first and second growing seasons

	O_3		WS		$O_3 \times$ WS	
	1st	2nd	1st	2nd	1st	2nd
Growth						
Leaf dry mass	↓	↓	–	↓	–	–
Bud dry mass	–	–	↓	↓	–	–
Stem dry mass	↓	↓	↓	↓	–	–
Root dry mass	↓	↓	–	↓	–	–
Whole-plant dry mass	↓	↓	↓	↓	–	–
Leaf area	–	↓	–	↓	–	–
Leaf number	–	↓	–	↓	–	–
Bud number	–	↓	↓	↓	–	–
Leaf number per bud	×	↓	×	–	×	–
S/R	–	↑	–	–	–	–
Leaf traits						
A_{sat}	↓	↓	↓	↓	–	Antag.
G_s	–	–	↓	↓	–	–
E	–	–	↓	↓	–	–
A_{max}	↓	↓	–	↓	–	Antag.
CE	↓	↓	–	–	–	Antag.
QY	–	–	–	↓	–	–
F_v/F_m	↓	↓	–	–	–	–
R_d	–	–	–	–	–	–
Water potential	–	–	↓	↓	–	–
Chl concentration	–	↓	–	–	–	–
Rubisco concentration	↓	↓	–	↓	–	–
TSP concentration	–	–	–	–	–	–
Carb. concentration	↓	↓	–	↑	–	–
Starch concentration	↓	↓	–	↑	–	–

After Yonekura (2006)

See text for the details of the gas and soil water treatments

↓: significant decrease, ↑: significant increase, –: not significant, ×: not determined, antag.: antagonistic effects of O_3 and WS

S/R shoot-to-root biomass ratio, A_{sat} light-saturated net photosynthetic rate, G_s stomatal conductance, E transpiration rate, A_{max} light and CO_2 saturated net photosynthetic rate, CE carboxylation efficiency, QY quantum yield, F_v/F_m maximum quantum yield, R_d dark respiration, Chl chlorophyll, Rubisco ribulose-1,5-bisphosphate carboxylase/oxygenase TSP total soluble protein, Carb. water-soluble carbohydrates intervals (water-stressed treatment). There were group differences in the combined effects of O_3 and water stress (Table 6.1). Additive effects (no significant interaction) of O_3 and water stress were detected in the growth and leaf traits in the first growing season. In the second growing season, on the other hand, there were significant effects of interactions between O_3 and water stress on some photosynthetic parameters, indicating water stress-induced amelioration of the negative effects of O_3. Lower leaf water potential and stomatal conductance may induce lower stomatal O_3 uptake. Furthermore, as reported by Watanabe et al. (2005), soil water stress in *F. crenata* seedlings increased the concentration of glutathione, a representative antioxidant, in the leaves, and these authors raised the possibility that the increased glutathione under soil water stress may moderate the negative impacts of O_3 on photosynthetic functions in *F. crenata* seedlings. It should be noted that, although antagonistic effects of O_3 and water stress were found in leaf traits, there was no significant effect of water stress on the growth response to O_3. Additive effects of O_3 and water stress on *F. crenata* and *B. ermanii* were also reported by Shimizu and Ito (2013) and Shimizu and Feng (2007), respectively. The extent of O_3 and water stress may modulate the interactive effects of the two stress factors.

6.5 Future Perspectives

The results obtained from the experimental studies described above indicate that most of the environmental factors examined have the potential to change O_3 sensitivities in Japanese forest tree species. The physiological mechanisms of changes in O_3 sensitivity seem to differ between environmental factors, although some of the mechanisms are the same (e.g. both elevated CO_2 and soil water stress decrease stomatal conductance and therefore stomatal O_3 uptake). To predict the effects of O_3 on Japanese forests under changing environments, deeper clarification of the modification effects of other environmental factors on O_3 sensitivity is crucial. However, the number of these studies has not yet been sufficient to achieve comprehensive understanding. Although studies with mature trees under natural conditions, such as free-air enrichment studies, are, of course, important, we consider that the accumulation of data on the combined effects of O_3 and other environmental factors on Japanese forest tree species, even with studies of seedlings, is important at the present time.

References

Ainsworth EA, Rogers A (2007) The response of photosynthesis and stomatal conductance to rising [CO_2]: mechanisms and environmental interactions. Plant Cell Environ 30:258–270

Dizengremel P, Jolivet Y, Tuzet A, Ranieri A, Le Thiec D (2013) Integrative leaf-level phytotoxic ozone dose assessment for forest risk modeling. In: Matyssek R, Clarke N, Cudlin P, Mikkelsen TN, Tuovinen J-P, Wieser G, Paoletti E (eds) Climate change, air pollution and global challenges: understanding and perspectives from forest research. Elsevier, Oxford, pp 267–288

Eguchi N, Fukatsu E, Funada R, Tobita H, Kitao M, Maruyama Y, Koike T (2004) Changes in morphology, anatomy, and photosynthetic capacity of needles of Japanese larch (*Larix kaempferi*) seedlings grown in high CO_2 concentrations. Photosynthetica 42:173–178

Hoshika Y, Wanatabe M, Inada N, Koike T (2012) Ozone-induced stomatal sluggishness develops progressively in Siebold's beech (*Fagus crenata*). Environ Pollut 166:152–156

Kam DG, Shi C, Watanabe M, Kita K, Satoh F, Koike T (2015) Growth of Japanese and hybrid larch seedlings grown under free-air O_3 fumigation – an initial assessment of the effects of adequate and excessive nitrogen. J Agric Meteorol 71:239–244

Karonen M, Ossipov V, Ossipova S, Kapari L, Loponen J, Matsumura H, Kohno Y, Mikami C, Sakai Y, Izuta T, Pihlaja K (2006) Effects of elevated carbon dioxide and ozone on foliar proanthocyanidins in *Betula platyphylla*, *Betula ermanii* and *Fagus crenata* seedlings. J Chem Ecol 32:1445–1458

Kitao M, Koike T, Tobita H, Maruyama Y (2005) Elevated CO_2 and limited nitrogen nutrition can restrict excitation energy dissipation in photosystem II of Japanese white birch (*Betula platyphylla* var. *japonica*) leaves. Physiol Plant 125:64–73

Kitao M, Komatsu M, Yazaki K, Kitaoka S, Tobita H (2015) Growth overcompensation against O_3 exposure in two Japanese oak species, *Quercus mongolica* var. *crispula* and *Quercus serrata*, grown under elevated CO_2. Environ Pollut 206:133–141

Koike T, Mao Q, Inada N, Kawaguchi K, Hoshika Y, Kita K, Watanabe M (2012) Growth and photosynthetic responses of cuttings of a hybrid larch (*Larix gmelinii* var. *japonica* × *L. kaempferi*) to elevated ozone and/or carbon dioxide. Asian J Atmos Environ 6:104–110

Magnani F, Mencuccini M, Borghetti M, Berbigier P, Berninger F, Delzon S, Grelle A, Hari P, Jarvis PG, Kolari P, Kowalski AS, Lankreijer H, Law BE, Lindroth A, Loustau D, Manca G, Moncrieff JB, Rayment M, Tedeschi V, Valentini R, Grace J (2007) The human footprint in the carbon cycle of temperate and boreal forests. Nature 447:848–852

Matsumura H, Mikami C, Sakai Y, Murayama K, Izuta T, Yonekura T, Miwa M, Kohno Y (2005) Impacts of elevated O_3 and/or CO_2 on growth of *Betula platyphylla*, *Betula ermanii*, *Fagus crenata*, *Pinus densiflora* and *Cryptomeria japonica* seedlings. J Agric Meteorol 60: 1121–1124

Matyssek R, Sandermann H (2003) Impact of ozone on trees: an ecophysiological perspective. Prog Bot 64:349–404

Monastersky R (2013) Global carbon dioxide levels near worrisome milestone. Nature 497:13–14

Nakaji T, Izuta T (2001) Effects of ozone and/or excess soil nitrogen on growth, needle gas exchange rates and Rubisco contents of *Pinus densiflora* seedlings. Water Air Soil Pollut 130:971–976

Nakaji T, Kobayashi T, Kuroha M, Omori K, Matsumoto Y, Yonekura T, Watanabe K, Utriainen J, Izuta T (2004) Growth and nitrogen availability of red pine seedlings under high nitrogen load and elevated ozone. Water Air Soil Pollut Focus 4:277–287

Norby RJ, Warren JM, Iversen CM, Medlyn BE, McMurtrie RE (2010) CO_2 enhancement of forest productivity constrained by limited nitrogen availability. Proc Natl Acad Sci USA 107: 19368–19373

Paoletti E, Schaub M, Matyssek R, Wieser G, Augustaitis A, Bastrup-Birk AM, Bytnerowicz A, Günthardt-Goerg MS, Müller-Starck G, Serengil Y (2010) Advances of air pollution science: from forest decline to multiple-stress effects on forest ecosystem services. Environ Pollut 158:1986–1989

Pell EJ, Sinn JP, Johansen CV (1995) Nitrogen supply as a limiting factor determining the sensitivity of *Populus tremuloides* Michx. to ozone stress. New Phytol 130:437–446

Peltonen PA, Vapaavuori E, Julkunen-Tiitto R (2005) Accumulation of phenolic compounds in birch leaves is changed by elevated carbon dioxide and ozone. Glob Chang Biol 11:1305–1324

Riikonen J, Lindsberg M-M, Holopainen T, Oksanen E, Lappi J, Peltonen P, Vapaavuori E (2004) Silver birch and climate change: variable growth and carbon allocation responses to elevated concentrations of carbon dioxide and ozone. Tree Physiol 24:1227–1237

Shimizu H, Feng YW (2007) Ozone and/or water stresses could have influenced the *Betula ermanii* Cham. forest decline observed at Oku-Nikko, Japan. Environ Monit Assess 128:109–119

Shimizu H, Ito S (2013) Combined stress effects on *Fagus crenata*. For Sci 67:14–17 (in Japanese)
Stocker TF, Qin D, Plattner G-K, Tignor M, Allen SK, Boschung J, Nauels A, Xia Y, Bex V, Midgley PM (2013) Climate change 2013: the physical science basis, Contribution of working group I to the fifth assessment report of the intergovernmental panel on climate change. Cambridge University Press, Cambridge, p 1535
Wang X, Qu L, Mao Q, Watanabe M, Hoshika Y, Koyama A, Kawaguchi K, Tamai Y, Koike T (2015) Ectomycorrhizal colonization and growth of the hybrid larch F_1 under elevated O_3 and CO_2. Environ Pollut 197:116–126
Watanabe M, Yonekura T, Honda Y, Yoshidome M, Nakaji T, Izuta T (2005) Effects of ozone and soil water stress, singly and in combination, on leaf antioxidative systems of *Fagus crenata* seedlings. J Agric Meteorol 60:1105–1108
Watanabe M, Yamaguchi M, Iwasaki M, Matsuo N, Naba J, Tabe C, Matsumura H, Kohno Y, Izuta T (2006) Effects of ozone and/or nitrogen load on the growth of *Larix kaempferi*, *Pinus densiflora* and *Cryptomeria japonica* seedlings. J Jpn Soc Atmos Environ 41:320–334
Watanabe M, Yamaguchi M, Tabe C, Iwasaki M, Yamashita R, Funada R, Fukami M, Matsumura H, Kohno Y, Izuta T (2007) Influences of nitrogen load on the growth and photosynthetic responses of *Quercus serrata* seedlings to O_3. Trees 21:421–432
Watanabe M, Yamaguchi M, Matsumura H, Kohno Y, Izuta T (2008) Effects of ozone on growth and photosynthesis of *Castanopsis sieboldii* seedlings grown under different nitrogen loads. J Agric Meteorol 24:143–155
Watanabe M, Umemoto-Yamaguchi M, Koike T, Izuta T (2010) Growth and photosynthetic response of *Fagus crenata* seedlings to ozone and/or elevated carbon dioxide. Landsc Ecol Eng 6:181–190
Watanabe M, Yamaguchi M, Matsumura H, Kohno Y, Izuta T (2012) Risk assessment of ozone impact on *Fagus crenata* in Japan: consideration of atmospheric nitrogen deposition. Eur J For Res 131:475–484
Yamaguchi M, Watanabe M, Iwasaki M, Tabe C, Matsumura H, Kohno Y, Izuta T (2007a) Growth and photosynthetic responses of *Fagus crenata* seedlings to O_3 under different nitrogen loads. Trees 21:707–718
Yamaguchi M, Watanabe M, Matsuo N, Naba J, Funada R, Fukami M, Matsumura H, Kohno Y, Izuta T (2007b) Effects of nitrogen supply on the sensitivity to O_3 of growth and photosynthesis of Japanese beech (*Fagus crenata*) seedlings. Water Air Soil Pollut Focus 7:131–136
Yamaguchi M, Watanabe M, Matsumura H, Kohno Y, Izuta T (2010) Effects of ozone on nitrogen metabolism in the leaves of *Fagus crenata* seedlings under different soil nitrogen loads. Trees 24:175–184
Yonekura T (2006) Water stress and plant. In: Izuta T (ed) Plants and environmental stresses. Corona Publishing, Tokyo, pp 145–167 (in Japanese)
Yonekura T, Dokiya Y, Fukami M, Izuta T (2001a) Effects of ozone and/or soil water stress on growth and photosynthesis of *Fagus crenata* seedlings. Water Air Soil Pollut 130:965–970
Yonekura T, Honda Y, Oksanen E, Yoshidome M, Watanabe M, Funada R, Koike T, Izuta T (2001b) The influences of ozone and soil water stress, singly and in combination, on leaf gas exchange rates, leaf ultrastructural characteristics and annual ring width of *Fagus crenata* seedlings. J Jpn Soc Atmos Environ 36:333–351
Yonekura T, Yoshidome M, Watanabe M, Honda Y, Ogiwara I, Izuta T (2004) Carry-over effects of ozone and water stress on leaf phenological characteristics and bud frost hardiness of *Fagus crenata* seedlings. Trees 18:581–588

Chapter 7
Environmental Monitoring with Indicator Plants for Air Pollutants in Asia

Hiroyuki Sase

Abstract Air pollution is a regional issue in Asia. To assess air pollution levels and their actual impacts on plants, environmental monitoring using indicator plants has been conducted since the 1970s. In particular, in Japan, during the 1970s/1980s when air pollution levels were relatively high, various kinds of indicator plants were used for monitoring air pollution. Some vascular plants, such as morning glory and petunia, were used as indicator plants for monitoring the pollution of photochemical oxidants (Ox), including O_3 and peroxyacetyl nitrate (PAN). In the middle of the 1970s, nationwide surveys using morning glory revealed Ox pollution in 37 of the 47 prefectures in Japan. Epiphytes, such as bryophytes and lichens, were used as bio-indicators for SO_2 pollution. The "epiphyte desert" observed in urban and industrial areas showed severe SO_2 pollution in large cities in Japan during the 1970s. An improvement in the distribution of epiphytes resulted in reduced SO_2 pollution in Japan during the 1980s. Similar environmental monitoring using bryophytes and lichens as bio-indicators was conducted in other Asian countries. Common tree species, such as cedar, can also be used as indicator plants for environmental monitoring. Further application of such monitoring in Asian countries should be promoted.

Keywords Morning glory • Petunia • Bryophyte • Lichen • Cedar

7.1 Introduction

Air pollution is a regional issue in Asia because the emission levels of various pollutants remain high in the region. According to Ohara et al. (2007), emissions increased rapidly from 1980 to 2003, by 28 % for black carbon (BC), 30 % for organic carbon (OC), 108 % for non-methane volatile organic compounds (NMVOCs), 119 % for sulfur dioxide (SO_2), and 176 % for nitrogen oxides (NO_X).

H. Sase
Asia Center for Air Pollution Research, 1182 Sowa, Nishi-ku, Niigata City 950-2144, Japan
e-mail: sase@acap.asia

T. Izuta (ed.), *Air Pollution Impacts on Plants in East Asia*,
DOI 10.1007/978-4-431-56438-6_7

Although the emission of SO_2 in China peaked in 2006 before declining thereafter, the emissions of BC and OC were still increasing as of 2010 (Lu et al. 2011). Air pollution, as well as being a human health issue, is a serious threat to plants in Asia.

Many cases of environmental pollution in the past were discovered due to abnormal changes in trees and crops near smelters, thermal power plants, and oil petrochemical complexes (Kuno and Ohashi 1993a). Moreover, some types of plants are very sensitive to environmental changes, including air pollution. Therefore, air pollution monitoring using indicator plants (or bio-indicators), such as some sensitive vascular plants, bryophytes, and lichens, has been utilized to evaluate pollution levels and/or actual impacts on plants. This chapter introduces the past and current issues related to the application of indicator plants for monitoring air pollutants.

7.2 Japanese Experience in the Application of Indicator Plants

SO_2 emissions in Japan significantly increased during the 1950s/1960s with rapid industrialization, peaked in 1969, and then rapidly decreased, according to the emissions inventory performed by Smith et al. (2011). In Japan, the Basic Law for Environmental Pollution Control and the Air Pollution Control Law were enacted in 1967 and 1968, respectively. Environmental quality standards (EQSs) were also established based on these laws, although damage to humans and plants by photochemical oxidants was still being reported in the Tokyo metropolitan area during the summer of 1970. Concentrations of primary pollutants, such as SO_2 and CO, rapidly decreased during the 1970s/1980s after the implementation of regulations for the total pollution loads of SO_X and NO_X in 1974 and 1981, respectively. However, emissions of NO_2 and photochemical oxidants (Ox) remained steady (Wakamatsu et al. 2013). In the 1970s/1980s, when the air concentrations of pollutants were still relatively high, various kinds of indicator plants were used for monitoring air pollution in Japan.

Kuno and Ohashi (1993a) reviewed indicator plants used for air pollution assessments and classified the methods of application of indicator plants into four categories:

1. Use of native/living plants in target areas: e.g., spatial distribution of tree decline symptoms, foliar discoloration, and other indicators in relation to air pollutant concentrations. Chemical analysis of pollutants absorbed in plant bodies and tree-ring analysis are also useful to assess chronic or long-term effects of air pollutants.
2. Replacement of indicator plants in target areas: e.g., visible foliar injuries to sensitive plants are compared in various locations.
3. Use of (open-top) chambers: e.g., in addition to visible foliar injuries, suppression of growth and yield reduction can be evaluated using filtered and non-filtered chambers.

4. Observations of the succession of plant groups: e.g., the species composition or density of mosses is examined in relation to air pollutant concentrations and then succession can be discussed to assess the long-term effects of air pollution.

Kuno and Ohashi (1993a) compiled data on the sensitivity of plant species to Ox, such as O_3 and peroxyacetyl nitrate (PAN); Japanese morning glory, petunia, tobacco, spinach, chickweed, kidney bean, and rice were listed as the plants most sensitive to Ox. These plants were injured by daily maximum values of approximately 80–100 ppb, although the sensitivities were different in various cultivars even within the same plant species. Many authors (e.g. Hatta and Terakado 1975; Nouchi and Aoki 1979) have reported morning glory (*Pharbitis nil* var Scarlett O'Hara) as a useful indicator plant for Ox. The review by Kuno and Ohashi (1993b), based on nationwide surveys conducted from 1974 to 1976, reported that visible foliar injuries to morning glory were observed in 37 of the 47prefectures of Japan. In addition, high concentrations of Ox above the EQS levels (60 ppb for 1-h value) were suggested in these prefectures. As for petunia, which is sensitive to PAN, Nouchi et al. (1984) reported that the percentage of visible foliar injuries in a sensitive cultivar (White ensign) escalated with an increase in the daily maximum concentration or daily dose of PAN. This finding was based on surveys conducted in Yurakucho, Tokyo, from 1976 to 1983. By using both the sensitive (White ensign) and tolerant (Blue ensign) cultivars, Nouchi et al. (1984) clarified that PAN pollution occurred at five representative locations (Ohme, Yohga, Shakujii, Yurakucho, and Adachi) in Tokyo in the late spring and autumn in 1982 and 1983. They also pointed out that cultivars with different sensitivities would be useful to estimate concentration levels of PAN. The national/regional assessments of air pollution discussed above were conducted by placing indicator plants in target areas and observing visible foliar injuries that were classified as category (2) above.

As for category (3), Izuta et al. (1988) studied the effects of ambient air on the growth of radish (*Raphanus sativus* L. cv. Comet) in Tokyo by using small open-top chambers. Their studies verified that the leaf area of cotyledons could be used as an indicator for the effects of O_3. Other studies using open-top chambers are introduced in other chapters of this book. In addition to higher plants, bryophytes are also used for the chamber method, since bryophytes, such as mosses, are sensitive to air pollution. Taoda (1973) developed a pair of filtered/non-filtered small open-top chambers using bryophytes and named the instrument the "bryo-meter". According to the study of an application of the "bryo-meter" using a moss species (*Marchantia polymorpha*) around the Kashima industrial district carried out by Yokobori (1978), the phytotoxicity, based on the growth rate of the bryophyte, was severe in the leeward direction from pollution sources. This finding corresponded to findings on the concentrations of air pollutants such as SO_2 and Ox. Shimizu et al. (1988) improved the bryo-meter system and used a different moss species (*Plagiomnium maximoviczii* (Lindb.) Kop.). They suggested that the species were sensitive to mixed gaseous pollutants, including 0.05 ppm of SO_2, 0.1 ppm of NO_2, and 0.07 ppm of O_3, and they noted that the system could be used for assessing atmospheric environments in urban areas of Japan.

As for categories (1) and (4) above, the following sections introduce the relevant studies in detail.

7.3 Bryophyte and Lichen Communities

Bryophytes have historically been used as indicator plants for air pollution in Japan. Taoda (1972) used epiphytic bryophytes on broad-leaved trees for assessing air pollution in Tokyo. He found an "epiphyte desert", in which no epiphytic bryophyte was observed in industrial areas around Tokyo Bay where SO_2 concentrations were over 0.05 ppm. He also pointed out that the Tokyo metropolitan area could be classified into five zones based on the number of bryophyte species that increased with the decline of SO_2 concentration levels. On the other hand, Taoda (1996) examined the effects of simulated acid rain on epiphytic bryophytes and reported that no correlation was found between distribution patterns of the species in the polluted area of Tokyo and tolerance to the acidity of the simulated rainwater. Mitsugi et al. (1978) used the Index of Atmospheric Purity (IAP: De Sloover and LeBlanc 1968), which was calculated from the frequency, coverage, density, and fertility of epiphytic bryophyte and lichen species, for assessing air pollution in Hyogo Prefecture, in the Kansai region of Japan. Multiple regression analysis showed that the combined effects of SO_2 and the soluble components of rainwater explained the IAP values. In their study, concentrations of heavy metals, such as Cu^{2+} and Mn^{2+}, had significant effects on the IAP values (Mitsugi et al. 1978). Mitsugi (1994) developed another index, the Evaluation Index of Air Quality (EIAQ), based on the number of epiphytic bryophyte species and their dominance scale. He classified the area of Hyogo prefecture into five zones, from Zone 0 (highly disturbed zone: epiphyte desert) around Kobe city along the Seto Inland Sea to Zone IV (undisturbed zone) in 1988; most of the area was classified as Zone III (quasi-normal zone). He also pointed out that the EIAQ values were better than those in 1975/1976 due to the reduction of SO_2 (and probably Ox) concentrations, especially in the industrial area near the sea. Omura and Murata (1984) assessed improvements in air pollution levels from 1978/1979 to 1982 by using modified IAP values (IAP*) in industrial areas in Fukuoka prefecture. They reported that the IAP* values were generally larger in 1982 than in 1978/1979, a finding which corresponded well to the reduction of SO_2 concentrations during that period. Thus, environmental monitoring using epiphytic bryophytes (and lichens) showed improvements in air pollution in the industrial areas of Hyogo and Fukuoka prefectures. Umezu (1978) applied the Braun-Blanquet method, based on phytosociology, to identify associations of epiphytic bryophytes and lichens in the industrial area around Ube city in Yamaguchi Prefecture. They classified the area into five zones according to the normal environmental degree, based on the species composition and their richness, which corresponded well to SO_2 concentrations.

As shown in the studies above, lichens were also used together with bryophytes for assessing air pollution. Since lichens are not plants but are composite organisms of algae (and/or cyanobacteria) and fungi, they should be called "bio-indicators." Nakagawa and Kobayashi (1990) classified Hyogo prefecture into six zones according to the upgraded IAP values, which were calculated based on the coverage of lichen species and their sensitivities to SO_2. The map of the upgraded IAP values reflected the pollution level in the area, from the lichen desert in the industrial area near the sea to the normal zone in the clean inland area. They also pointed out that two indicator species, *Parmelia tinctorum* (currently reclassified as *Parmotrema tinctorum*) and *Parmelia caperata* (currently reclassified as *Flavoparmelia caperata*), could be used for the rough evaluation of pollution levels. The usefulness of *P. tinctorum* as a bio-indicator for SO_2 air pollution had already been suggested by Sugiyama et al. (1976), who surveyed the distribution of the species on tombstones in the cemeteries of Buddhist temples in five cities in Japan; namely, Sendai, Tokyo, Fuji, Shizuoka, and Yokkaichi, from 1972 to 1974. They reported that the species were absent in the area where SO_2 concentrations were higher than 0.020 ppm. Ohmura et al. (2008) resurveyed the distribution of *P. tinctorum* in Shizuoka city in 1978, 1994, and 2003. They found that the lichen desert observed in 1972 near the port areas had shrunk in 1978 with a decrease in SO_2 concentrations. However, a new lichen desert appeared in 1978 near the crossing point of the Tomei Expressway and Japan National Route 1. This desert expanded along Route 1 until 2003. Although the new lichen desert did not show clear correlation with pollutants, such as NO, NO_2, and NO_X, the effects of increasing traffic, including air pollutants, drought conditions, high temperatures, etc., might affect the lichen species (Ohmura et al. 2008). The studies on bryophytes and lichens in Japan are summarized in Table 7.1. Almost

Table 7.1 Spatial assessment of air pollution using distribution of bryophytes and lichens as bio-indicators

Area, prefecture	Year	Biological indicator	Evaluation method or index	Reference
Tokyo	1969–1971	Epiphytic bryophyte	By number and abundance	Taoda (1972)
Hyogo	1975/1976	Epiphytic bryophytes and lichens	Index of Atmospheric Purity (IAP)	Mitsugi et al. (1978)
	1988 (partly compared with 1975/1976)	Epiphytic bryophytes	Evaluation Index of Air Quality (EIAQ)	Mitsugi (1994)
Ube city, Yamaguchi	1972–1976	Epiphytic bryophytes and lichens	Braun-Blanquet method based on phytosociology	Umezu (1978)
Omuta and Kitakyushu cities, Fukuoka	1978/1979 and 1982	Epiphytic bryophytes and lichens	Modified IAP (IAP*, as the mean of the top four values among the IAP values of five neighboring plots)	Omura and Murata (1984)
Hyogo	1984–1988	Epiphytic lichens	Upgraded IAP (based on coverage and sensitivity to SO_2 of lichens)	Nakagawa and Kobayashi (1990)
Sendai, Tokyo, Fuji, Shizuoka, and Yokkaichi	1972–1974	Lichen (*Parmotrema tinctorum*) on tombstones in the cemeteries of Buddhist temples	Presence or absence of species	Sugiyama et al. (1976)
Shimizu city, Shizuoka	1978, 1993, 2003	Lichen (*Parmotrema tinctorum*)	Presence or absence of the species	Ohmura et al. (2008)

all these studies were conducted in the 1970s/1980s, when SO_2 concentrations were still high. The current SO_2 concentration level in Japan is significantly lower than it was in the 1970s/1980s. However, lichens, such as *P. tinctorum*, may reflect changes in some more general atmospheric conditions, as suggested by Ohmura et al. (2008). Therefore, similar long-term surveys should be conducted to assess the improvement (or deterioration) of bryophyte and/or lichen communities in Japan.

The use of lichen communities as bio-indicators for air pollution was also applied in other Asian countries, including Hong Kong (Thrower 1980), Republic of Korea (Ahn et al. 2011), and Thailand (Saipunkaew et al. 2007). Thrower (1980) surveyed Hong Kong in the late 1970s in cooperation with secondary school students and classified the area into four zones based on the shapes of lichen species, including crustose, foliose, and fruticose lichens. Lichen deserts were found near power stations and industrial areas, and the distribution of crustose, foliose, and fruticose lichens was related to topography as well as the distance from major pollution sources (Thrower 1980). In Seoul, Ahn et al. (2011) compared the distribution of epiphytic macrolichens in 2010 with recorded data from 1975. They reported that the change in lichen distribution for that 35-year period was related to concentrations of air pollutants, such as SO_2, NO_2, and O_3, and the species richness tended to increase with distance from the city center. They also pointed out that the appearance of nitrophilic species might reflect increases of NO_2 concentrations. In northern Thailand, the distribution of epiphytic macrolichens corresponded to population density and was related to particulate matter less than 10 μm in diameter (PM_{10}), rather than SO_2, based on surveys in 2002 and 2004 (Saipunkaew et al. 2007). Therefore, not only the effects of SO_2 but also the effects of other pollutants, such as NO_2 (and/or nitrogen [N] deposition), and PM, should be taken into consideration as possible factors regulating lichen communities. In fact, in Europe and the United States, the relationship between the distribution of lichens and N deposition has been discussed recently, and critical N loads for lichen communities were proposed (e.g. Giordani et al. 2014).

7.4 Chemical Analysis of Bryophytes and Lichens

The studies discussed in the previous section assessed the effects of air pollution according to observations of bryophyte and lichen communities. In addition to field observations, some studies utilized chemical analysis of bryophytes and lichens to elucidate more direct relationships with pollution. Kobayashi et al. (1986) assessed atmospheric mercury (Hg) pollution around a steelworks facility in Hyogo prefecture in western Japan by using epiphytic lichens (*P. tinctorum*). They reported that the Hg concentration in lichen thallus decreased with the distance from the steelworks facility and noted that the Hg concentration in the thallus was significantly correlated with the level of atmospheric Hg. Moreover, Kobayashi and Nakagawa (1990) showed that the Hg concentration was positively correlated with the S concentration in lichen thallus, while a similar correlation was observed in atmospheric concentrations of

Hg. They pointed out that the Hg concentration could be used as an indicator of air pollution, since the upgraded IAP values (Nakagawa and Kobayashi 1990) were negatively correlated with the concentration of Hg. Nogami et al. (1987) suggested the Hg concentration in a moss species (*Bryum argenteum* Hedw.) could also be an indicator, based on surveys undertaken in 1986 in Okayama city in western Japan. They reported that the Hg concentration in the moss body was significantly correlated with the concentration of antimony (Sb; which showed a high enrichment factor), and was mainly derived from anthropogenic sources. They also pointed out that the moss species had an advantage over the sensitive lichen species, such as *P. tinctorum*, for assessing the conditions of air pollution over a wide area, since *P. tinctorum* could not be found in the industrial and urban areas (Nogami et al. 1987). Ishibashi et al. (1982) compared heavy metal concentrations in epiphytic moss bodies in three polluted areas, near a zinc smelter and cement factories, with concentrations in three control areas. They reported that cadmium (Cd), zinc (Zn), and/or lead (Pb) concentrations in the epiphytic moss bodies were significantly higher in the polluted areas than in the control areas. The heavy metal concentrations in moss bodies decreased with the distance from the zinc smelter (Ishibashi et al. 1982).

In addition to elemental analysis, isotopic analyses of epilithic mosses have also been applied to assess the atmospheric deposition of N and S in southwest China; e.g., the N isotopic ratio ($\delta^{15}N$) has been used for the assessment of N deposition (Liu et al. 2008), the S isotopic ratio ($\delta^{34}S$) for the assessment of S deposition (Liu et al. 2009), and the carbon isotopic ratio ($\delta^{13}C$) for the assessment of N deposition (Liu et al. 2010). In these cases, epilithic lichens, which live on rocks in open fields, were used for assessing the deposition of these elements, since the effects of the substratum could be regarded as negligible.

7.5 Use of Common Tree Species as Bio-indicators

This chapter has mainly concentrated on the use of sensitive plant and epiphyte species as bio-indicators. However, native/living plants in target areas can also be used as indicator plants, as Kuno and Ohashi (1993a) suggested (category 1 in Sect. 7.2). Common tree species, such as pine and cedar, are extensively planted for afforestation. Some of these trees could be used as bio-indicators, even if they are not as sensitive to air pollution as bryophytes and lichens. By using trees, not only can the current condition of air pollution be assessed but also the past condition or chronological change over time can be assessed. Several studies traced past air pollution conditions by utilizing tree rings. Katoh et al. (1988a) detected signs of air pollution effects from the ring width of Japanese cedar (*Cryptomeria japonica*) in Fukui prefecture. Conspicuous inhibition of growth in the species was observed within approximately an 8-km radius of the power station, and annual mean concentrations of SO_2 and NO_2 were closely related to the tree growth (Katoh et al. 1988b). Kawamura et al. (2006) analyzed sulfur isotopic ratio ($\delta^{34}S$) values in the tree rings of *C. japonica* and Japanese cypress (*Chamaecyparis obtusa*) collected in Fukuoka prefecture. They found that the profile of the $\delta^{34}S$ values significantly decreased in the 1960s/1970s, when high SO_2 concentrations were recorded, and increased gradually thereafter. Ishida et al. (2015) also analyzed tree rings of *C. japonica*, collected from stumps near Nagoya. They found a similar profile with lower $\delta^{34}S$ values in the 1960s/1970s. The profile was significantly correlated with anthropogenic SO_2 emissions in Japan. In contrast to tree rings, the outer bark, which is exposed to atmospheric concentrations, may reflect current air pollution conditions. In China, Kuang et al. (2007) clarified heavy metal pollution near a Pb-Zn smelter in Qujiang, Guangdong province, by analyzing the heavy metal concentrations in the outer bark of trees. Satake et al. (1996) clarified the progression of Pb pollution due to leaded gasoline used rom 1949 to 1987 in Japan – by comparing Pb concentrations in the outer bark and "bark pocket." The bark pocket is the outer bark from the past left within tree rings during the injury-repair process, and it preserves historical environmental information. Thus, the past and current conditions of air pollution have been clarified by using the tree rings and/or bark of these tree species.

The leaves/needles of these tree species have also been used as bio-indicators for air pollution. Sase et al. (1998a, b) clarified changes in the carbon and oxygen concentrations and carbon/oxygen (C/O) ratios in epicuticular wax in *C. japonica* due to natural and anthropogenic environmental factors. The amount of wax was higher both with the decrease of the C/O ratio caused by exposure to volcanic gas in Osorezan, Aomori prefecture in northern Japan (Sase et al. 1998a) and with the air pollutants from an electrochemical plant in Yakushima Island in southern Japan (Sase et al. 1998b). Takamatsu et al. (2000) analyzed heavy metals in the particulate matter strongly adsorbed on the leaf surfaces of *C. japonica* collected from Kanto and Tohoku, in eastern Japan, and Yakushima Island. They showed that Sb concentrations in the leaf-surface particles were correlated with population density, and NO_X and could be a good indicator of air pollution. Leaf surfaces can trap air pollutants effectively and may reflect air pollution conditions. Therefore, utilization of the leaf surface condition as a bio-indicator for air pollution should also be promoted.

7.6 Summary and Future Perspectives

Environmental monitoring, using indicator plants to assess air pollution levels and actual impacts on plants in Japan and other countries, has been conducted since the 1970s. In particular, in Japan, during the 1970s/1980s when air pollution levels were relatively high, various kinds of indicator plants and approaches were applied for monitoring air pollution. A list of sensitive plant species and the relevant air pollutants has been compiled in Japan. Some vascular plants, such as Japanese morning glory and petunia, were used as indicator plants for pollution by photochemical oxidants (Ox) including O_3 and PAN. Epiphytes, such as bryophytes and lichens, were also used as bio-indicators, for SO_2 pollution. Similar environmental monitoring using bio-indicators, such as bryophytes and lichens, was conducted in other Asian countries. Not only the sensitive plant species noted above but also common tree species, such as cedar, can be utilized as bio-indicators by performing analyses of their bark, tree rings, and leaf surface conditions.

With the development of an emission inventory for pollutants, and with simulations using chemical-transport models, the current and past conditions of air pollution may be reproduced precisely. However, the actual conditions for plants, and their reactions, should be proven in the field. Environmental monitoring using indicator plants is a strong tool for showing actual ecological conditions. Various approaches can be considered, as shown in this chapter. Further application in Asian countries should be promoted.

References

Ahn et al (2011) Epiphytic macrolichens in Seoul: 35 years after the first lichen study in Korea. J Ecol Field Biol 34:381–391

De Sloover J, LeBlanc F (1968) Mapping of atmospheric pollution on the basis of lichen sensitivity. In: Misra R, Gopal B (eds) Proceeding of the symposium on recent advances in tropical ecology. Banaras Hindu University, Varanasi, pp 42–56

Giordani P et al (2014) Detecting the nitrogen critical loads on European forests by means of epiphytic lichens. A signal-to-noise evaluation. For Ecol Manag 311:29–40

Hatta H, Terakado K (1975) Effect of photochemical oxidant on morning glory and its validity as indicator plant, I. Characteristics of oxidant damage on morning glory and correlation between leaf damage index and oxidant dose. J Jpn Soc Air Pollut 9:722–728 (in Japanese with English summary)

Ishibashi R et al (1982) Heavy metal determination of epiphytic mosses as an attempt to indicate industrial pollution of environment. J Jpn Soc Air Pollut 17:63–69 (in Japanese with English summary)

Ishida T et al (2015) Characterization of sulfur deposition over the period of industrialization in Japan using sulfur isotope ratio in Japanese cedar tree rings taken from stumps. Environ Monit Assess 187:459

Izuta T et al. (1988) An evaluation of atmospheric environment by a plant indicator based on the growth of radish plants grown in small open-top chambers. J Jpn Soc Air Pollut 23:284–292 (in Japanese with English summary)

Katoh T et al (1988a) An assessment of effects of air pollution on the ring width of Japanese cedars (Cryptomeria japonica D. Don) (I) –relationships between the operation of thermoelectric power stations and fluctuations of the standardized ring index. J Jpn Soc Air Pollut 23:311–319 (in Japanese with English summary)

Katoh T et al (1988b) An assessment of effects of air pollution on the ring width of Japanese cedars (Cryptomeria japonica D. Don) (II) –relationships between the standardized ring index and the pollution levels of SO2 and NO2 in the ambient air at the surrounding area of thermoelectric power stations. J Jpn Soc Air Pollut 23:320–328 (in Japanese with English summary)

Kawamura H et al (2006) Isotopic evidence in tree rings for historical changes in atmospheric sulfur sources. Environ Sci Technol 40:5750–5754

Kobayashi T, Nakagawa Y (1986) Estimation of atmospheric mercury pollution by using epiphytic lichen. J Jpn Soc Air Pollut 21:151–155 (in Japanese with English summary)

Kobayashi T, Nakagawa Y (1990) Estimation of atmospheric mercury pollution by using epiphytic lichens. –Mapping the distribution of mercury content in Parmelia tinctorum over the Hyogo Prefecture. J Jpn Soc Air Pollut 25:206–211 (in Japanese with English summary)

Kuang YW et al (2007) Heavy metals in bark of Pinus massoniana (Lamb.) as an indicator of atmospheric deposition near a smeltery at Qujiang, China. Environ Sci Pollut Res 14:270–275

Kuno H, Ohashi (1993a) Indicator plants for air pollution (1). J Jpn Soc Air Pollut 28:A45–A52 (in Japanese)
Kuno H, Ohashi (1993b) Indicator plants for air pollution (2). J Jpn Soc Air Pollut 28:A65–A76 (in Japanese)
Liu XY et al (2008) Tissue N content and ^{15}N natural abundance in epilithic mosses for indicating atmospheric N deposition in the Guiyang area, SW China. Appl Geochem 23:2708–2715
Liu XY et al (2009) Assessment of atmospheric sulfur with the epilithic moss Haplocladium microphyllum: evidences from tissue sulfur and $\delta^{34}S$ analysis. Environ Pollut 157:2066–2071
Liu XY et al (2010) Response of stable carbon isotope in epilithic mosses to atmospheric nitrogen deposition. Environ Pollut 158:2273–2281
Lu Z et al (2011) Sulfur dioxide and primary carbonaceous aerosol emissions in China and India, 1996–2010. Atmos Chem Phys 11:9839–9864
Mitsugi H (1994) Mapping of environmental aspect based on epiphytic bryophyte vegetation in Hyogo Prefecture. Environ Sci 7:313–323 (in Japanese with English summary)
Mitsugi et al (1978) Epiphytic bryophytes and lichens as the indicator of air pollution, -correlation between some air pollutants and IAP values. J Jpn Soc Air Pollut 13:26–32 (in Japanese with English summary)
Nakagawa Y, Kobayashi T (1990) Estimation of air pollution based on the distribution and the component of epiphytic lichens by means of the modified IAP method. J Jpn Soc Air Pollut 25:233–241 (in Japanese with English summary)
Nogami et al (1987) An evaluation of the air pollution by using heavy metals accumulation in *Bryum argenteum* Hedw. J Jpn Soc Air Pollut 22:347–354 (in Japanese with English summary)
Nouchi I, Aoki K (1979) Morning glory as a photochemical oxidant indicator. Environ Pollut 18:289–303, 1970
Nouchi I et al (1984) Atmospheric PAN concentrations and foliar injury to petunia indicator plants in Tokyo. J Jpn Soc Air Pollut 19:392–402, in Japanese with English summary
Ohara T et al (2007) An Asian emission inventory of anthropogenic emission sources for the period 1980–2020. Atmos Chem Phys 7:4419–4444
Ohmura Y et al (2008) Long-term monitoring of *Parmotrema tinctorum* and qualitative changes of air pollution in Shimizu Ward, Shizuoka City, Japan. J Jpn Soc Atmos Environ 43:47–54 (in Japanese with English summary)
Omura M, Murata A (1984) Epiphytic lichen and bryophyte vegetation as a bioindicator of air pollution, – the recent transition in industrial areas of Fukuoka Prefecture. J Jpn Soc Air Pollut 19:462–472 (in Japanese with English summary)
Saipunkaew et al (2007) Epiphytic macrolichens as indicators of environmental alteration in Northern Thailand. Environ Pollut 146:366–374
Sase H et al (1998a) Variation in amount and elemental composition of epicuticular wax in Japanese cedar (*Cryptomeria japonica*) leaves associated with natural environmental factors. Can J For Res 28:87–97
Sase H et al (1998b) Changes in properties of epicuticular wax and the related water loss in Japanese cedar (*Cryptomeria japonica*) affected by anthropogenic environmental factors. Can J For Res 28:546–556
Satake K et al (1996) Accumulation of lead in tree trunk bark pockets as pollution time capsules. Sci Total Environ 181:25–30
Shimizu H et al (1988) Studies on the evaluation of atmospheric quality by bryophytes -development and characteristics of a method using twin chambers. Proc Bryol Soc Jpn 4:155–161
Smith SJ et al (2011) Anthropogenic sulfur dioxide emissions: 1850–2005. Atmos Chem Phys 11:1101–1116
Sugiyama K et al (1976) Studies on lichens as a bioindicator of air pollution. I. Correlation of distribution of *Parmelia tinctorum* with SO2 air pollution. Jpn J Ecol 26:209–212
Takamatsu T et al (2000) Aerosol elements on tree leaves – antimony as a possible indicator of air pollution. Global Environ Res 4:49–60
Taoda H (1972) Mapping of atmospheric pollution in Tokyo based upon epiphytic bryophytes. Jpn J Ecol 22:125–133

Taoda H (1973) Bryo-meter, an instrument for measuring the phytotoxic air pollution. HIKOBIA 6:224–228

Taoda H (1996) Tolerance of some epiphytic bryophytes to simulated acid rain. Proc Bryol Soc Jpn 6:237–243

Thrower (1980) Air pollution and lichens in Hong Kong. Lichenologist 12:305–311

Umezu Y (1978) Mapping of air pollution intensity by epiphytic bryophyte and lichen communities in heavy industry region. Jpn J Ecol 28:143–154 (in Japanese with English summary and German captions)

Wakamatsu et al (2013) Air pollution trends in Japan between 1970 and 2012 and impact of urban air pollution countermeasures. Asian J Atmos Environ 7:177–190

Yokobori (1978) Measuring of phytotoxic air pollution based upon response of bryophytes using a filtered-air growth chamber. Jpn J Ecol 28:17–23 (in Japanese with English summary)

Part III
Case Studies in Japanese Forests

Chapter 8
Flux-Based O_3 Risk Assessment for Japanese Temperate Forests

Mitsutoshi Kitao, Yukio Yasuda, Masabumi Komatsu, Satoshi Kitaoka, Kenichi Yazaki, Hiroyuki Tobita, Kenich Yoshimura, Takafumi Miyama, Yuji Kominami, Yasuko Mizoguchi, Katsumi Yamanoi, Takayoshi Koike, and Takeshi Izuta

Abstract Ground-level ozone (O_3) levels are expected to increase over the twenty-first century, particularly in the region of East Asia. We performed an O_3 flux-based risk assessment of C sequestering capacity in an old cool temperate deciduous forest, consisting of O_3-sensitive Japanese beech (*Fagus crenata*), and in a warm temperate deciduous and evergreen forest dominated by O_3-tolerant Konara oak (*Quercus serrata*), based on long-term CO_2 flux observations. Light-saturated

---

M. Kitao (✉) • M. Komatsu • S. Kitaoka • K. Yazaki • H. Tobita
Department of Plant Ecology, Forestry and Forest Products Research Institute,
Matsunosato 1, Tsukuba 305-8687, Japan
e-mail: kitao@ffpri.affrc.go.jp; kopine@ffpri.affrc.go.jp; skitaoka3104@gmail.com; kyazaki@ffpri.affrc.go.jp; tobi@ffpri.affrc.go.jp

Y. Yasuda
Tohoku Research Center, Forestry and Forest Products Research Institute,
Nabeyashiki 92-25, Morioka 020-0123, Japan
e-mail: yassan@ffpri.affrc.go.jp

K. Yoshimura • T. Miyama • Y. Kominami
Kansai Research Center, Forestry and Forest Products Research Institute,
Nagaikyutaroh 68, Kyoto 612-0855, Japan
e-mail: mt_everysashi@hotmail.com; tmiyama@affrc.go.jp; kominy@ffpri.affrc.go.jp

Y. Mizoguchi • K. Yamanoi
Hokkaido Research Center, Forestry and Forest Products Research Institute,
Hitsujigaoka 7, Sapporo 062-8516, Japan
e-mail: pop128@affrc.go.jp; yamanoi@ffpri.affrc.go.jp

T. Koike
Department of Forest Science, Hokkaido University, Sapporo 060-8589, Japan
e-mail: tkoike@for.agr.hokudai.ac.jp

T. Izuta
Institute of Agriculture, Tokyo University of Agriculture and Technology,
Fuchu, Tokyo 183-8509, Japan
e-mail: izuta@cc.tuat.ac.jp

T. Izuta (ed.), *Air Pollution Impacts on Plants in East Asia*,
DOI 10.1007/978-4-431-56438-6_8

gross primary production, as a measure of C sequestering capacity, declined earlier in the late-growth season with increasing cumulative O_3 uptake, suggesting an earlier autumn senescence in the O_3-sensitive beech forest, but not in the O_3-tolerant oak forest.

Keywords C sequestration • O_3 flux • O_3-sensitive forest • O_3-tolerant forest • Deciduous temperate forests

8.1 Introduction

In Japan, which is located on the edge of East Asia, a continuous increase in O_3 concentration has been observed since the latter half of the 1980s, partly due to pollutants advected from foreign sources, particularly in East Asia (Richter et al. 2005; Ohara et al. 2008; Nagashima et al. 2003; Akimoto 2003). Since 2000, the flux tower sites of the Forestry and Forest Products Research Institute (FFPRI) have monitored CO_2, energy, and water vapor fluxes in several Japanese forests with different tree species. Although O_3 is known to be a detrimental air pollutant for trees, as it reduces the photosynthetic rate, increases the respiration rate, and accelerates leaf senescence (Matyssek et al. 2010), few studies have been conducted to assess the O_3 risk in forests directly. Thus, we have a unique chance to assess the O_3 risk in Japanese forests based on the long-term CO_2 flux observations.

O_3 flux-based risk assessment is more essential than O_3 exposure-based risk assessment for evaluating the physiological effects of O_3 on plants (Emberson et al. 2000; Matyssek et al. 2007). To estimate O_3 fluxes at the forest level, canopy-level stomatal conductance is required, as well as O_3 concentration over the forest, which is generally estimated by the Penman–Monteith (P–M) equation, based on energy and water flux over a forest (Gerosa et al. 2005; Monteith 1981). However, the use of this approach is valid only when the entire evaporation process in the canopy takes place through stomatal transpiration. In this context, it is difficult to conduct a continuous estimation of the canopy-level stomatal conductance using the P–M approach in temperate deciduous forests, which have regular rainfall and a period when the canopy is not closed, in spring and autumn (Biftu and Gan 2000).

To solve this problem, we have developed a novel approach (Kitao et al. 2014) combining the P–M approach with a semi-empirical photosynthesis-dependent stomatal model (Ball–Woodrow–Berry [BWB] model; Ball et al. 1987), where photosynthesis, relative humidity, and CO_2 concentration are assumed to regulate stomatal conductance. Based on the BWB model, leaf-level stomatal conductance (g_s) is estimated as:

$$g_s = g_{min} + a A_g \mathrm{rh} / C_s \tag{8.1}$$

where g_{min} denotes the minimum conductance in the dark, a is an empirical scaling parameter, A_g the gross photosynthetic rate, rh the relative humidity, and C_s the leaf surface CO_2 concentration.

In the present study, we applied the modified BWB model to estimate canopy-level stomatal conductance (G_s) taking into account the non-linear response of canopy stomatal conductance to relative humidity (Fares et al. 2013) as:

$$G_s = G_{min} + ab^{rh}\text{GPP} / C_s \tag{8.2}$$

where G_{min} denotes the canopy-level minimum conductance in the dark, a and b are empirical scaling parameters, and GPP is the gross primary production (= net ecosystem exchange [NEE] – respiration of ecosystem [R_{eco}]). We first determined the coefficients in the equation, using G_s derived by the (P–M) equation when the canopy was closed from June to August for the beech forest, and G_s derived by the (P–M) equation from June to September for the oak forest without rain (>1 mm within 24 h) for each year. Then we estimated the stomatal conductance over the canopy of temperate deciduous forests continuously, using this novel approach combining the P–M approach with a photosynthesis-dependent stomatal model (Kitao et al. 2014).

Based on the above approach, we performed flux-based assessments of O_3 effects on photosynthetic CO_2 uptake in a cool temperate deciduous forest, consisting of O_3-sensitive deciduous broadleaf tree species, Japanese beech (*Fagus crenata* Blume; 70–80 years old), and a warm temperate mixed deciduous and evergreen broadleaf forest, dominated by O_3-tolerant deciduous broadleaf tree species, Konara oak (*Quercus serrata* Thunb. ex. Murray; approximately 30 years old) (Yasuda et al. 2012; Kominami et al. 2012; Yamaguchi et al. 2011; Kitao et al. 2015).

8.2 Study Sites

The Appi forest meteorology research site (40° 00′ N, 140° 56′ E, 825 m above sea level) is located on the Appi highland in Iwate Prefecture, Japan (details of which are described in Yasuda et al. 2012) (Fig. 8.1). The site is located in a secondary cool temperate deciduous broadleaf forest primarily consisting of the Japanese beech (*Fagus crenata* Blume), which was approximately 80 years old. The canopy height was measured to be 19–20 m in 2009. There is not much vegetation on the forest floor, and evergreen trees are rarely observed. It snows heavily from November to May, with the snow depth reaching 2 m. The annual mean temperature was 5.9 °C (in 2000–2006), annual precipitation was 1,869 mm (in 2007–2009), and annual mean solar radiation was 12.7 MJ m^{-2} day^{-1} (in 2000–2006). The soil was classified as moderately moist brown forest soil.

The Yamashiro forest hydrology research site (34° 47′ N, 135° 50′ E, 220 m above sea level) is situated in the southern part of Kyoto Prefecture, Japan (details of which are described in Kominami et al. 2012) (Fig. 8.2). The site is located in a warm temperate mixed deciduous and evergreen broadleaf forest, which is built upon weathered granite. After an invasion by pine wilt disease in the 1980s, Konara oak (*Quercus serrata* Thunb. ex. Murray) has taken over and the forest is now

Fig. 8.1 Cool temperate deciduous forest consisting of Japanese beech at the Appi tower site

Fig. 8.2 Warm temperate deciduous and evergreen mixed forest, predominantly consisting of Konara oak, at the Yamashiro tower site

regenerated. The tree biomass (diameter at breast height [DBH] ≥3 cm) was estimated at 51 Mg C ha^{-1} in 1999, dominated by Konara oak, classified as a deciduous broadleaf tree species (66 % of biomass), and *Ilex pedunculosa* Miq. (an evergreen broadleaf tree species; 28 % of biomass) (Goto et al. 2003). The canopy height ranged from 6 to 20 m with an average of 12 m. The annual mean temperature was 14.7 °C (in 2000–2002), annual precipitation was 1,095 mm (in 2000–2002), and annual mean solar radiation was 11.9 MJ m^{-2} day^{-1} (in 2000–2002).

8.3 Estimation of Light-Saturated GPP

Light-saturated GPP was derived from the relationship between GPP and photosynthetic photon flux density (PPFD). The data were rejected when the friction velocity (u*) was below 0.25 m s^{-1}, but in cases of precipitation, the data were included. We regressed the relation between GPP and PPFD as follows:

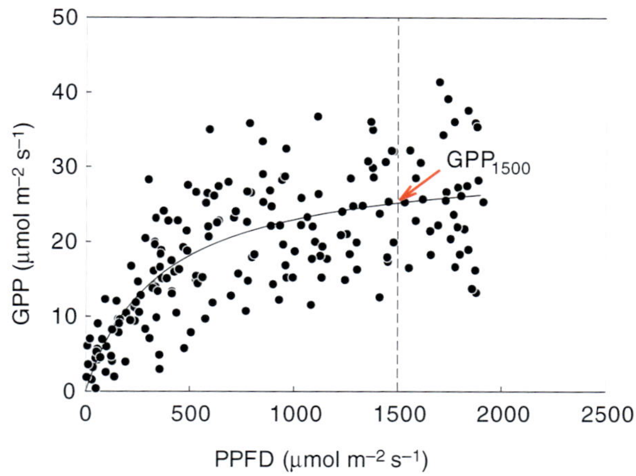

Fig. 8.3 Gross primary production (GPP) as a function of photosynthetic photon flux density (PPFD) in the beech forest, based on the pooled data of 9–10 weeks after the budbreak. Light-saturated GPP was determined as GPP at PPFD of 1,500 μmol m^{-2} s^{-1}

$$GPP = \alpha GPP_{max} PPFD / (GPP_{max} + \alpha PPFD) \tag{8.3}$$

where α denotes the ecosystem quantum yield and GPP_{max} the maximum GPP. We derived α and GPP_{max} for the pooled data at 2-week intervals from the onset of budbreak. We set the GPP at a PPFD of 1,500 μmol m^{-2} s^{-1} based on the equation as the light-saturated GPP (Fig. 8.3).

The maximum light-saturated GPP of the beech forest varied among the years, but the environmental factors determining the inter-annual variation have not yet been fully identified (Yasuda et al. 2012). A survey of each individual tree suggested that the oak forest was still growing, as the total biomass increased from 2004 to 2009 (Kominami et al. 2012). Therefore, to investigate the O_3 effects on the seasonal changes in foliar photosynthetic maturation and senescence, we used a relative unit of GPP (GPP_rel), which is calculated as:

$$\begin{aligned} GPP\_rel = & (\text{light} - \text{saturated GPP}) \\ & /(\text{the maximum light} - \text{saturated GPP during} \\ & \text{the growth period for each year}) \end{aligned}$$

The light-saturated GPP was estimated as the GPP at PPFD of 1,500 μmol m^{-2} s^{-1} based on the GPP light-response curves derived from the pooled data for 2-week intervals from the budbreak. We categorized the growth season as spring–summer (April or May to July) for leaf maturation and summer–autumn (August to October or November) for leaf senescence stages.

8.4 Estimation of Cumulative O_3 Uptake (COU)

The approach used to estimate stomatal O_3 fluxes involves several steps, according to Cieslik (2004) and Gerosa et al. (2003). The aerodynamic resistance (R_a) is calculated from measured micrometeorological parameters such as friction velocity and sensible heat flux by using the Monin–Obukhov similarity theory (see Gerosa et al. 2003, for calculation details), while the quasi-laminar layer resistance (R_b) is calculated by using the parameterization proposed by Hicks et al. (1987). We calculated the surface resistance (R_c) from the stomatal (R_{ST}) and non-stomatal resistance to O_3 (R_{NS}) as follows:

$$R_c = 1/\left(1/R_{ST} + 1/R_{NS}\right) \tag{8.4}$$

$$R_{NS} = 1/\left(\text{LAI}/r_{ext} + 1/\left(R_{inc} + R_{gs}\right)\right) \tag{8.5}$$

where LAI is the leaf area index ($m^2\ m^{-2}$), r_{ext} denotes the external leaf resistance, R_{inc} the in-canopy resistance, and R_{gs} the ground surface resistance. The maximum LAI in the beech forest was estimated from the amounts of leaf litter (Yasuda et al. 2012). Based on the field observations, we assumed that the LAI of the beech forest increased from 0 to the maximum within 1 month from the budbreak, and then decreased from the maximum to 0, also within 1 month, before the end of the foliage period. In contrast, seasonal changes in the LAI in the oak forest were measured using a plant canopy analyzer (LAI-2000; Li-Cor, Lincoln, NE, USA). The external leaf resistance (r_{ext}) is set at 2,500 s m^{-1}, the in-canopy resistance (R_{inc}) is defined as b LAI h/u*, where h is the canopy height and b an empirical constant taken as 14 m^{-1}, and R_{gs} is set at 200 s m^{-1} (Erisman et al. 1994; Simpson et al. 2012).

The stomatal O_3 flux (F_{ST}) is obtained as follows:

$$F_{ST} = C_m * R_c / \left[\left(R_a + R_b + R_c\right) R_{ST}\right] \tag{8.6}$$

where C_m denotes the ozone concentration at the measurement height. The stomatal resistance (R_{ST}) is calculated by multiplying the stomatal resistance for water vapor flux (noted as R_s) by 1.65 (Gerosa et al. 2003). R_s is expressed as $1/G_s$, where G_s is determined from the modified BWB model described above (Eq. 8.2). We estimated C_m at the flux sites using the O_3 concentration data monitored by the nearest air pollution stations (Komatsu et al. 2015).

The COU in the forest ecosystem was calculated by summing up the ozone stomatal flux (F_{ST}) during the daytime (PPFD >0) from the budbreak to a given period as follows:

$$\text{COU} = \sum F_{ST} \Delta t, \tag{8.7}$$

where the mean daily sum of F_{ST} for the days of each month when F_{ST} data were available was substituted for the rest days when F_{ST} data were unavailable.

8.5 Effects of Cumulative O_3 Uptake (COU) on C Sequestering Capacity in the Beech and Oak Forests

To quantitatively evaluate the influence of the major explanatory factor(s) on the photosynthetic capacity, multiple regression analysis was used. We initially set four explanatory factors affecting potential photosynthetic performance

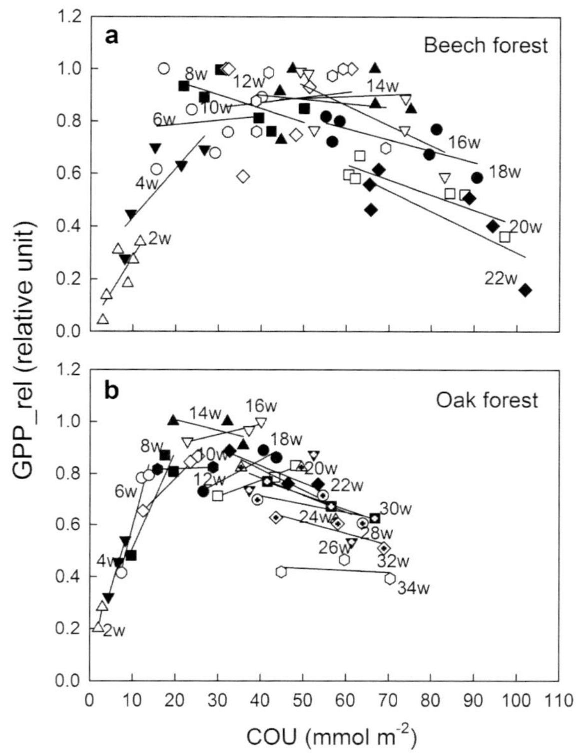

Fig. 8.4 Relationship between the relative unit of light-saturated GPP (shown as relative unit) and cumulative O_3 uptake (COU) at the end of each period from the budbreak in the beech (**a**) and oak (**b**) forests. Data are grouped by 2-week intervals from the budbreak, indicated by different symbols. Budbreaks in the beech forest occurred during April 30 to May 16, while those in the oak forest occurred during March 30 to April 8 (after Kitao et al. 2016)

(GPP_rel): leaf age, photoperiod, air temperature, and COU (Bauerle et al. 2012). GPP_rel in the spring–summer period could be explained by three explanatory factors, i.e., leaf age, photoperiod, and COU, in both the beech and oak forests. Increased COU, in addition to the photoperiod and leaf age, showed a positive effect on leaf maturation (Fig. 8.4a, b). As young leaves generally have less sensitivity to O_3 than old leaves (Kitao et al. 2015; Bohler et al. 2010), a stimulating effect of O_3 on photosynthesis at the early stage of leaf development needs to be further investigated. In contrast, GPP_rel in the summer–autumn period in the beech forest could be explained by two environmental factors, i.e., leaf age and COU. Thus, increased COU may accelerate age-dependent leaf senescence in the O_3-sensitive beech forest (Fig. 8.4a), whereas leaf senescence was not influenced by COU but was primarily influenced by air temperature in the O_3-tolerant oak forest (Fig. 8.4b). The lower O_3 sensitivity during leaf senescence in the oak forest may be attributed to the lower COU (due to the lower G_s) than in the beech forest (data not shown), as reported for the seedlings of Konara oak grown under free-air O_3 fumigation (Kitao et al. 2015).

8.6 Conclusion

In summary, our findings, based on long-term CO_2 flux observations, indicate that the photosynthetic C sequestering capacity of the forest ecosystem in an O_3-sensitive forest is potentially affected by the present-level O_3. O_3-induced earlier leaf senescence has been reported in several previous studies using open-top chambers and free-air fumigation systems (Karnosky et al. 2005; Kitao et al. 2009; Yamaji et al. 2003; Calatayud et al. 2011). In the present study, it is noteworthy that such accelerated leaf senescence induced by O_3 was apparently detected in the O_3-sensitive beech species at the real-world forest level. As the atmospheric O_3 concentrations are predicted to increase, particularly in East Asia, including Japan (Akimoto 2003; Karnosky et al. 2005; Ashmore 2005; Koike et al. 2013), earlier leaf senescence induced by elevated O_3 could cause further adverse effects on forest C sequestering in the future, particularly in forests consisting of O_3-sensitive species.

Acknowledgments Japan's Ministry of the Environment financially supported this study under a program of the Environment Research and Technology Development Fund (5B-1105, 2011–2013). We greatly appreciate Iwate prefecture and Kyoto prefecture for providing ground-based ozone data.

References

Akimoto H (2003) Global air quality and pollution. Science 302:1716–1719
Ashmore MR (2005) Assessing the future global impacts of ozone on vegetation. Plant Cell Environ 28:949–964
Ball JT et al (1987) A model predicting stomatal conductance and its contribution to the control of photosynthesis under different environmental conditions. In: Biggens J (ed) Progress in photosynthesis research. Martinus-Nijhoff Publishers, Dordrecht, pp 221–224
Bauerle WL et al (2012) Photoperiodic regulation of the seasonal pattern of photosynthetic capacity and the implications for carbon cycling. PNAS 109:8612–8617
Biftu GF, Gan TY (2000) Assessment of evapotranspiration models applied to a watershed of Canadian Prairies with mixed land-uses. Hydrol Proc 14:1305–1325
Bohler S et al (2010) Differential impact of chronic ozone exposure on expanding and fully expanded poplar leaves. Tree Physiol 30:1415–1432
Calatayud V et al (2011) Responses of evergreen and deciduous *Quercus* species to enhanced ozone levels. Environ Pollut 159:55–63
Cieslik SA (2004) Ozone uptake by various surface types: a comparison between dose and exposure. Atmos Environ 38:2409–2420
Emberson LD et al (2000) Modelling stomatal ozone flux across Europe. Environ Pollut 109: 403–413
Erisman JW et al (1994) Parametrization of surface resistance for the quantification of atmospheric deposition of acidifying pollutants and ozone. Atmos Environ 28:2595–2607
Fares S et al (2013) Testing of models of stomatal ozone fluxes with field measurements in a mixed Mediterranean forest. Atmos Environ 67:242–251
Gerosa et al (2003) Micrometeorological determination of time-integrated stomatal ozone fluxes over wheat: a case study in Northern Italy. Atmos Environ 37:777–788
Gerosa G et al (2005) Ozone uptake by an evergreen Mediterranean Forest (*Quercus ilex*) in Italy. Part I: micrometeorological flux measurements and flux partitioning. Atmos Environ 39: 3255–3266

Goto Y et al (2003) Aboveground biomass and net primary production of a broad-leaved secondary forest in the southern part of Kyoto prefecture, central Japan. Bull FFPRI 387:115–147 (in Japanese with English summary)

Hicks BB et al (1987) A preliminary multiple resistance routine for deriving dry deposition velocities from measured quantities. Water Air Soil Pollut 36:311–330

Karnosky DF et al (2005) Scaling ozone responses of forest trees to the ecosystem level in a changing climate. Plant Cell Environ 28:965–981

Kitao M et al (2009) Effects of chronic elevated ozone exposure on gas exchange responses of adult beech trees (*Fagus sylvatica*) as related to the within-canopy light gradient. Environ Pollut 157:537–544

Kitao M et al (2014) Seasonal ozone uptake by a warm-temperate mixed deciduous and evergreen broadleaf forest in western Japan estimated by the Penman-Monteith approach combined with a photosynthesis-dependent stomatal model. Environ Pollut 184:457–463

Kitao M et al (2015) Growth over-compensation against O_3 exposure in two Japanese oak species, *Quercus mongolica* var. *crispula* and *Quercus serrata*, grown under elevated CO_2. Environ Pollut 206:133–141

Kitao et al (2016) Increased phytotoxic O_3 dose accelerates autumn senescence in an O_3-sensitive beech forest even under the present-level O_3. Sci Rep 6:32549

Koike T et al (2013) Effects of ozone on forest ecosystems in East and Southeast Asia. Elsevier Dev Environ Sci 13:371–390

Komatsu M et al (2015) Estimation of ozone concentrations above forests using atmospheric observations at urban air pollution monitoring stations. J Agric Meteorol 71:202–210

Kominami Y et al (2012) Heterotrophic respiration causes seasonal hysteresis in soil respiration in a warm-temperate forest. J For Res 17:296–304

Matyssek R et al (2007) Promoting the O_3 flux concept for European forest trees. Environ Pollut 146:587–607

Matyssek R et al (2010) Advances in understanding ozone impact on forest trees: message from novel phytotron and free-air fumigation studies. Environ Pollut 158:1990–2006

Monteith JL (1981) Evaporation and surface temperature. Q J R Meteorol Soc 107:1–27

Nagashima T et al (2003) The relative importance of various source regions on East Asian surface ozone. Atmos Chem Phys 10:11305–11322

Ohara T et al (2008) Long-Term simulations of surface ozone in East Asia during 1980–2020 with CMAQ. In: Borrego C, Miranda AI (eds) NATO science for peace and security series – C: environmental security, air pollution modelling and its application XIX. Springer, Dordrecht, pp 136–144

Richter A et al (2005) Increase in tropospheric nitrogen dioxide over China observed from space. Nature 437:129–132

Simpson D et al (2012) The EMEP MSC-W chemical transport model – technical description. Atmos Chem Phys 12:7825–7865

Yamaguchi M et al (2011) Experimental studies on the effects of ozone on growth and photosynthetic activity of Japanese forest tree species. Asian J Atmos Environ 5:65–78

Yamaji K et al (2003) Ozone exposure over two growing seasons alters root-to-shoot ratio and chemical composition of birch (*Betula pendula* Roth). Glob Chang Biol 9:1363–1377

Yasuda Y et al (2012) Carbon balance in a cool-temperate deciduous forest in northern Japan: seasonal and interannual variations, and environmental controls of its annual balance. J For Res 17:253–267

Chapter 9
Tree Decline at the Somma of Lake Mashu in Northern Japan

Takashi Yamaguchi, Makoto Watanabe, Izumi Noguchi, and Takayoshi Koike

Abstract Decline of the mountain birch (*Betula ermanii*) has been observed at the somma of Lake Mashu in northern Japan. To clarify its causes, ozone (O_3) and fog chemistry were evaluated. O_3 concentration was lower in summer, but higher in spring, with a monthly mean of 50–60 ppbv. The results of on-site open-top-chamber (OTC) experiments using birch seedlings suggested the possibility that O_3 decreased the leaf photosynthetic capacity and shifted biomass allocation to the roots. The results of fog measurements showed that the fog acidity was not serious; however, nitrogen input via fog deposition was estimated to be 30 meq m^{-2} in the plant-growing season, and this could be the main nitrogen source for trees in this area. These results indicated that anthropogenic air pollutants have influenced plant growth in this area. However, there is still great uncertainty about O_3 effects on the mountain birch, such as its effects causing stomatal sluggishness, water stresses, and these effects combined with nitrogen input. Further evaluation of these factors is necessary to assess the tree decline.

Keywords Lake Mashu • Mountain birch • Tropospheric ozone • Open-top chambers • Fog deposition • Nitrogen deposition

---

T. Yamaguchi (✉) • I. Noguchi
Institute of Environmental Sciences, Environmental and Geological Research Department, Hokkaido Research Organization, N19 W12, Kita-ku, Sapporo 060-0819, Japan
e-mail: t-yamaguchi@hro.or.jp; izumi@hro.or.jp

M. Watanabe
Institute of Agriculture, Tokyo University of Agriculture and Technology, Fuchu, Tokyo 183-8509, Japan
e-mail: nab0602@cc.tuat.ac.jp

T. Koike
Department of Forest Science, Hokkaido University, Sapporo, Hokkaido 060-8589, Japan
e-mail: tkoike@for.agr.hokudai.ac.jp

T. Izuta (ed.), *Air Pollution Impacts on Plants in East Asia*,
DOI 10.1007/978-4-431-56438-6_9

9.1 Introduction

Lake Mashu, located in northeastern Japan (Fig. 9.1) is a caldera lake and is a popular sightseeing spot because of its transparency, recorded at 41.6 m in 1931, the world record at that time. There are many studies of the geology and water quality of this lake because no surface streams flow into it. This lake is also famous for the dense fog that flows in from the Pacific Ocean in the warm season, mainly from May to September.

The somma of Lake Mashu was formed about 7,000 years ago. The soil material is composed of nutrient-poor volcanic eruption products (Katsui 1955) and the humus layer in the somma is shallow, around 20 cm at most (Sakuma et al. 2014).

The dominant tree species is the Japanese mountain birch (*Betula ermanii*) and its growing season is almost same as the foggy season, from late May to September. The forest floor is tightly covered by dwarf bamboo (*Sasa senanensis*) (Igarashi 1986).

Recently, the decline of the Japanese mountain birch has been reported around Lake Mashu (Yamaguchi et al. 2012b; Sakuma et al. 2014).

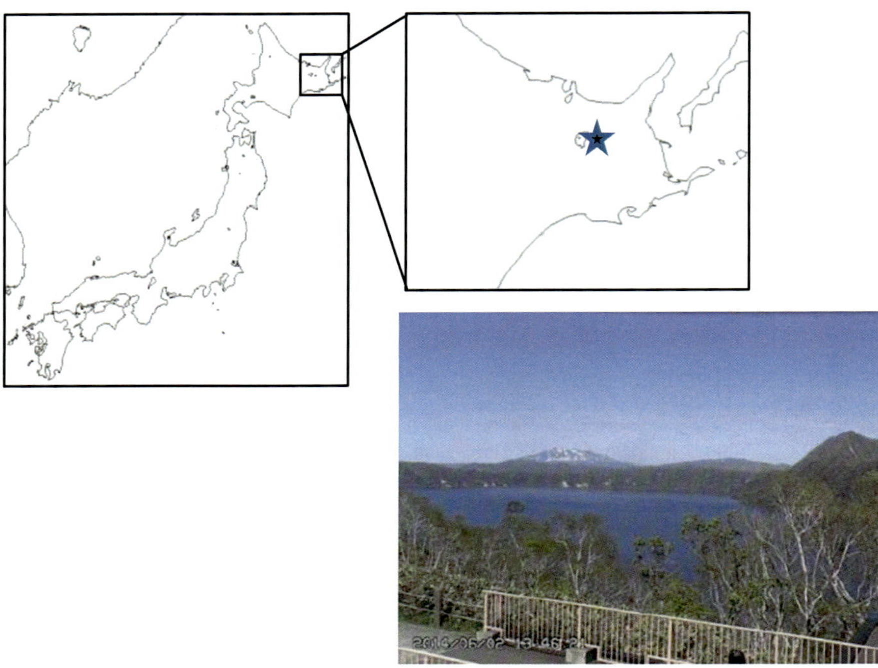

Fig. 9.1 Location of Lake Mashu

9.2 Stresses on Trees at the Somma of Lake Mashu

Tree decline areas, estimated from two aerial photographs, taken in 1977 and 2008, and an example of landscape change are shown in Fig. 9.2 (Yamaguchi and Noguchi, 2012). Trees and shrubs were growing around the path to the observatory in 1965. After 50 years, all of them were wiped out and the area had turned into grassland. The tree decline observed at the somma of Lake Mashu is still occurring now (Fig. 9.3). A healthy-looking tree died between 2005 and 2010, and then fell down in 2015. Some other dead trees also fell down in the same period; the landscape change is now being observed.

The estimated tree decline areas are distributed mainly on the southern and western outside slopes of the somma. On the western slope, road construction could be a reason; however, there is no construction on the southern slope, indicating that the tree decline is not induced by artificial disturbances or plant disease/insect damage. A field investigation revealed the influence of Sika deer (*Cervus nippon yesoensis*) overbrowsing on woody species in eastern Hokkaido, including the Lake Mashu region (Maeda Ippoen Foundation 1994). The results suggested that the damage to vegetation was serious; however, the impact on the Japanese mountain birch at the somma of Lake Mashu was relatively small because of the animals' feeding preference.

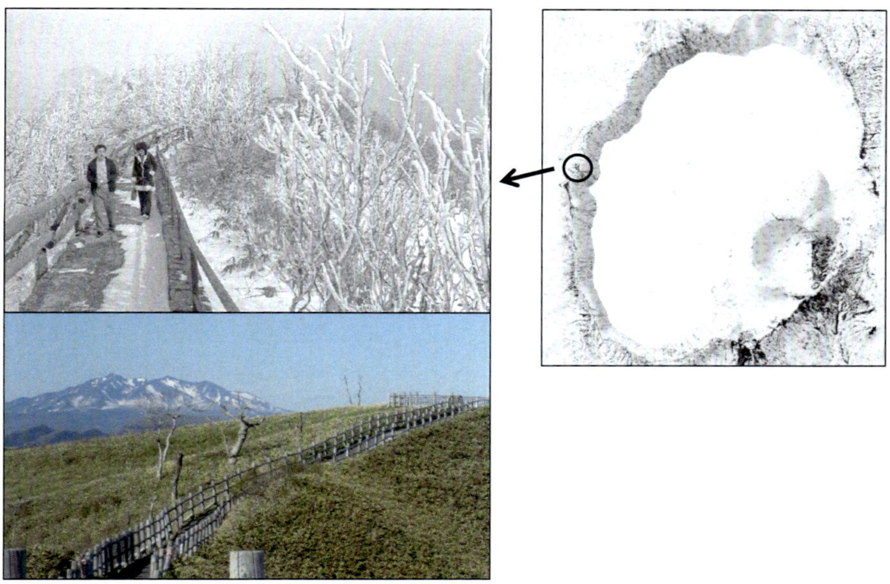

Fig. 9.2 Estimated tree decline areas (*right, black areas*) and an example of landscape change at Lake Mashu from 1965 to 2015 (Photo from 1965 provided by Fuji Taito)

Regarding anthropogenic pollution, SO_2, ozone (O_3), nitrogen oxides (NOx), acid rain, and fog acidity have been discussed as the causes of tree decline (Izuta 2001). Even Hokkaido, located at the northeastern edge of Japan, is under the influence of O_3 production in the Asian continent (Sudo and Akimoto 2007; Koike et al. 2013).

In Hokkaido, long-term monitoring of acid deposition, including rain and gases, has been carried out since the 1980s. The results show that rain acidity and the deposition of air pollutants in Hokkaido were relatively low compared with findings in other regions of Japan (EANET 2013). However, acid fog has been detected over the northwestern Pacific Ocean (Sasakawa and Uematsu 2002; Yoshida et al. 2007) and in coastal regions in eastern Hokkaido (Nishio et al. 1995; Tanaka et al. 1998). Previous studies have pointed out that the deposition of air pollutants via fog could be the cause of tree decline in central Japan near the Tokyo metropolitan area (Igawa et al. 1998, 2002).

Against this background, O_3 concentration measurements and on-site experiments with open-top chambers (OTCs) have been carried out (Tatsuda et al. 2010; Hoshika et al. 2013). Soil nutrients and water condition were also examined as predisposing factors for tree decline (Sakuma et al. 2013, 2014) , and fog chemistry and deposition have also been investigated (Yamaguchi et al. 2013, 2015). In this chapter, we will introduce the results of these studies.

Fig. 9.3 An example of tree decline in the past decade

9.3 Studies at Lake Mashu

9.3.1 Ozone Concentration Measurements

9.3.1.1 Passive Sampler

The measurement of O_3 started from 2006, using the Ogawa passive sampler because of power supply difficulties. At the beginning of the study, SO_2 and NOx were also measured; however, their concentrations were low enough not to affect vegetation. The O_3 concentration was higher in spring, with 50–60 ppbv as the monthly mean (Fig. 9.4) (Yamaguchi et al. 2010).

The monthly mean O_3 concentration showed clear seasonal changes. It was higher in spring, at approximately 60 ppbv, and the highest concentration, 75 ppbv, was recorded in March 2007. The concentration in summer after the foliation stage, May to June, ranged from 20 to 40 ppbv. For a more detailed evaluation, we started continuous monitoring at the site.

9.3.1.2 O_3 Monitoring

Since 2009, the concentration of O_3 has been measured by a continuous ozone monitor (Model 1150, Dylec Co. Ltd., Ibaraki, Japan); (Yamaguchi et al. 2010). The results with this monitor showed the well known seasonal change of O_3 concentration (e.g. Oltmans 1981), high in spring and low in summer (Fig. 9.5), as shown by the passive sampler. In addition, some high-concentration O_3 events, ranging fom 60 to 80 ppbv, were observed, mainly in spring, with the highest concentration, 84 ppbv, observed in April 2009. After the foliation period, the highest concentration was 63 ppbv, in June. These short-term high-concentration events were also

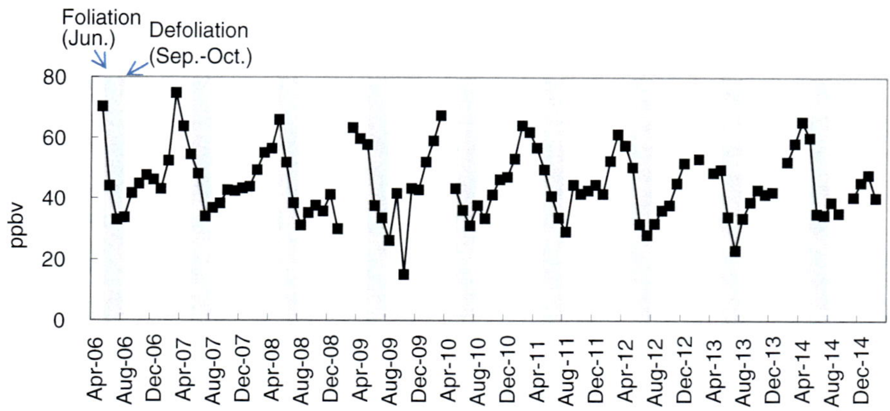

Fig. 9.4 Monthly mean ozone concentration measured with passive sampler. *Shaded areas* indicate foliation period

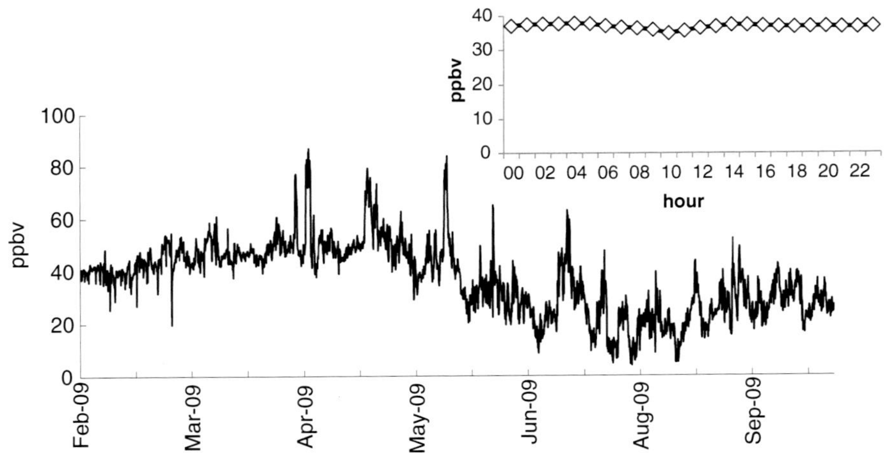

Fig. 9.5 The concentration change of O_3 and its diurnal variation (*right top*) in 2009

observed at other monitoring stations in Hokkaido, indicating that long-range transport of air pollution had occurred.

AOT40 (Accumulated Ozone Exposure over a threshold of 40 ppbv) in 2012 was calculated to be 1.2 ppm h (Ministry of the Environment, Japan 2014). This value seemed much lower than the threshold level (see Chap. 5); however, the small diurnal variation of O_3 indicated that night-time O_3 concentration was relatively high; Matyssek et al. (1995) pointed out that nighttime exposure of O_3 to plants should be considered.

Considering these results, it was thought that O_3 could cause damage to the vegetation at Lake Mashu. Consequently on-site OTC experiments began.

9.4 On-Site Experiments with Open-Top Chambers

As mentioned above, the O_3 concentration at the somma of Lake Mashu seems to be high enough to induce growth reduction in birch species. Therefore, we carried out OTC experiments to examine the effects of O_3 at the somma of Lake Mashu on the growth and photosynthetic traits of two common birch species in Japan (mountain birch and white birch) (Hoshika et al. 2013) (Fig. 9.6).

Two-year-old seedlings of the two birch species were grown in OTCs ($1.2 \times 1.2 \times 1.2$ m) for 3 months from June 2 to September 1, 2009. The OTCs were located in the observatory. We set up two gas treatments, charcoal-filtered air (CF) and non-filtered air (NF). Average concentrations of O_3 in the OTCs with CF and NF treatment during the experiment were 14.1 and 18.5 nmol mol^{-2}, respectively.

On 1 September, 2009, we evaluated the photosynthetic traits of the birch seedlings. The intercellular CO_2 concentration (C_i) response curve of the net photosynthetic rate (A), i.e., the A/C_i curve, was analyzed (Farquhar et al. 1980, see Chap. 5). We deter-

Fig. 9.6 The open-top chambers located at the somma of the Lake Mashu

Table 9.1 Gas exchange traits of leaves of mountain birch seedlings grown in charcoal-filtered air (CF) and non-filtered air (NF) open-top chambers at the somma of Lake Mashu measured on September 1, 2009

	CF	NF	*t*-test
A380	9.3 (0.1)	9.1 (0.4)	n.s.
G_s	0.25 (0.03)	0.21 (0.04)	n.s.
V_{cmax}	45.4 (2.7)	46.9 (2.4)	n.s.
J_{max}	107.8 (3.8)	99.5 (3.1)	*
A_{max}	21.2 (0.6)	19.2 (0.2)	**

After Hoshika et al. (2013)
Each value is the mean (SD) of three replications rate at 1700 mol mol^{-1} CO_2
A380 net photosynthetic rate at 380 mol mol^{-1} CO_2, *Gs* stomatal conductance to water vapor, *Vcmax* maximum rate of carboxylation, *Jmax* maximum rate of electron transport, *Amax* net photosynthetic
t-test: *$P<0.05$; **, $P<0.01$; *n.s.* not significant

mined *A* at 380 and 1,700 μmol mol^{-1} CO_2 (A_{380} and A_{max}, respectively), and determined the stomatal conductance at 380 μmol mol^{-1} CO_2 (G_s), the maximum rate of carboxylation (V_{cmax}), and the maximum rate of electron transport (J_{max}). At the end of the experiment (2 September, 2009), all seedlings were harvested and the dry mass of each organ (i.e., stems, branches, leaves, and roots) and the leaf area were determined.

Significantly lower J_{max} and A_{max} values in the mountain birch were found for NF treatment as compared with CF treatment (Table 9.1), although there was no significant difference between CF and NF for the other gas exchange parameters. These results suggest that O_3 may inhibit ribulose-1,5-bisphosphate (RuBP) regeneration in the Calvin cycle in the mountain birch; this cycle depends on the electron transport rate in the thylakoid membrane and/or the supply or utilization of inorganic phosphate (e.g. Farquhar et al. 1980; Sharkey 1985). White birch seedlings had lower G_s values than the mountain birch seedlings (Table 9.1), thus possibly leading to lower stomatal O_3 uptake. Similar findings were reported in another OTC experiment with

Table 9.2 Leaf area and dry mass of plant organs of mountain birch and white birch seedlings grown in charcoal-filtered air (CF) and non-filtered air (NF) open-top chambers at the somma of Lake Mashu measured on September 2, 2009

	Mountain birch		White birch		Two-way ANOVA		
	CF	NF	CF	NF	O_3	Species	O_3 × species
Leaf area (cm^2)	154.5 (11.5)a	187.7 (24.6)a	345.2 (41.9)b	421 (46.5) b	*	***	n.s.
Dry mass (g)							
Leaf	0.88 (0.03)a	1.01 (0.07)a	1.97 (0.20)b	2.31 (0.26)b	*	***	n.s.
Branch	0.54 (0.01)a	0.52 (0.10)a	0.97 (0.14)b	1.26 (0.11)c	0.053	***	*
Stem	0.78 (0.10)a	0.76 (0.03)a	2.12 (0.15)b	2.37 (0.34)b	n.s.	***	n.s.
Root	3.33 (0.14)a	2.97 (0.20)a	6.27 (0.22)b	7.43 (0.63)c	0.087	***	**
Whole plant	5.42 (0.12)a	5.21 (0.33)a	11.33 (0.58)b	13.38 (1.17)c	*	***	*
T/R ratio	0.66 (0.04)a	0.76 (0.02)b	0.82 (0.04)b	0.80 (0.02)b	0.069	***	*

After Hoshika et al. (2013)
Standard deviation is shown in parentheses (n = 3).
Two-way analysis of variance (ANOVA): * $P < 0.05$; ** $P < 0.01$; *** $P < 0.001$; n.s. not significant. Actual P values were shown when $0.05 < P < 0.1$
Different letters show significant differences among values in each row (Tukey HSD test, $p < 0.05$).
T/R ratio: ratio of above-ground dry mass to below-gorund dry mass

the Japanese birch species (Hoshika et al. 2012), in which the authors reported that O_3 exposure of 60 nmol mol^{-1} caused a reduction in branch biomass and the net photosynthetic rate and an increase in the number of defoliated leaves in the mountain birch, while no adverse effect caused by O_3 was recorded in the white birch. The limitation of stomatal O_3 uptake in the white birch may lead to a low level of O_3 stress. Actually, it was found that the mountain birch has the highest susceptibility to elevated O_3 among three species of birch native to Hokkaido (Mao et al. 2012).

Ozone affected the growth and allocation pattern of the two species of birch seedlings in our study. The leaf area of birch seedlings in the NF group was significantly larger than that in the CF group (Table 9.2). A significant increase in leaf dry mass was found in the NF group compared with the CF group. Although higher whole-plant dry mass was observed in the white birch under NF treatment as compared with CF treatment, the whole-plant dry mass of the mountain birch did not significantly differ between the gas treatments. An interactive effect of O_3 and species on the ratio of above-ground dry mass to below-ground dry mass (T/R ratio) was found. The T/R ratio in the CF group was lower than that in the NF group for mountain birch seedlings, while O_3 did not affect the T/R ratio of the white birch seedlings. Figure 9.7 shows the ratio of leaf dry mass or root dry mass to whole-plant dry mass in mountain birch seedlings. Higher ratios of the leaf dry mass and

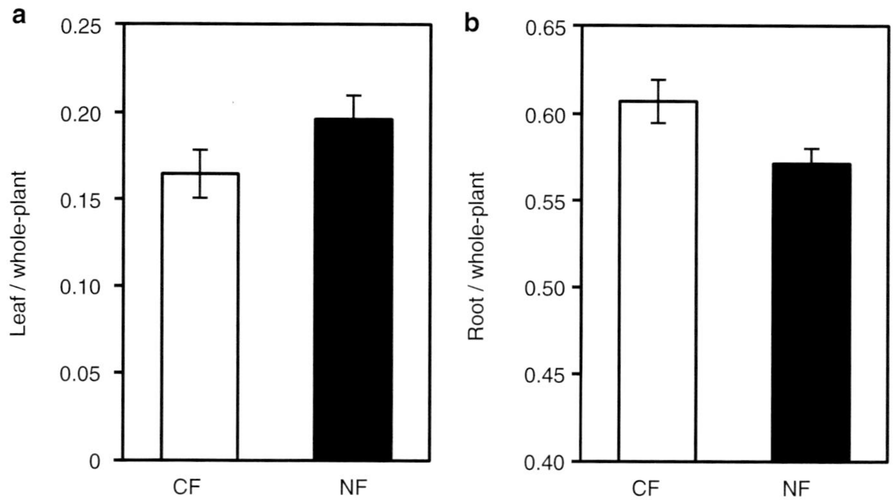

Fig. 9.7 The ratios of leaf dry mass (**a**) and root dry mass (**b**) to whole-plant dry mass of mountain birch seedlings grown in charcoal-filtered air (CF) and non-filtered air (NF) in open-top chambers at the somma of Lake Mashu, measured on September 1, 2009. There was a significant difference between CF and NF treatment in both ratios ($P<0.05$)

smaller ratios of the root dry mass to the whole-plant dry mass were recorded for the NF condition as compared with CF. Studies of the European birch (*Betula pendula*) also reported that O_3 induced similar growth enhancement in leaves, considered as a compensation response to elevated O_3 (Karlsson et al. 2003; Wittmann et al. 2007).

From the OTC experiments, we conclude that ambient O_3 at the somma of Lake Mashu has the potential to decrease leaf photosynthetic capacity and may shift the allocation of biomass to above-ground rather than below-ground in the mountain birch (Hoshika et al. 2013). The change in allocation pattern may adversely affect water and nutrient status in mountain birches. In general, the leaves have a higher demand for water and nutrients to maintain their high metabolic activities, while roots have the function of water/nutrient uptake from soil. Birches are infected with symbiotic ectomycorrhiza (ECM), and Wang et al. (2012) reported lower ECM infection rates under NF treatment in another OTC experiment carried out at the same place in 2011–2012. Similar results have been reported in hybrid larch seedlings in an artificial O_3 fumigation study (Wang et al. 2015). Wang et al. (2015) also reported less infection by ECM and different ectomycorrhizal fungal species composition in larch roots at elevated O_3 compared with findings at ambient O_3. ECM fungi help the host plant obtain water and nutrient uptake from soil in exchange for carbohydrate supplied by the host plant. According to Sakuma et al. (2014), the soil at the somma of Lake Mashu is shallow, especially in the area with declining mountain birch. Also, lower leaf water potential in mountain birch grown in the field has been observed even in relatively healthy trees (Sakuma et al. 2013). There is a possibility that O_3 at the somma of Lake Mashu may adversely affect the health of mountain birch through an increase of water/nutrient demand with a decrease of water/nutrient supply by roots, including symbiosis with ECM fungi.

9.5 Fog

The northern North Pacific region is an area with one of the most frequent sea fog occurrences in the world, and eastern Hokkaido is very frequently covered by advected sea fog during spring and summer (Hori 1953; Yoshihiro et al. 1985), and acid fog has been detected in the coastal areas (Nishio et al. 1995; Tanaka et al. 1998; Honma et al. 2000).

Generally, fog contains higher concentrations of air pollutants than rain; accordingly, some previous studies have pointed out that fog could be a cause of tree decline. In Japan, some previous studies about the relationship between fog and tree decline have pointed out that the deposition of air pollutants via fog could be the cause of tree decline near urban areas (Igawa et al. 1998, 2002) (see Chap. 10).

Thus, fog acidity was thought to be a cause of tree decline, and fog studies have been carried out since 2006 (Yamaguchi et al. 2012a, b, 2013, 2015) (Fig. 9.8).

9.5.1 *Fog Acidity*

Fog water sampling was conducted at the southern part of the somma using an automated active-string sampler during the plant-growing season (June–October), and samples were recovered weekly in 2007–2010 and daily in 2011 and 2012. Major inorganic ion components and the acidity of fog water were analyzed. The frequencies of acidic fog, below pH 4, were 0–3 in the weekly sampling periods and 3–7 in the daily sampling periods (Fig. 9.9). The acidification was thought to be caused by anthropogenic pollutants transported from domestic/foreign regions, and a volcanic plume (Yamaguchi et al. 2012b). However, the mean fog acidity in the two sampling periods was pH 4.6, indicating that the fog acidity at Lake Mashu was relatively neutral compared with the fog acidity in urban areas. Considering previous studies about the relationship of fog/cloud acidity to plant injury (see Chap. 15), the frequency of acidic fog/acidity of fog water at Lake Mashu was thought not to be the main reason for the tree decline.

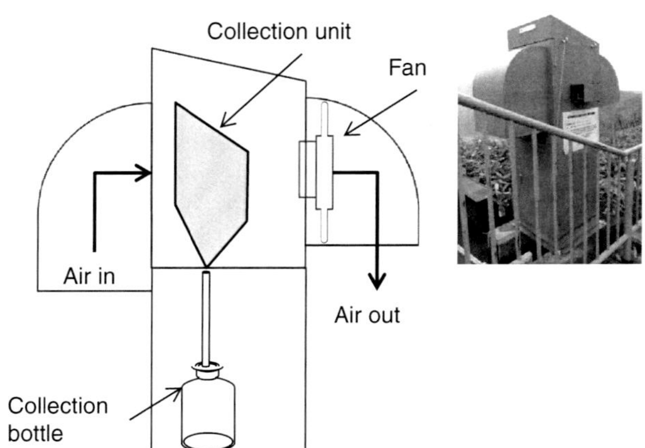

Fig. 9.8 Active fog sampler

9.5.2 Nitrogen Ions in Fog Water

The monthly mean concentrations of major ions in fog water at Lake Mashu are shown in Table 9.3 (summarized from Yamaguchi et al. 2015). NH_4^+ and SO_4^{2-} were the most abundant cations and anions, and their mean concentrations were 175 and 147 μeq L^{-1}, respectively. The mean NH_4^+ concentration was higher than those reported in previous studies in rural areas. These results indicate that NH_3 gas emitted from agricultural sources, such as fertilizer and manure, plays a role as a neutralizer of fog acidity at Lake Mashu. NH_3 gas and NH_4^+ particle

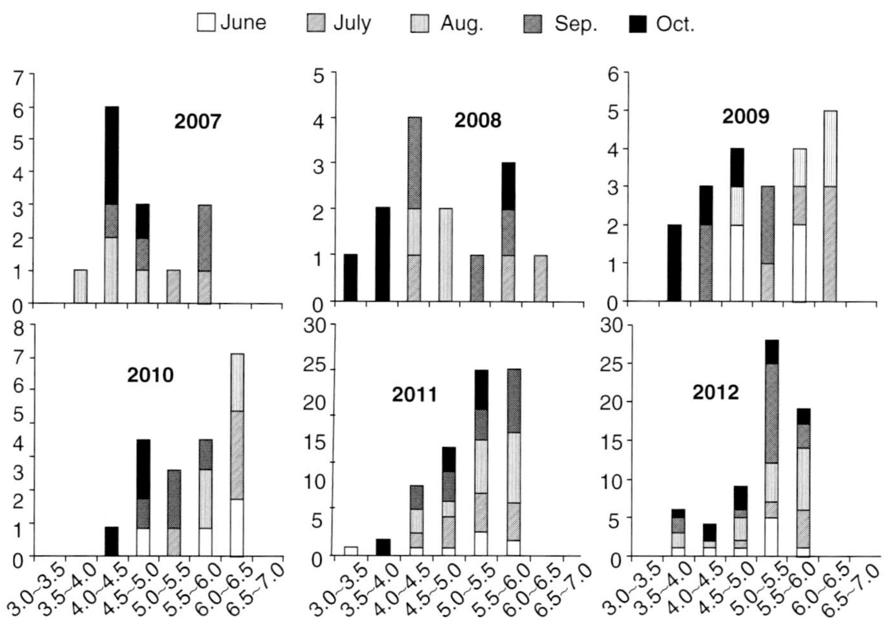

Fig. 9.9 pH frequency distributions of fog water

Table 9.3 Summary of fog acidity and the concentrations of major inorganic ions at the somma of Lake Mashu from 2006 to 2012

Year	pH	H^+	NH_4^+	Na^+	K^+	Mg^{2+}	Ca^{2+}	Cl^-	NO_3^-	SO_4^{2-}
		(μeq L^{-1})								
2006	4.48	33	205	127	14	28	14	144	89	174
2007	4.60	25	182	94	8.7	21	12	109	66	153
2008	4.29	51	205	74	13	20	14	87	76	206
2009	4.53	30	159	77	7	16	10	90	53	139
2010	5.07	8.5	169	70	7.1	16	11	82	51	129
2011	4.79	16	200	52	8.9	14	12	74	58	160
2012	4.66	22	156	61	7.2	13	9.9	72	51	123
2006–2012 mean	4.61	24	175	76	8.8	17	11	90	60	147

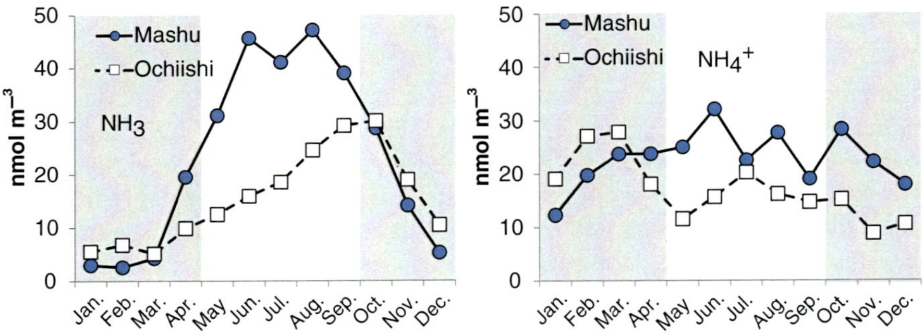

Fig. 9.10 The annual concentration changes of NH_3 gas and NH_4^+ particles at Lake Mashu and at the Ochiishi coastal site

concentrations at Lake Mashu were larger than these concentrations in the examined coastal area (unpublished data, Yamaguchi and Noguchi, Fig. 9.10), also suggesting a contribution from NH_3 emissions in the agricultural area in eastern Hokkaido.

9.5.3 Fog and Nitrogen Deposition

The nitrogen load decreases the diversity of plant communities. Previous studies of fog deposition in mountainous areas have suggested that the depositions of fog water and its contained air pollutants were non-negligible (Klemm and Wrzesinsky 2007), besides, these depositions were especially important at the forest edge (Weathers et al. 2001). Considering the high concentration of NH_4^+ in fog water at Lake Mashu, fog deposition could play a major role there as a nitrogen source. Therefore, fog-water deposition and nitrogen deposition via fog water were measured, by a throughfall method (Fig. 9.11), and were estimated by a model calculation based on Katata et al. (2008) and Yamaguchi et al. (2013, 2015).

The fog-water deposition in the plant-growing season ranged from 100 to 160 mm as precipitation, corresponding to approximately 20 % of the rainfall in the same period. Nitrogen deposition via fog water (NH_4^+ and NO_3^-) was estimated to be approximately 30 meq m^{-2} in 5 months (4 kg-N/ha) (Fig. 9.12). At rural sites in Hokkaido, annual nitrogen input via rainfall in recent years was approximately 20–36 meq m^{-2} (EANET 2013) and the nitrogen amount in snow cover at those sites ranged from 1 to 5 meq m^{-2} (published data on the Web, Hokkaido Research Organization, Institute of Environmental Sciences 2012). Nitrogen in dry deposition was estimated to be 5 meq m^{-2} in a rural area in Hokkaido where NH_3 and NH_4^+ concentrations were similar to those at Lake Mashu (EANET 2013). Considering these results, fog deposition could be the main nitrogen source for the trees at Lake Mashu.

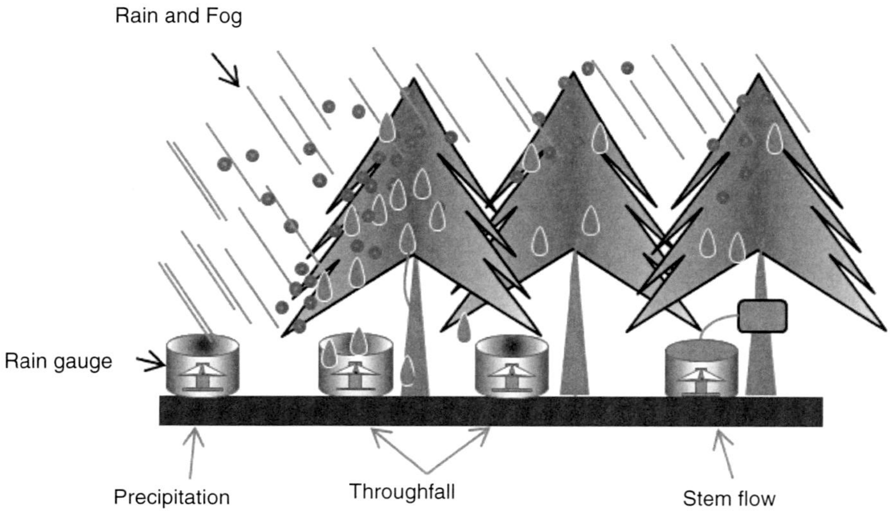

Fig. 9.11 Throughfall method

Fig. 9.12 Fog deposition and nitrogen deposition

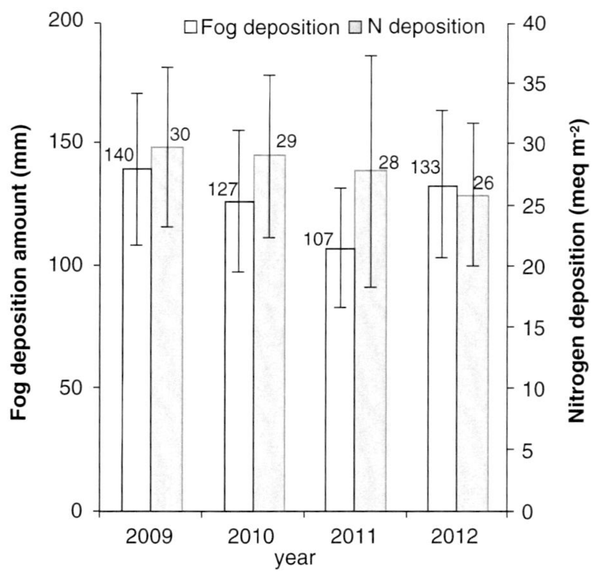

9.6 Conclusion

Ozone concentration was higher in spring at Lake Mashu, with a monthly mean of 60–70 ppbv, and high-concentration events were confirmed. The on-site OTC experiments suggested adverse influences on mountain birch seedlings, and these might be connected with the water/nutrient demand and changes in ECM fungi.

However, there are large uncertainties surrounding O_3 effects on the mountain birch, such as O_3 effects causing stomatal sluggishness, water stresses, and these effects combined with nitrogen input. More evaluation of these factors is needed.

Acidified fog was confirmed, but the frequency was low. Fog acidity itself was thought not to be the main cause of the tree decline, while nitrogen deposition via fog was estimated to be comparable to that via annual rainfall. Furthermore, increasing trends of nitrogen deposition in the snow cover in Hokkaido (Yamaguchi and Noguchi 2015) have been reported. It is necessary to conduct more comprehensive evaluation of nitrogen deposition and its effect on vegetation around Lake Mashu.

There are few records about vegetation at Lake Mashu in the early twentieth century, and formation processes or species changes have been reported in later twentieth century studies, although only dominant tree species were reported (Shibakusa and Matushita 1962; Igarashi 1986). Consequently, the forest development process for these 50 years in this region is unknown, and it should be noted that the "tree decline" phenomenon might be part of natural succession or transition.

To clarify the causes of tree decline at the somma of Lake Mashu, more long-term monitoring of tree growth is needed, including the above- and below-ground processes in the mountain birch, in addition to continuous measurements of air pollutants such as O_3 and nitrogen deposition,.

References

EANET (2013) Data report on the acid deposition in the east Asian region 2012. http://www.eanet.asia/product/datarep/datarep12/datarep12.pdf. Accessed 1 Dec 2015

Farquhar GD, von Caemmerer S, Berry JA (1980) A biochemical model of photosynthetic CO_2 assimilation in leaves of C_3 species. Planta 149:78–90. doi:10.1007/BF00386231

Hokkaido Research Organization, Institute of Environmental Sciences (2012) Ion components in snowcover. http://www.hro.or.jp/list/environmental/research/ies/seisakuka/acid_rain/acid_snow_data.xls. Accessed 25 Dec 2015

Honma S, Tanaka H, Teramoto S et al (2000) Effects of naturally-occurring acid fog on inflammatory mediators in airway and pulmonary functions in asthmatic patients. Respir Med 94: 935–942

Hori T (1953) Studies on fogs in relation to fog-preventing forest. Tanne Trading Company, Sapporo

Hoshika Y, Watanabe M, Inada N, Koike T (2012) Growth and leaf gas exchange in three birch species exposed to elevated ozone and CO_2 in summer. Water Air Soil Pollut 223:5017–5025. doi:10.1007/s11270-012-1253-y

Hoshika Y, Tatsuda S, Watanabe M et al (2013) Effect of ambient ozone at the somma of Lake Mashu on growth and leaf gas exchange in *Betula ermanii* and *Betula platyphylla var. japonica*. Environ Exp Bot 90:12–16

Igarashi T (1986) Forest Vegetation of the Akan National Park, Hokkaido, Japan. Res Bull Coll Exp For Hokkaido Univ 43:335–494

Igawa M, Tsutsumi Y, Mori T, Okochi H (1998) Fogwater chemistry at a mountainside forest and the estimation of the air pollutant deposition via fog droplets based on the atmospheric quality at the mountain base. Environ Sci Technol 32:1566–1572

Igawa M, Matsumura K, Okochi H (2002) High frequency and large deposition of acid fog on high elevation forest. Environ Sci Technol 36:1–6

Izuta T (2001) Forest decline. In: Nouchi I (ed) Change of atmospheric environment and plant responses. Yokendo Ltd, Tokyo, pp 168–208

Karlsson PE, Uddling J, Skärby L et al (2003) Impact of ozone on the growth of birch (*Betula pendula*) saplings. Environ Pollut 124:485–495

Katata G, Nagai H, Wrzesinsky T et al (2008) Development of a land surface model including cloud water deposition on vegetation. J Appl Meteorol Climatol 47:2129–2146. doi:10.1175/2008JAMC1758.1

Katsui Y (1955) Geology and petrology of the Volcano Mashu Hokkaido, Japan. J Geol Soc Jpn 61:481–495. doi:10.5575/geosoc.61.481

Klemm O, Wrzesinsky T (2007) Fog deposition fluxes of water and ions to a mountainous site in Central Europe. Tellus B 59:705–714

Koike T, Watanabe M, Hoshika Y et al (2013) Effects of ozone on forest ecosystems in East and Southeast Asia. In: Matyssek R, Clarke N, Cudlín P et al (eds) Climate change, air pollution and global challenges: understanding and solutions from forest research. Elsevier, Oxford, pp 371–390

Maeda Ippoen Foundation (1994) Assessment report of the vegetation damages and landscape changes caused by the population increase of the sika deer in Akan National Park, Japan

Mao Q, Hoshika Y, Watanabe M, Koike T (2012) Symptom of ozone injured leaves in 3 kinds of birch species in Hokkaido. Boreal For Res 60:35–38

Matyssek R, Gunthardt-Goerg MS, Maurer S, Keller T (1995) Nighttime exposure to ozone reduces whole-plant production in *Betula pendula*. Tree Physiol 15:159

Ministry of the Environment, Japan (2014) Long-term acid deposition monitoring report (2008–2012). http://www.env.go.jp/air/acidrain/monitoring/rep3.html. Accessed 1 Dec 2015

Nishio F, Oda N, Tomisawa H (1995) Acid fog in Kiritappu marsh, Hokkaido, Japan. J Hokkaido Univ Educ Kushiro 27:169–184

Oltmans SJ (1981) Surface ozone measurements in clean air. J Geophys Res Oceans 1978–2012 86:1174–1180

Sakuma A, Watanabe M, Wakamatsu A et al (2013) Relationship between the vigor of mountain birch and soil factors at Lake Mashu somma. Boreal For Res 61:105–106

Sakuma A, Watanaabe M, Wakamatsu A et al (2014) Soil properties and moisture characteristics of leaves of *Betula ermanii* decline in Lake Mashu somma. Boreal For Res 62:61–64

Sasakawa M, Uematsu M (2002) Chemical composition of aerosol, sea fog, and rainwater in the marine boundary layer of the northwestern North Pacific and its marginal seas. J Geophys Res Atmos 1984–2012 107:ACH 17–1–ACH 17–9

Sharkey T (1985) Photosynthesis in intact leaves of C_3 plants: physics, physiology and rate limitations. Bot Rev 51:53–105. doi:10.1007/BF02861058

Shibakusa R, Matushita A (1962) Vegetation on the islet of Kamuishu, Lake Mashu, Prov Kushiro, Hokkaido in Japan. Jpn J Ecol 12:89–94

Sudo K, Akimoto H (2007) Global source attribution of tropospheric ozone: long-range transport from various source regions. J Geophys Res 112:D12302

Tanaka H, Honma S, Nishi M et al (1998) Acid fog and hospital visits for asthma: an epidemiological study. Eur Respir J 11:1301–1306

Tatsuda S, Watanabe M, Imori M et al (2010) Effects of ambient ozone in somma of lake Mashu on seedlings of two *Betula* species evaluating by Open-top Chamber method. Trans Meet Hokkaido Branch Jpn For Soc 58:21–22

Wang Y, Yu W, Pan Y, Wu D (2012) Acid neutralization of precipitation in Northern China. J Air Waste Manag Assoc 62:204–211

Wang X, Qu L, Mao Q et al (2015) Ectomycorrhizal colonization and growth of the hybrid larch F 1 under elevated CO_2 and O_3. Environ Pollut 197:116–126

Weathers KC, Cadenasso ML, Pickett ST (2001) Forest edges as nutrient and pollutant concentrators: potential synergisms between fragmentation, forest canopies, and the atmosphere. Conserv Biol 15:1506–1514

Wittmann C, Matyssek R, Pfanz H, Humar M (2007) Effects of ozone impact on the gas exchange and chlorophyll fluorescence of juvenile birch stems (*Betula pendula Roth.*). Environ Pollut 150:258–266

Yamaguchi T, Noguchi I (2015) Long-term trends for nitrate and sulfate ions in snowcover on Hokkaido, northern Japan. J Agric Meteorol 71:196–201. doi:10.2480/agrmet.D-14-00056
Yamaguchi T, Noguchi I, Eguchi M (2010) Ambient ozone concentration around Lake Mashu, Hokkaido, Japan. Trans Meet Hokkaido Branch Jpn For Soc 58:123–124
Yamaguchi T, Noguchi I, Kouda H (2012a) Temporal and spatial change of tropospheric ozone in Hokkaido. Rep Inst Environ Sci 2:29–34
Yamaguchi T, Sakai S, Noguchi I, et al (2012b) Estimation of tree decline area and fogwater chemistry around Lake Mashu. Boreal For Res 60:45–46
Yamaguchi T, Noguchi I, Watanabe Y et al (2013) Aerosol deposition and behavior on leaves in cool-temperate deciduous forests-Part 2: characteristics of fog water chemistry and fog deposition in Northern Japan. Asian J Atmos Environ 7:8–16
Yamaguchi T, Katata G, Noguchi I et al (2015) Long-term observation of fog chemistry and estimation of fog water and nitrogen input via fog water deposition at a mountainous site in Hokkaido, Japan. Atmos Res 151:82–92. doi:10.1016/j.atmosres.2014.01.023
Yoshida K, Narita Y, Griessbaum F et al (2007) Chemical composition and size distribution of sea fog over the northern North Pacific. Geochemistry 41:165–172
Yoshihiro M, Taniguchi T, Kikuchi K (1985) Observations on the atmospheric electrical environment of sea fog in the Pacific Coast of Eastern Hokkaido, Japan. Geophys Bull Hokkaido Univ 46:13–31. doi:10.14943/gbhu.46.13

Chapter 10
Decline of *Fagus crenata* in the Tanzawa Mountains, Japan

Yoshihisa Kohno

Abstract The Tanzawa Mountains are located in the southwestern part of the Kanto District of Japan, and many hikers and climbers visit the area from several gateways, leading to the overuse of trails and the retreat of vegetation. Siebold's beech (*Fagus crenata*) forests are distributed in the high-elevation areas. Beech growth on the southern slopes, along the ridgeline, and around peaks has declined significantly. Recent ozone monitoring data suggest that high ozone concentrations may be a possible chronic cause of the loss of beech vitality. Outbreaks of sawfly and repeated sawfly attacks are fatal for the weakened beech trees. Another indirect biotic factor is the increased population of sika deer (*Cervus nippon*), which destroy ground vegetation and the related community balance. The factors affecting beech decline in the Tanzawa Mountains are complicated, and further scientific research activities in various fields are required to understand the phenomena and to recover the beech forest vegetation.

Keywords Tanzawa Mountains • Forest decline • *Fagus crenata* • Ozone • Beech sawfly • Sika deer

10.1 Background

The Tanzawa Mountains are a mountain range of about 40 kha in the Kanto District of Japan. The mountain range covers the northwestern part of Kanagawa Prefecture and borders Yamanashi and Shizuoka Prefectures.

The highest peak is Mt. Hirugatake (1,673 m) and some other peaks are Hinokiboramaru (1,601 m), Tanzawayama (1,567 m), and Tonodake (1,491 m). Siebold's beech (also known as Japanese beech; *Fagus crenata*) forests are located at about 1,000 m above sea level, and below are *Quercus* forests. A fir (*Abies firma*)

Y. Kohno
Central Research Institute of Electric Power Industry,
1646 Abiko, Abiko City, Chiba 270-1194, Japan
e-mail: kohno@peach.ocn.ne.jp

T. Izuta (ed.), *Air Pollution Impacts on Plants in East Asia*,
DOI 10.1007/978-4-431-56438-6_10

Fig. 10.1 Fir forest at Mt. Oyama

forest is distributed in the zone between the *Fagus* and *Quercus* forests. The Tanzawa Mountain range was registered as a Kanagawa Prefectural natural park in 1960 and as a National park in 1965. As many hikers and climbers are visiting, trails are overused and vegetation along the trails has retreated (Koshiji et al. 1996; Konohira 2007).

In the 1970s, firs standing on the southeastern slope of Mt. Oyama (1,252 m) showed a decline and in the 1980s, a decline in Japanese beech (*Fagus crenata*) became apparent in the ridge between Mt. Tonodake and Hirugatake, and around the peak of Mt. Hinokiboramaru. Also, natural vegetation and the regeneration of trees are suffering from feeding damage caused by sika deer (*Cervus nippon*) due to their increased population (Koshiji et al. 1996; RGTM 2007).

10.2 Fir Decline

The natural fir forest on the southeastern slope of Mt. Oyama, in the eastern part of the Tanzawa Mountains, has an area of about 100 ha and is distributed between 400 and 1,100 m above sea level (Fig. 10.1). Dieback of trees estimated to be 200 years old became apparent in the 1970s. Analysis of aerial photographs and tree rings suggested that fir decline had occurred before 1954 and the number of dead fir trees increased significantly from the 1960s to the 1970s. However, since the 1970s new decline or dieback was not observed (Suzuki 1992). Suzuki (1992) suggested that air pollutants, especially sulfur dioxide, rather than extreme meteorological factors, might have a possible link to the fir decline.

Concentrations of air pollutants at Mt. Oyama in the 1960s–1970s, caused by the emissions of primary pollutants and meteorological conditions, were estimated, and the annual mean concentration of sulfur dioxide was 0.01–0.02 ppm in 1965 and S deposition was about 60 kg S/ha. The concentration of O_3 was about 30 % higher than the O_3 concentrations in 1990–1992. The estimated annual mean pH of precipitation was 4.1 in the 1970s and fog below pH 3.0 occurred frequently and continued for a long time. However, soil pH ranged from 5.64 to 6.12. From this information, it seemed that air pollutants might have caused the fir decline. However, Japanese cedar (*Cryptomeria japonica*) and other trees co-standing with firs did not show any dieback (Fig. 10.1). Infestation of a host-specific insect, *Lymantria fumida* or infection with *Armillaria mellea* might be a possible cause of such fir decline under subtle abiotic stress conditions (Aso et al. 2001).

Hosono et al. (1994) conducted fog observations on the mountainside of Mt. Oyama and their recorded pH ranged from 2.61to 7.00. Igawa et al. (1997) exposed fir seedlings to simulated acid fog of pH 3.0 and reported that the fir leaves became reddish-brown, and thereafter defoliated, and new leaves did not develop. Such experimental results suggested that acid fog had a possible and significant role in inducing fir decline.

The dominant component of fog at Mt. Oyama was nitrate and its wet deposition of nitrogen might cause nitrogen saturation in fir forests (Igawa et al. 2001; Okochi and Igawa 2001). Such acidic wet deposition might induce accelerated leaching of Ca and Mg from leaves. As the leaching of boron associated with Ca and sugars in the cell wall also occurred, it seems possible that acid fog might have an important role in the development of fir decline (Igawa et al. 2002a, b; Shigihara et al. 2008a, b).

Japanese cedar in fir forests did not show any decline syndromes. Levels of membrane-bound Ca (mCa) in cedar exposed to acid fog of pH 3.0 did not show significant changes; however, the levels in fir were reduced. This suggested that fir is more sensitive to acid fog than Japanese cedar and this sensitivity might be a possible cause of the fir decline (Shigihara et al. 2009).

As described above, a direct cause of the fir decline at Mt. Oyama could be a biotic factor, an infection of *Armillaria mellea* or infestation of *Lymantria*. Accelerated leaching of Ca and B from fir needles could be due to chronic effects of acidic fog. Concentrations of Ca in the stem flow and throughfall increased with increasing acidity of precipitation. As leachate elements could be resupplied within the plant tissues, elemental deficiency would not develop. If reduced supply from the rhizosphere occurred, reduced levels of basic elements and signs of acidification in soil would be apparent. However, field surveys did not show such significant changes.

Further, the simulated acid-rain exposure experiment suggested that acidic precipitation around pH 3–4 would not be a possible direct cause of fir decline (Kohno et al. 1994 and Kohno 2001). Thus the decline or dieback of fir trees on Mt. Oyama in the 1970s could be directly induced by an infestation of a host-specific insect *Lymantria*.

10.3 Beech Decline

Fir decline at Mt. Oyama, on the eastern side of the Tanzawa Mountains, was distributed in a relatively limited area. However, beech decline is widely distributed on the south to southeastern slopes of ridgelines and around the peaks of Mts. Tanzawa, Hirugatake, Hinokiboramaru, and others.

Koshiji et al. (1996) analyzed aerial photographs of the Tanzawa Mountains after World War II. Trees showing decline were distributed in about 30 % of the natural forests in the area and concentrated around the peaks. These trees were old and had large trunks. The decline possibly started in the 1970s–1980s and has continued to date. Sasakawa et al. (2005) revealed that about 17 ha of forested area, including beech forests, in ridgelines from Mt. Hinokiboramaru to Mt. Tonodake had disappeared. Intensive field surveys in progressively declining areas of Mts. Tanzawa and Hinokiboramaru revealed that the trees showing decline were beech, but no other deciduous broad-leaved tree species showed declines. Such decline were located on the south to southwestern slopes.

10.3.1 Ozone Concentration Around Tanzawa Mountains

In the early stage of the above phenomena, the direct impacts of high concentrations of primary air pollutants, such as sulfur dioxide, might have been a potential causal factor. However, after air quality improvements in industrial areas, concentrations of sulfur dioxide and nitrogen dioxide in a remote mountain area are unlikely to be an important causal factor in tree decline, although ozone could have a possible link with the decline.

Aso et al. (2007) measured ozone concentrations in the Tanzawa Mountains with passive samplers and reported a monthly mean value of 46.8 ppb from May to September in 2005. However, there were no available hourly data for calculating an ozone exposure index. Kohno and collaborators set up an active monitoring station for ozone at an elevation of 1,540 m near the peak of Mt. Hinokiboramaru in 2004 (Kohno, 2005). Two years of monitoring results are summarized in Table 10.1. The mean concentration of ozone at Mt. Hinokiboramaru was comparable with that reported by Aso et al. (2007); however, ozone doses, expressed as AOT40 (Accumulated exposure Over a Threshold of 40 ppb) values (April–September) were 46 ppmh for 24 h and 20 ppmh for daytime 12 h. There are several monitoring stations around the Tanzawa Mountains, as shown in Table 10.2. Low elevation sites had lower ozone concentrations than those of the mountain sites: Inukoeji and Hinokiboramaru, and AOT40 (12 h) values at the low elevation sites were similar to the AOT40 (24 h), as the night-time ozone concentration was low. In contrast, AOT40 (24 h) values at mountain sites were about twice AOT40 (12 h), as the night-time ozone concentration was high.

The seasonal mean concentrations of ozone calculated from values for 3 months showed quite different patterns at the low and high elevation sites, as shown in Fig. 10.2; the low elevation site at Isehara showed a high concentration at midday and low concentration in the night to early morning hours. In contrast, at Hinokiboramaru, differences between night and daytime hours were relatively small; however, the baseline concentration was higher than that at Isehara. Ozone concentration was higher in the spring time and low in the fall and winter season. Beech trees developed young leaves in the period with high concentrations of ozone.

Table 10.1 Results of active monitoring of ozone concentration at Mt Hinokiboramaru

			Max	Mean			AOT40		
Year	Days	Hours	(ppb)	(24 h, ppb)	(12 h, ppb)	(8 h, ppb)	(24 h, ppb · h)	(12 h, ppb · h)	(8 h, ppb · h)
2004	365	7,973	111	43.1	41.5	41.8	68,371	29,635	19,969
2005	365	7,601	119	42.4	41.8	42.2	61,115	27,779	18,736
Mean				42.7	41.7	42.0	64,743	28,707	19,353
2004 (April–September)				46.3	44.5	44.8	46,814	19,959	13,725
2005 (April–September)				44.2	43.4	44.2	45,949	20,538	14,158
Mean				45.2	44.0	44.5	46,381	20,249	13,942

Elevation: 1,540 m
Period: 2004/08/01–2006/07/31; 2004/08/01–2005/07/31 for 2004 and 2005/08/01–2006/07/31 for 2005
8 h: 09:00–16:59
12 h: 06:00–17:59
24 h: 00:00–23:59
AOT40(8, 12 and 24h) indicated accumulated hours for 8 hours (09:00–16:59), 12h (06:00–17:59) and 24h (00:00–23:59), respectively, from April to September.

Table 10.2 Comparison of mean AOT40 values around the Tanzawa Mountains in 2004–2005

Station (elevation, m)									
	Odawara	Isehara	Minamiashigara	Ebina	Hadano	Sagamihara	Tsukui	Inukoeji	Hinokiboramaru
AOT40	(15)	(25)	(30)	(35)	(110)	(125)	(170)	(920)	(1,540)
12 h, ppbh	13,852	14,939	10,324	8,242	8,545	8,498	6,307	22,292	20,249
24 h, ppbh	16,040	18,381	11,912	8,919	10,834	9,813	7,238	36,169	46,381

Period: April–September
Values, except for those at Hinokiboramaru, were calculated from Kanagawa Environmental Research Center data

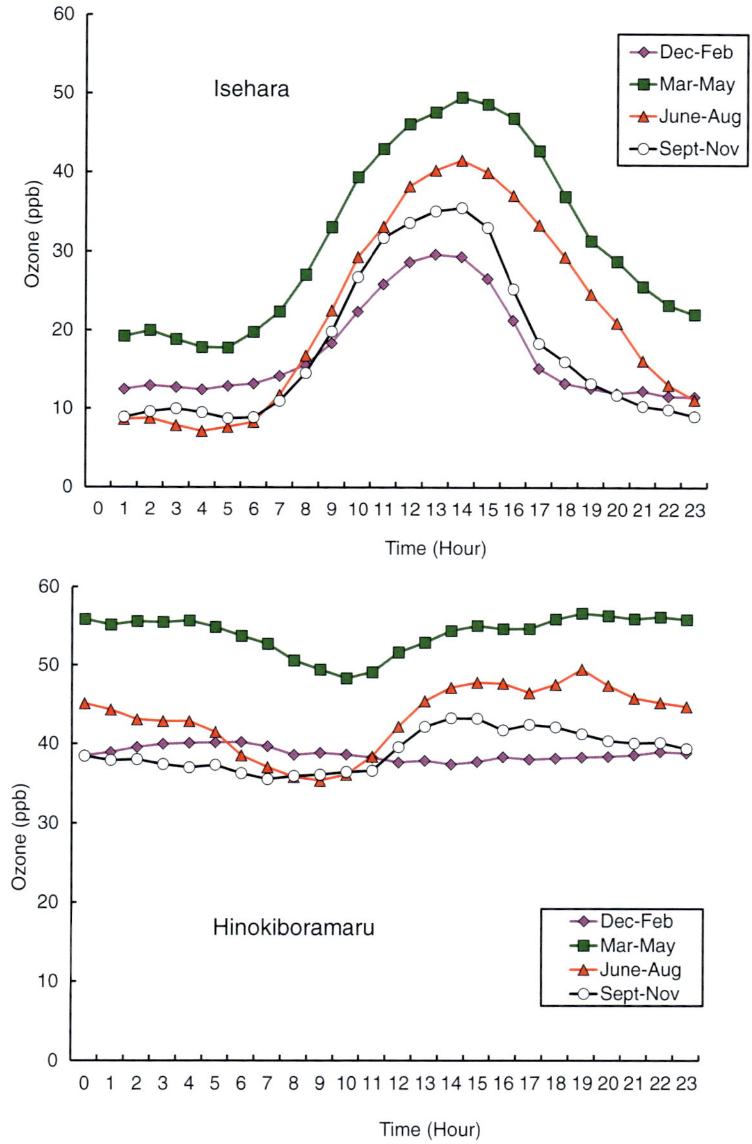

Fig. 10.2 Seasonal changes in ozone concentration at Isehara and Hinokiboramaru in 2004–2005

Based on the ozone exposure experiments in tree seedlings performed by Kohno et al. (2005), beech was found to be one of the ozone-sensitive species among broad-leaved trees, and the suggested critical level was 20 ppmh (24 h). Matyssek et al. (1995) reported that ozone reduced tree growth not only by daytime exposure but also by night-time exposure, and growth reduction increased proportionately with the ozone concentration. Therefore, it seems that beech stands at high elevation sites such as in the Tanzawa Mountains may be receiving extremely high ozone stress.

Izuta et al. (1996) suggested that ambient levels of ozone detrimentally affected the growth, photosynthetic rate, and chlorophyll content of *F. crenata*. Takeda and Aihara (2007) also reported that chlorophyll content in the beech leaves exposed to ambient air at the Inukoeji site in the Tanzawa Mountains was significantly decreased. Early defoliation and growth reduction in beech seedlings were observed. Further, a reduction of bud numbers in the winter suggested that beech growth would be greatly reduced in the next year.

Simulated acid-rain experiments suggested that the current acidity, frequency, and amount of precipitation did not directly affect beech trees. However, an increase of nitrogen input by acid deposition might induce a phenological response in trees under ozone stress, such as that caused by the combined exposure to acid rain and ozone, evidenced by an increased top/root ratio (Kohno and Matsumura 1999). Increased above-ground biomass with less biomass under the ground may induce a condition sensitive to water stress under drought and/or high-temperature stress. Yonekura et al. (2001) experimentally demonstrated that photosynthesis and the structural characteristics of *F. crenata* were affected by ambient levels of ozone and long-term mild water stress, and these two stresses were related to the beech decline.

Watanabe et al. (2011, 2012) suggested that nitrogen deposition-induced change in the ozone sensitivity of *F. crenata* could be an important factor in the discordance between areas with high risk and those with high ozone exposure.

10.3.2 Wind and Ozone in the Tanzawa Mountains

Suto et al. (2008) developed a dry-deposition model for ozone to estimate O_3 deposition in complex terrains, and the qualitative validity of the predicted O_3 concentration in the field was confirmed by a comparison with observed data collected by passive samplers in the Tanzawa Mountains. The simulation revealed that wind velocity near the ground increased around the ridgelines and peaks of the mountains. The areas with strong wind corresponded well with the beech decline sites at high altitudes, suggesting that strong wind is one of the important factors in the localization of areas of beech decline. A direct relationship was not observed between the forest decline and O_3 concentration. The O_3 concentration, however, tends to increase as wind velocity becomes higher; thus, the O_3 concentration itself could be a possible secondary factor in the local decline phenomena in the region when the O_3 concentration level is high. While the diffusion flux did not clearly correspond with the distributions of areas of beech decline, a large advection flux of O_3 showed a distribution pattern similar to that of high wind velocity corresponding with an area of beech decline. Considering the impact of ozone, strong wind with a large advection flux of O_3 may play a key role in the localized development of tree/forest decline in the high mountain ridges and peaks.

Takeda et al. (2011) developed a flux-type passive sampler from a diffusive-type one and showed the applicability of this sampler to mountain sites without any available electricity. Their results suggested that ozone concentration did not vary among the sites; however, flux data reflecting wind speed indicated locality.

10.3.3 Outbreak of Beech Sawfly

Yamagami et al. (2007) reported that beech decline in the ridgelines and peaks in Tanzawa Mountains became evident after the 1980s; however, in those days there was no evidence about outbreaks of beech sawfly, *Fagineura crenativora*. An outbreak of sawfly was first recorded in 1993, and thereafter sawfly feeding damage to beech trees expanded in the mountains. Trees damaged by repeated feeding attacks from sawfly larvae died within a couple of years (Fig. 10.3). Severe damage and leaf loss in the beech trees due to sawfly larvae feeding may reduce tree vitality. Considering the ozone exposure conditions in the Tanzawa Mountains, it is possible that the beech trees might had reduced physiological activity caused by the chronic exposure to high ozone concentrations and they had become sensitive to water stress, and their potentiality for recovery was reduced by the feeding attacks. Trees with lost vitality showed delayed leaf development and had received intensive feeding attacks. These trees developed secondary leaves in summer, being spent energy for reproducing leaves, instead of storing carbohydrates for preparing for winter season. Such growth responses could have resulted in their decline.

10.3.4 Sika Deer Impacts

Generally, an increasing deer population causes critical changes in the composition and structure of forest ecosystems. In the Tanzawa Mountains, grazing pressure from sika deer has caused a decline in vegetation over 20 years (Tamura 2016). As a result, it is difficult to find young seedlings in the beech forests in the Tanzawa Mountains. Forest floor vegetation was grazed and had disappeared, as shown in Fig. 10.4, and this process resulted in the acceleration of surface soil drying and a reduction in the variety of the forest floor community.

Yamagami et al. (2007) pointed out that when sawfly larvae made cocoons in the surface soil layer, they were attacked by parasites and the sawfly population was maintained at a low density in a healthy system. However, the disappearance of

Fig. 10.3 Healthy beech (*left*) and damaged beech (*right*) in the Tanzawa Mountains

Fig. 10.4 Degradation of forest floor vegetation by sika deer grazing, and recovered vegetation inside protecting fence

floor vegetation, owing to grazing caused by increasing sika deer density, could change the natural environmental conditions. Such a significant environmental change could disturb co-relationships among the members of the forest floor community. Controlling sika deer density may lead not only to the recovery of forest floor vegetation but also to the maintenance of biodiversity.

10.4 Conclusion

Takeda et al. (1999) suggested that accelerated leaching from plant tissues due to simulated acid precipitation might have an indirect impact on plant growth. Shigihara et al. (2008b) suggested that reduced physiological activities due to chronic exposure to acid fog resulted in growth reduction. Such acidic precipitation might have indirect effects on vegetation; however, it could hardly explain the specific tree decline phenomena.

At high elevation sites such as the Tanzawa Mountains, the concentrations and doses of ozone are higher than those at the low elevation sites. Beech, which is sensitive to ozone, grows under a crucial condition in the Tanzawa Mountains. Global warming-related air temperature rises and surface soil drying can have negative effects on environmental conditions for beech trees. Furthermore, outbreaks of beech sawfly, *Fagineura crenativora*, and its repeated feeding on the young leaves of beech can have a significant and direct impact on beech decline (RGTM 2007). Further, high populations of sika deer disturb the surface vegetation community including young beech seedlings. It will be resulted in difficulties of rehabilitation of beech forests.

References

Aso T, Takeda M, Aihara K (2001) Observational study of acid rain and air pollutants in the western part of Mt. Tanazawa (1995–2000). Bull Kanagawa Environ Res Cent 24:82–89 (in Japanese)

Aso T, Utiyama Y, Yamane M, Koshiji M, and Aihara K (2007) The ozone distribution in Tanzawa Oyama mountainous area at foliage leaf duration of beech, p.396-399, Results of the Scientific Research on the Tanzawa Mountains, Hiraoka Environmental Science Laboratory, Sagamihara, Kanagawa, Japan. (in Japanese)

Hosono T, Okochi H, Igawa M (1994) Fog water chemistry at a mountainside in Japan. Bull Chem Soc Jpn 67:368–374

Igawa M, Kameda H, Maruyama F, Okochi H, Otsuka I (1997) Effect of simulated acid fog on needles of fir seedlings. Environ Exp Bot 38:155–163

Igawa M, Matsumura K, Okochi H (2001) Fog water chemistry at Mt. Oyama and its dominant factors. Water Air Soil Pollut 130:607–612

Igawa M, Kase T, Satake K, Okochi H (2002a) Severe leaching of calcium ions form fir needles caused by acid fog. Environ Pollut 119:375–382

Igawa M, Okuyama K, Okochi H, Sakurai N (2002b) Acid fog removes calcium and boron from fir tree: one of the possible causes of forest decline. J For Res 7:213–215

Izuta T, Umemoto M, Horie K, Aoki M, Totsuka T (1996) Effects of ambient levels of ozone on growth, gas exchange rates and chlorophyll contents of *Fagus crenata* seedlings. J Jpn Soc Atmos Environ 31(2):95–105

Kohno Y (2001) A review on effects of acidic deposition on trees by long-term exposure experiments. J Jpn Soc Atmos Environ 36(2):47–59 (in Japanese with English summary)

Kohno Y (2005) Study on the impacts of acidic and oxidative substances on vegetation and establishment of tentative critical levels for protecting East Asian vegetation. p 279–284, Summary report of research results under the GERF(Global Environmental Research Fund) in FY2004, Ministry of the Environment, Japan

Kohno Y, Matsumura H, Kobayashi T (1994) Effect of simulated acid rain on the development of leaf injury in tree seedlings. J Jpn Soc Air Pollut 29(4):206–219 (in Japanese with English summary)

Kohno Y, and Matsumura H (1999) Combined effect of simulated acid rain and ozone on the growth of Japanese conifer seedlings, J Jpn Soc Atmos Environ 34(2):74–85 (in Japanese with English summary)

Kohno Y, Matsumura H, Ishii T, and Izuta T (2005) Establishing critical levels of air pollutants for protecting East Asian vegetation – A challenge, p.243–250, In: Omasa K, Nouchi I, De Kok L J (eds) Plant responses to air pollution and global change, Springer-Verlag Tokyo.

Konohira Y (2007) The scientific research on theTanzawa Mountains (1995) and the conservation plan. p 1–10, Results of the Scientific Research on the Tanzawa Mountains, Hiraoka Environmental Science Laboratory, Sagamihara, Kanagawa, Japan. (in Japanese)

Koshiji M, Suzuki K, Suga K (1996) Investigation of forest decline in the Tanzawa Mountain (1) Distribution of decline of *Fagus crenata*, *Abies firma* and other tree species. Bull Kanagawa Prefecture For Res Inst 22:7–18 (in Japanese)

Matyssek R, Gunthardt-Goerg M S, Maurer S, and Keller T (1995) Nighttime exposure to ozone reduces whole-plant production in *Betula pendula*, Tree Physiol 15:159–165.

Okochi H, Igawa M (2001) Elevational patterns of acid deposition into a forest and nitrogen saturation on Mt. Oyama. Japan. Water Air Soil Pollut 130:1091–1096

Research Group of the Tanzawa Mountains (RGTM) (2007) Results of the Scientific Research on the Tanzawa Mountains, pp.794, Hiraoka Environmental Science Laboratory, Sagamihara, Kanagawa, Japan. (in Japanese)

Sasakawa et al (2005) Identifying declining forests – a case of beech forests in Tanzawa Mountain- Proceedings of ACRS (Asian Conference on Remote Sensing) 2005, D2-P27

Shigihara A, Matsumoto K, Sakurai N, Igawa M (2008a) Leaching of cell wall components caused by acid deposition on fir needles and trees. Sci Total Environ 398:185–195

Shigihara A, Matsumoto K, Sakurai N, Igawa M (2008b) Growth and physiological responses of beech seedlings to long-term exposure of acid fog. Sci Total Environ 391:124–131

Shigihara A, Matsumura Y, Matsumoto K, Igawa M (2009) Effect of simulated acid fog on membrane-bound calcium (mCa) in fir (*Abies firma*) and cedar (*Cryptomeria japonica*) mesophyll cells. J For Res 14:188–192

Suto H, Hattori Y, Tanaka N, Kohno Y (2008) Effects of strong wind and ozone on localized tree decline in the Tanzawa Mountains of Japan. Asian J Atmos Environ 2(2):81–89

Suzuki K (1992) Fluctuation of Momi (*Abies firma*) dead standing trees and change of annual ring width at Mt. Ohyama and the around areas in Kanagawa Pref. Bull Kanagawa Prefecture For Res Inst 19:23–42 (in Japanese)

Takeda M, Aihara K (2007) Effect of ambient ozone concentrations on beech (*Fagus crenata*) seedlings in the Tanzawa Mountains, Kanagawa Prefecture, Japan. J Jpn Soc Atmos Environ 42(2):107–117 (in Japanese with English summary)

Takeda M, Aihara K, Aso T (1999) Leakage of cation species from leaves by exposure of simulated acidic fog. Bull Kanagawa Environ Res Center 22:22–25 (in Japanese)

Takeda M, Komatsu H, Iida N (2011) Evaluations of ozone-flux and wind-speed by using flux-type passive sampler in Mt. Tanzawa. Bull Kanagawa Environ Res Cent 34:14–20 (in Japanese)

Tamura A (2016) Potential of soil seed banks in the ecological restoration of overgrazed floor vegetation in a cool-temperate old-growth damp forest in eastern Japan. J For Res 21:43–56

Watanabe M, Yamaguchi M, Matsumura H, Kohno Y, Koike Y, Izuta T (2011) A case study of risk assessment of ozone impact on forest tree species in Japan. Asian J Atmos Environ 5(4): 205–215

Watanabe M, Yamaguchi M, Matsumura H, Kohno Y, Izuta T (2012) Risk assessment of ozone impact on *Fagus crenata* in Japan: consideration of atmospheric nitrogen deposition. Eur J For Res 131:475–484

Yamagami A, Tani S, Banno H (2007) Death of Siebold's beech trees, *Fagus crenata*, and decline of the beech forest damaged by feeding of a saw fly, *Fagineura crenativora* (Hymenoptera, Tenthredinidae), in the Tanzawa Mountains, Central Japan. p 256–268, Results of the Scientific Research on the Tanzawa Mountains, Hiraoka Environmental Science Laboratory, Sagamihara, Kanagawa, Japan. (in Japanese)

Yonekura T, Honda Y, Oksanen E, Yoshidome M, Watanabe M, Funada R, Koike T, Izuta T (2001) The influences of ozone and soil water stress, singly and in combination, on leaf gas exchange rates, leaf ultrastructural characteristics and annual ring width of *Fagus crenata* seedlings. J Jpn Soc Atmos Environ 36(6):333–351

Chapter 11
Reactions Between Ozone and Terpenoids Within a Forest on Mt. Fuji

Akira Tani

Abstract Biogenic hydrocarbons react with ozone in forest ecosystems to produce oxygenated compounds, which are partially incorporated into secondary organic aerosols (SOAs). The hydrocarbons also react with OH radicals, controlling the levels of photochemical oxidants. These reactions occur simultaneously, and the concentrations of ozone and SOAs depend on the abundances of the relevant components, such as biogenic hydrocarbons. In this chapter, results of observations of these compounds in a Japanese larch forest (*Larix kaempferi*) in the foothills of Mt. Fuji are provided as a case study. Ozone concentrations increased with the height measured within the canopy, suggesting that ozone is decomposed within the forest and/or is absorbed by the forest canopy. The biogenic hydrocarbons isoprene and monoterpenes are emitted from the forest ecosystem and the monoterpene flux can be explained by the temperature and soil water content. The annual emission of monoterpenes plus isoprene was determined to be 0.93 % of the net ecosystem exchange. Oxygenated compounds, originating from isoprene and α-pinene, were found in the suspended particulate matter. The concentrations of these compounds and ozone were well correlated, suggesting that, along with terpenoid emissions, the anthropogenic inflow and subsequent oxidation of the terpenoids also promote biogenic SOA formation within the forest canopy.

Keywords Isoprene • Monoterpene • α-Pinene • Secondary organic aerosols (SOAs) • Ozone

---

A. Tani
School of Food and Nutritional Science, University of Shizuoka,
52-1 Yada, Shizuoka 422-8526, Japan
e-mail: atani@u-shizuoka-ken.ac.jp

T. Izuta (ed.), *Air Pollution Impacts on Plants in East Asia*,
DOI 10.1007/978-4-431-56438-6_11

11.1 Ozone-Related Reactions Within the Forest

In forest ecosystems, ozone not only degrades the air quality and negatively affects the plant growth, but also acts as an oxidizing agent (Fig. 11.1). Reactive hydrocarbons from both anthropogenic and biogenic sources are oxidized by ozone in the atmosphere to produce oxygenated compounds, which are precursors of secondary organic aerosols (SOAs). In the forest atmosphere, biogenic hydrocarbons are emitted from leaves, stems, roots, and soils. Among their major components are highly reactive terpenoids, including isoprene, monoterpenes, and sesquiterpenes. The lifetimes of the major monoterpenes, α-pinene and limonene, in reactions with a typical concentration of ozone (30 ppb) are 4.6 h and 2.0 h, respectively, which are much shorter than those of ubiquitous anthropogenic hydrocarbons such as benzene and toluene (>4.5 years for both) (Atkinson 2000).

The SOAs may scatter sunlight, and through reactions in the atmosphere, form cloud condensation nuclei (CCN). SOAs may have direct and indirect cooling effects on global climatic conditions, and therefore, are considered to be important components of the earth's atmosphere. Recent studies indicate that the production of biogenic SOAs is substantially greater than that of anthropogenic SOAs (Kanakidou et al. 2005).

The hydrocarbons also react with the OH radical, yielding peroxides. The peroxides oxidize nitrogen monoxide (NO) to nitrogen dioxide (NO_2) in the atmosphere, and this process can reduce the ozone-driven oxidation of NO to NO_2, suppressing ozone decomposition in the forest atmosphere.

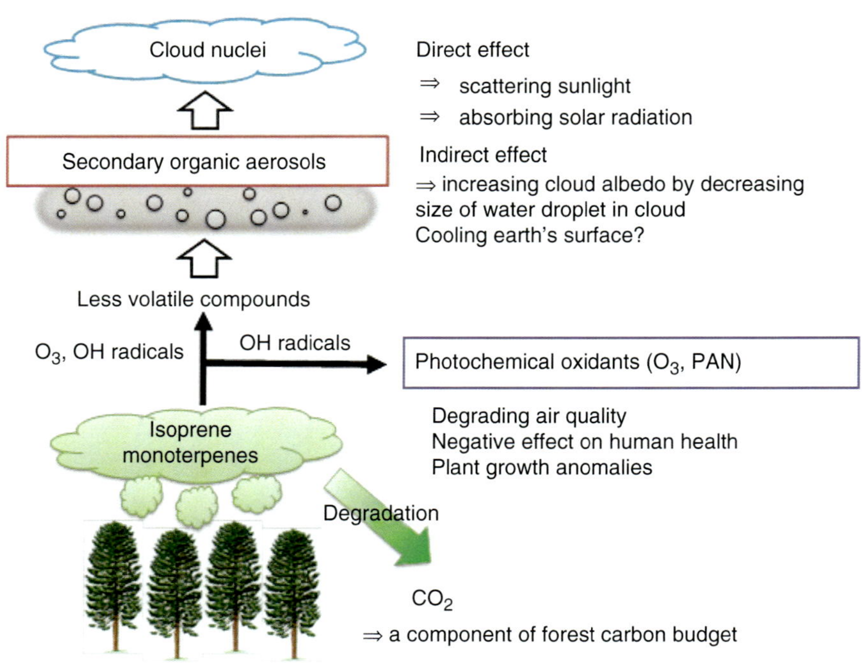

Fig. 11.1 Ozone-related reactions within the forest

The reactions mentioned above occur simultaneously, and ozone and SOA concentrations depend on the abundances of the relevant components. In this chapter, the results of observations of these compounds in a Japanese larch forest (*Larix kaempferi*) are provided as a case study (Mochizuki et al. 2014, 2015).

11.2 Research Site

The study site is a *L. kaempferi* plantation at the Fuji-Hokuroku Flux Research Site (35° 26′ N, 138° 45′ E), which is located at an elevation of 1100 m (slope: 3–4°) in the foothills of Mt. Fuji in Yamanashi Prefecture, Japan. *L. kaempferi* is commonly found in Japan and occupies 5 % of Japan's total forest area. Its congeneric species *Larix sibirica* and *Larix gmelinii* occur widely in boreal coniferous forests in Siberia and northern Europe. At the study site, the larch trees were planted uniformly 55 years ago, over an area of 150 ha. Tree height in the study period was 20–25 m. The forest floor is covered with a fern species, *Dryopteris crassirhizoma*. The annual mean air temperature and annual precipitation measured at the meteorological flux tower are approximately 9 °C and 2,300 mm, respectively. Two meteorological towers (32 m and 20 m high) situated at the center of the forest site were used for observation. We report the results of the study measurements of ozone, terpenoids, oxygenated volatile organic compounds (OVOCs), and SOA tracers, and the results of 1-year flux measurements of terpenoids.

11.3 Ozone Concentration

Figure 11.2 shows the vertical profile of the ozone concentrations measured above and within the larch forest (at 2, 10, 16, and 28 m) on July 10 and 16, 2012. The O_3 concentration was higher during the daytime than during the nighttime. In

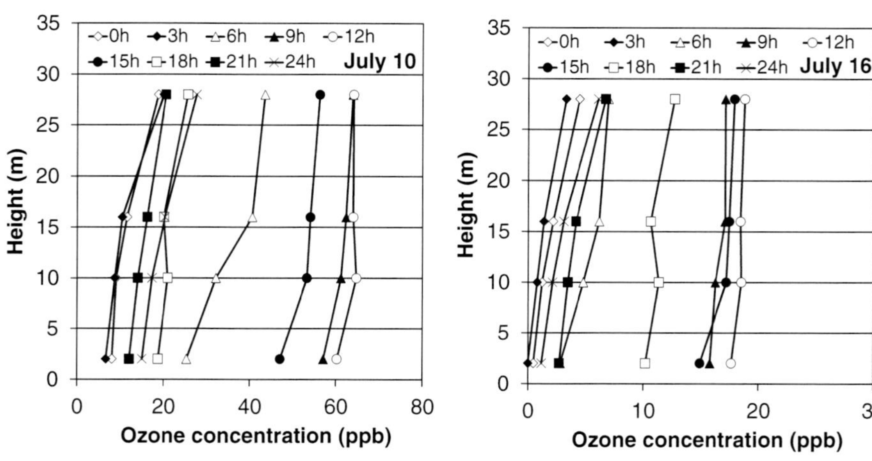

Fig. 11.2 *Vertical* profile of ozone concentrations measured above and within the larch forest

particular, the concentration recorded on July 10 reached a daily maximum of more than 60 ppb around noon. On that day, we assumed that the polluted air masses had been transported from the nearby urban area by northeasterly and easterly winds. On other days, the forest air was assumed to have been transported from the nearby forests and mountain areas by southerly and southwesterly winds. This assumption was supported by the findings of lower concentrations of toluene, benzene, and nitrogen oxides (NO_x) on July 5 and July 15–17.

In most cases, the ozone concentration decreased when the measurement height was lowered, and the lowest ozone concentration was observed near the ground surface. This means that ozone is decomposed within the forest and/or absorbed by forest leaves. Ozone decomposition may have been caused by the reaction between ozone and hydrocarbons, and by physical contacts of ozone with the forest canopy. As this reaction might be significant in this forest, we studied the oxygenated products of the reaction in the forest atmosphere.

11.4 Components of Ozone Reactions: Isoprene and Monoterpenes

As terpenoids are components of ozone reactions, it is necessary to determine the amounts of these compounds that are emitted from the forest ecosystem and to determine how the emissions are affected by environmental factors. Isoprene and monoterpene fluxes above the larch (*L. kaempferi*) forest were measured across a year (May 2011–2012) by using a relaxed eddy accumulation method. High isoprene and monoterpene fluxes were observed in July and August. α-Pinene was observed to be the dominant compound, and the total monoterpene flux was dependent on the temperature, regardless of the season (Fig. 11.3). This exponential relationship of monoterpene flux with temperature has been observed for a number of tree species, and followed the G93 model – named after its developer, Alex Guenther, and year of publication of the study (Guenther et al. 1993). At our study site, isoprene is emitted from a species of fern, *D. crassirhizoma*, found on the forest floor,

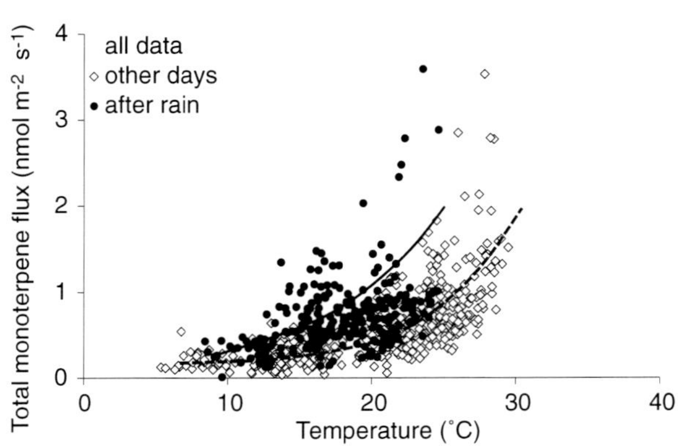

Fig. 11.3 Relationship between temperature and total monoterpene flux measured throughout a year, using a relaxed eddy accumulation sampling technique

and monoterpenes are emitted from *L. kaempferi*. These might be the major compounds that react with ozone in the forest.

However, several unusually high positive fluxes were observed after rainfall events. A good correlation (r = 0.88) was obtained between the total monoterpene flux and the volumetric soil water content (Fig. 11.4), and this correlation was used to estimate the monoterpene flux after rain events and to calculate the annual terpenoid emissions. The G93 model was improved upon by accounting for this relationship (improved G93 model); this model can better estimate the monoterpene emissions after rainfall events (Fig. 11.5a). Monthly monoterpene flux was calculated by using the improved G93 model (Fig. 11.5b). The annual carbon emission in the form of total monoterpenes was estimated to be 1,671 mg C m^{-2} $year^{-1}$. Total carbon emission in the form of terpenoids (monoterpenes plus isoprene) was

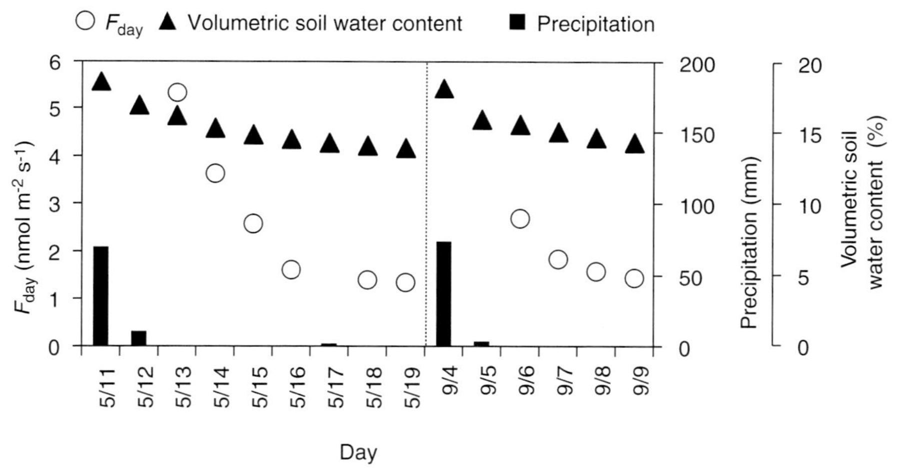

Fig. 11.4 Change in the daily monoterpene flux (F_{day}) and soil water content after rain events

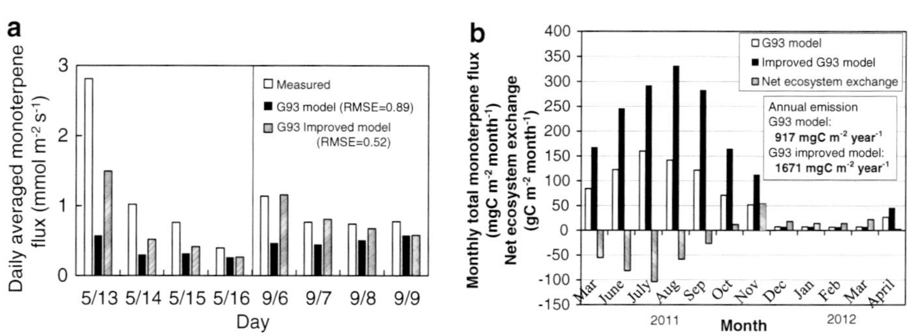

Fig. 11.5 Comparison of the monoterpene fluxes measured and estimated using the G93 model – named after its developer, Alex Guenther, and year of publication of the study (Guenther et al. 1993), and improved G93 models after rainfall events (panel **a**) and the estimated monoterpene flux during four seasons (panel **b**)

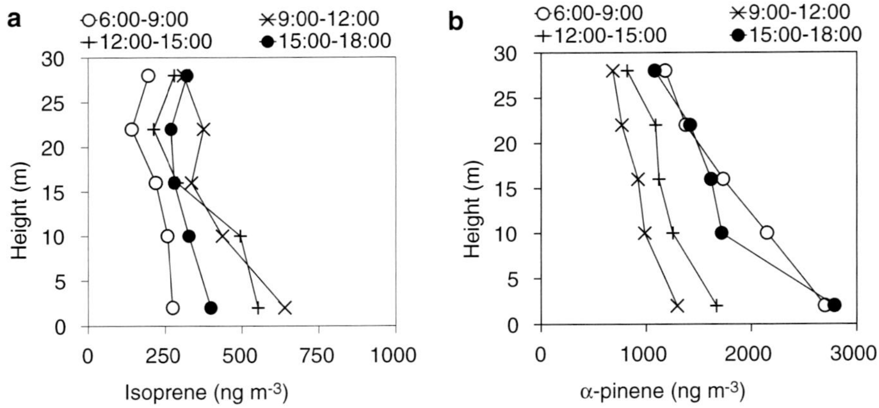

Fig. 11.6 *Vertical* profile of the isoprene (**a**) and α-pinene (**b**) concentrations measured above and within the larch forest. Gas samples were collected for 3 h. *Values* represent the means of measurements carried out over 5 days

determined to be 0.93 % of the net ecosystem exchange (189 g C m^{-2} $year^{-1}$). If we do not consider the effect of rainfall, the carbon emissions may be underestimated by about 50 %. The results here suggest that the moisture content of forest soils is one of the key factors that control monoterpene emissions from the forest ecosystem.

In regard to the isoprene profile, higher isoprene concentration was observed to be restricted to lower forest levels during the daytime (9:00–15:00), because of the existing isoprene emitter *D. crassirhizoma* near the forest floor (Fig. 11.6a). As isoprene emission is increased by sunlight, its concentration near the forest floor tended to be high during daytime. Monoterpene concentrations were also higher at a lower forest level, and the highest value was recorded near the ground surface (for α-pinene, see Fig. 11.6b). This might be caused by monoterpenes being emitted not only from the larch leaves, but also from the forest floor, which includes soil, litter, and roots (e.g. Hayward et al. 2001), and by weaker vertical mixing of air near the forest floor. The α-pinene concentration tended to be high in the early morning regardless of the measurement height, because of less vertical mixing of the forest air during the night. This monoterpene storage within the forest has been reported to be significant when forest monoterpene emission rates are estimated using flux data (Tani et al. 2002). This so-called storage effect has been reported to be indispensable when considering CO_2 exchange between the forest ecosystem and the atmosphere (Papale et al. 2006).

11.5 Oxidation Products: Secondary Organic Aerosols

Total suspended particulate matter was collected on quartz-fiber filters (8 × 10 in.), using a high-volume air sampler at a height of 16 m, just beneath the larch canopy (Mochizuki et al. 2015). We determined the concentrations of the oxidation

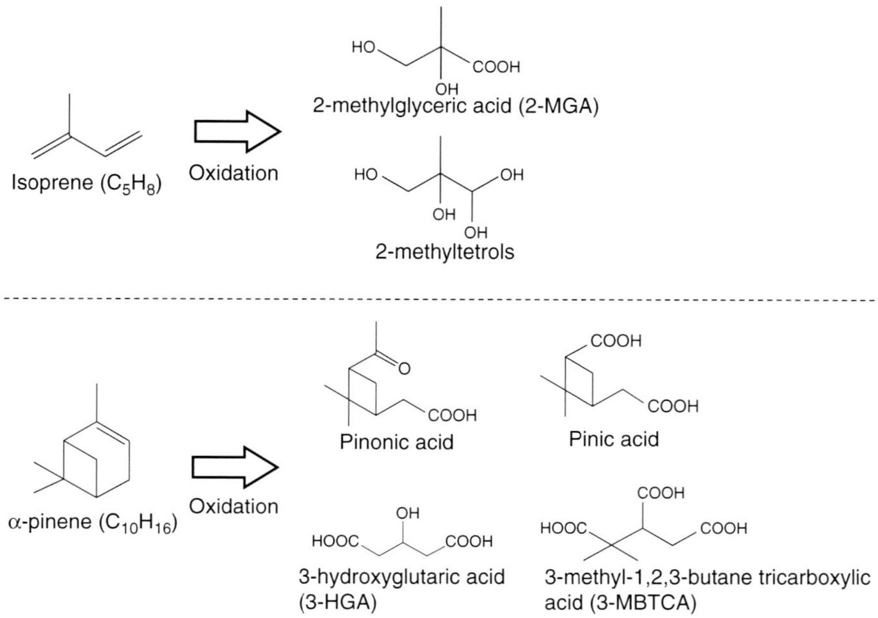

Fig. 11.7 Chemical structures of biogenic secondary organic aerosol (SOA) tracers formed from isoprene and α-pinene

products of isoprene and monoterpenes included in the particles. The oxidation products were 2-methylglyceric acid (2-MGA) and 2-MTLs (the sum of 2-methylerythritol and 2-methylthreitol), which are isoprene-derived SOA tracers; as well as pinic acid, pinonic acid (PA), 3-methyl-1, 2, 3-butanetricarboxylic acid (3-MBTCA), and 3-hydroxyglutaric acid (3-HGA), which are α-pinene-derived SOA tracers (Fig. 11.7).

The concentrations of 2-MGA and 2-MTLs had higher values (up to ~10 ng m^{-3} and ~30 ng m^{-3}, respectively) on July 10 and 17 than on the other days. The concentrations of PA and 3-MBTCA on July 10 and 17 were also higher. To elucidate the reactivity of isoprene and monoterpenes to form SOAs, the ratios 2-MGA/isoprene, 2-MTLs/isoprene, PA/α-pinene, and 3-MBTCA/α-pinene were plotted against the O_3 concentration (Fig. 11.8). These ratios increased with increasing O_3 concentrations. Although the time delay in the formation of these oxygenates should be considered, these relationships indicate that the concentration of the oxidizing agent (ozone) is important for the formation of these oxygenates.

Anthropogenic influence on the observed aerosols was significant on July 10, the day on which the highest concentrations of the biogenic SOA tracers and ozone were observed. This finding is also supported by the substantially large concentrations of SO_4^{2-} (~6,400 ng m^{-3}) and NO_x (~3.5 ppbv) observed on that day. The present results suggest that the anthropogenic inflow and subsequent oxidation of isoprene and α-pinene promote biogenic SOA formation within the forest canopy.

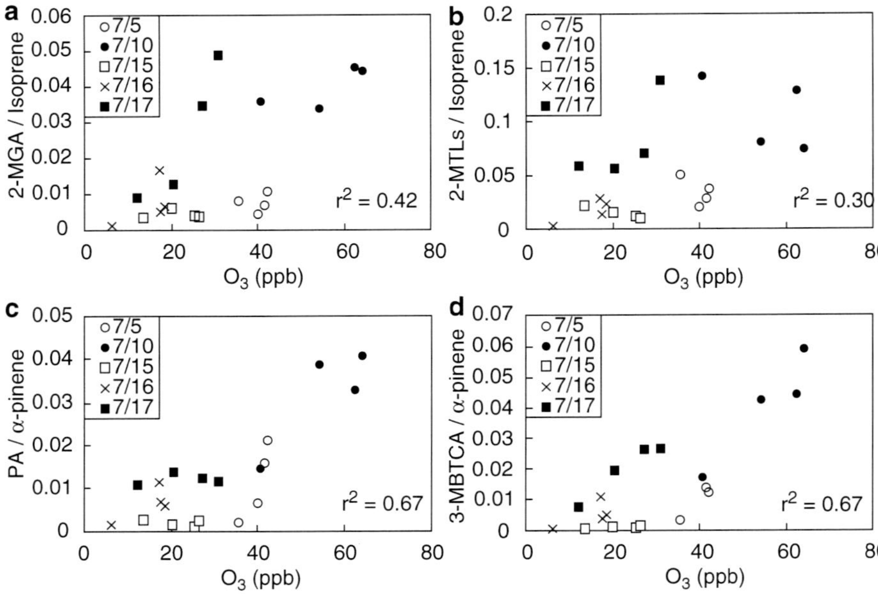

Fig. 11.8 Relationship between ozone concentration and the ratios (**a**) 2-methylglyceric acid (2-MGA)/isoprene, (**b**) the sum of 2-methylerythritol and 2-methylthreitol (2-MTLs)/isoprene, (**c**) pinonic acid (PA)/α-pinene, and (**d**) 3-methyl-1, 2, 3-butanetricarboxylic acid (3-MBTCA)/ α-pinene

11.6 Concluding Summary

At the *L. kaempferi* plantation of the Fuji-Hokuroku Flux Research Site, the ozone concentration decreased with decreasing measurement height, suggesting that ozone is decomposed within the forest and/or is absorbed by the forest canopy. Isoprene and monoterpenes – the major components of the ozone-related reactions – were emitted from the forest ecosystem, and the monoterpene flux could be explained by the temperature and soil water content. The annual emission of terpenoids (monoterpenes plus isoprene) was determined to be 0.93 % of the net ecosystem exchange. Oxygenated compounds, produced by the reactions between terpenoids and oxidizing agents such as ozone, were incorporated in the suspended particulate matter collected under the forest canopy. Good correlations were observed between the concentrations of these biogenic SOA tracers and ozone, suggesting that not only terpenoid emission, but also the anthropogenic inflow and subsequent oxidation of the terpenoids, promote biogenic SOA formation within the forest canopy.

References

Atkinson R (2000) Atmospheric chemistry of OVOCs and NOx. Atmos Environ 34:2063–2101

Guenther AB, Zimmerman PR, Harley PC (1993) Isoprene and monoterpene emission rate variability: model evaluations and sensitivity analysis. J Geophys Res 98(D7):12609–12617

Hayward S, Muncey RJ, James AE, Halsall CJ, Hewitt CN (2001) Monoterpene emissions from soil in a Sitka spruce forest. Atmos Environ 35:4081–4087

Kanakidou M, Seinfeld JH, Pandis SN, Barnes I, Dentener FJ, Facchini MC, Van Dingenen R, Ervens B, Nenes A, Nielsen CJ, Moortgat GK, Winterhalter R, Myhre CEL, Tsigaridis K, Vignati E, Stephanou EG, Wilson J (2005) Organic aerosol and global climate modeling: a review. Atmos Chem Phys 5:1053–1123

Mochizuki T, Tani A, Takahashi Y, Saigusa N, Ueyama M (2014) Long-term measurement of terpenoid flux above a *Larix kaempferi* forest using a relaxed eddy accumulation method. Atmos Environ 83:53–61

Mochizuki T, Miyazaki Y, Ono K, Wada R, Takahashi Y, Saigusa N, Kawamura K, Tani A (2015) Emissions of biogenic volatile organic compounds and subsequent formation of secondary organic aerosols in a *Larix kaempferi* forest. Atmos Chem Phys 15:12029–12041

Papale D, Reichstein M, Aubinet M, Canfora E, Bernhofer C, Kutsch W, Longdoz B, Rambal S, Valentini R, Vesala T, Yakir D (2006) Towards a standardized processing of Net Ecosystem Exchange measured with eddy covariance technique: algorithms and uncertainty estimation. Biogeosciences 3(4):571–583

Tani A, Nozoe S, Aoki M, Hewitt CN (2002) Monoterpene fluxes measured above a Japanese red pine forest at Oshiba plateau, Japan. Atmos Environ 36:3391–3402

Part IV
Effects of Gaseous Air Pollutants on Plants in China

Chapter 12
Effects of Ozone on Crops in China

Zhaozhong Feng, Haoye Tang, and Kazuhiko Kobayashi

Abstract Current ambient O_3 concentrations in China are high, as shown by observations of typical O_3 symptoms in some plant species and crop yield losses, as detected by the use of chemical protectants against O_3. Experiments with artificially elevated O_3 concentrations have shown the effects of O_3 on growth processes, grain yield, grain quality, CH_4 emissions, and soil microbiology. The experiments have facilitated estimations of the yield losses in wheat and rice caused by current and future O_3 concentrations at the national scale. Further studies are warranted on the interactions between O_3 and other environmental changes, such as increasing CO_2 concentrations, increased nitrogen deposition, aerosol loading, and climatic changes. Future needs for research include improvement of O_3 impact models and the development of an O_3 monitoring network to cover the vast areas of crop production in China. The establishment of an air quality standard for protecting crops from O_3 damage is of critical importance for food security in China.

Keywords China • CH_4 emission • Rice • Wheat • O_3 • Yield

Z. Feng (✉)
State Key Laboratory of Urban and Regional Ecology,
Research Center for Eco-Environmental Sciences, Chinese Academy of Sciences,
Shuangqing Road 18, Haidian District, Beijing 100085, China
e-mail: zhzhfeng201@hotmail.com; fzz@rcees.ac.cn

H. Tang
State Key Laboratory of Soil and Sustainable Agriculture, Institute of Soil Sciences,
Chinese Academy of Sciences, Nanjing 210008, People's Republic of China
e-mail: hytang@issas.ac.cn

K. Kobayashi
Graduate School of Agricultural and Life Sciences, The University of Tokyo,
1-1-1 Yayoi, Bunkyo-ku, Tokyo 113-8657, Japan
e-mail: k.kobayashi.ut@gmail.com

T. Izuta (ed.), *Air Pollution Impacts on Plants in East Asia*,
DOI 10.1007/978-4-431-56438-6_12

12.1 Current O_3 Concentrations in China

Until 1990, ozone (O_3) concentrations ([O_3]) in China were lower than those in the United States and European cities, but they have increased quite rapidly since then, due to the increased emissions of O_3 precursors arising from automobile transportation, electricity generation, and other industrial activities. Current [O_3] is rising at a higher rate in China than in other countries, and the daily 24 h mean [O_3] often exceeds 50 ppb on average across the crop growing season in some regions (Tang et al. 2013; Zhao et al. 2009). Regional O_3 pollution has become one of the top environmental concerns in China, especially in economically vibrant and densely populated regions. Some major cities in China, such as Beijing, Shanghai, Jinan, Hong Kong, and Guangzhou, are faced with photochemical threat (Feng et al. 2015a). In recent years, ambient [O_3] has often reached 150 ppb in the afternoons from May to August in farmlands near Beijing city.

The variability of [O_3] is a function of geographic location. In the central and northern parts of China, [O_3] reaches a maximum in summer. In southern China, by comparison, [O_3] is generally characterized by a peak in fall and a trough in summer (Feng et al. 2015a). On a monthly-mean basis, surface [O_3] peaks in May in the Yangtze River Delta, in June in the North China Plain, and in October in the Pearl River Delta (Wang et al. 2011). The peak hourly [O_3] reaches as high as 316 ppb in the North China Plain (Feng et al. 2015a). The high surface [O_3] can therefore be regarded as a present threat to food security.

12.2 O_3 Injuries Observed in the Field

O_3 concentrations in the Beijing area are high enough to induce foliar symptoms in plants. During our survey around Beijing in 2013, we found 12 species, including food crops, vegetables, and fruit trees, showing typical O_3 symptoms (Table 12.1). Among the crops, different types of beans belonging to the genera *Canavalia*, *Vigna*, and *Phaseolus* showed distinctive and severe O_3 symptoms at most of the places examined. These species have strong potential to be used as O_3 bio-indicators in China, given that they are O_3-sensitive and commonly planted. The observed O_3 symptoms are described in Table 12.1. During the survey in 2014, however, we found fewer plant species showing O_3 injury in the field, presumably due to the weather conditions not being conducive to photochemical O_3 production. Furthermore, no more plant species, other than those shown in the list in Table 12.1, were found to show O_3 injury at sites situated in Hebei Province or Tianjin.

Table 12.1 O_3-sensitive species and their common names, and description of the O_3 symptoms observed in the field (Feng et al. 2014)

Species	Common name	Symptom description
Abelmoschus esculentus	Okra plant	Fine light brown and purple interveinal stippling on the adaxial side of older leaves
Abutilon theophrasti	Velvetleaf, China jute, Buttonweed	Fine yellow to light brown stippling between main and secondary veins, which remain green
Arachis hypogaea	Peanut	Very fine light brown interveinal stippling, finally also with dark purple stippling
Benincasa pruriens	Winter gourd	Fine light brown diffuse stippling between main and secondary veins, which remain green
Canavalia gladiata	Sword bean	Distinctive purple-brown or dark purple stippling that can cover a large portion of the interveinal areas of leaves. Sometimes with associated chlorosis
Citrullus lanatus	Watermelon	Dark brown stippling on interveinal parts of older leaves. Stippling evolves to form very distinctive brown necrotic areas, sometimes with associated white flecking. Chlorosis is frequently observed together with stippling
Humulus scandens	Hops	Very fine light brown interveinal stippling
Luffa cylindrica	Courgette	Fine yellow to light brown stippling between main and secondary veins, which remain green
Phaseolus vulgaris	Common bean	Distinctive purple-brown or dark purple stippling that can cover a large portion of the interveinal areas of leaves. Sometimes with associated chlorosis. Severely affected leaves finally become dry
Prunus persica f. duplex	Peach f. duplex	Dark purple stippling, usually more intense towards the leaf margins
Vigna unguiculata subsp. sesquipedalis	Chinese long bean, cowpea	Distinctive purple-brown or dark purple stippling that can cover a large portion of the interveinal areas of leaves. Sometimes with associated chlorosis. Severely affected leaves finally become dry
Vitis vinifera	Grapevine	Brown and purple-brown interveinal stippling on the adaxial side of older leaves

12.3 Effects of Ambient O_3 on Crops as Detected by Using a Plant Protectant

Ethylenediurea (N-[2-(2-oxo-1-imidazolidinyl)ethyl]-N-phenylurea), abbreviated as EDU, has protective effects against O_3 damage, and, hence, has been used extensively to assess the O_3-induced impact on many crops, e.g., wheat, potato, bean, and tomato (Feng et al. 2010a). The application of EDU as a 'control' for ambient O_3 is useful to determine ambient O_3 effects on field-grown plants, particularly in remote and under-developed regions where the availability of electricity and funding is limited (Feng et al. 2010a; Manning et al. 2011). So far, only two EDU studies have been conducted in China, focusing on the effects of ambient O_3 on one cultivar of wheat and rice (Wang et al. 2007) and various cultivars of snap bean (Yuan et al. 2015).

Yuan et al. (2015) investigated the effects of ambient O_3 (accumulated hourly O_3 concentration over a threshold of 40 ppb during daytime (AOT40) of 29.0 ppm·h) on four genotypes of snap bean (*Phaseolus vulgaris* L.) in a cropland area around Beijing by using 450 ppm of EDU as a protectant. All genotypes showed foliar injuries, but O_3-sensitive genotypes exhibited much more injury than O_3-tolerant ones. In the O_3-sensitive genotypes, EDU significantly alleviated foliar injuries, increased the photosynthesis rate and chlorophyll *a* fluorescence, and alleviated the O_3 effects on photosynthetic parameters (maximum carboxylation efficiency (V_{cmax}) and maximum rate of electron transport (J_{max})) and seed and pod weights. These effects of EDU were not observed in the O_3-tolerant genotypes. EDU did not, however, significantly affect antioxidant content in any of the genotypes.

Wang et al. (2007) applied EDU, at four concentrations of 0, 150, 300, or 450 ppm, as a foliar spray to field-grown rice (*Oryza sativa* L.) and wheat (*Triticum aestivum* L.) in the Yangtze Delta in China. They found that rice and wheat responded differently to ambient O_3 and EDU applications. In wheat, some growth characteristics, such as yield, seed number per plant, seed set rate, and harvest index, were increased significantly with 300 ppm EDU treatment. In rice, by comparison, no parameters measured were significantly different among the various EDU application levels. It is, therefore, suggested that EDU is effective in demonstrating O_3 effects on wheat, but not on rice. The different response to EDU between wheat and rice could be attributed to the facts that the wheat cultivar used is more sensitive to O_3 than the rice cultivar and that $[O_3]$ was lower during the rice season than during the wheat season.

From these two studies, it can be concluded that EDU can be regarded as a useful tool in the risk assessment of ambient O_3 on O_3-sensitive crop species. Currently, high ambient $[O_3]$ occurs in summer throughout China, which suggests the threat of ground-level O_3 to food security. Breeding O_3-tolerant cultivars will be needed to protect the crops from the threat, considering the significant varietal difference in O_3 sensitivity, as found in the O_3 exposure studies noted later in this chapter.

12.4 Effects of Elevated O_3 Concentration on Crops Studied with Artificial O_3 Exposures

Artificial O_3 exposure experiments with agricultural crops: winter wheat, rice, and oil-rape (*Brassica napus* L.) have been conducted at five different sites in China, using open-top chamber (OTC) or free-air O_3 concentration enrichment (FACE-O_3) facilities (Tables 12.2 and 12.3). Experiments in both Jiaxing and Jiangdu, which

Table 12.2 Summary of O_3 impacts on wheat yield reported in studies conducted in China compared with findings in European studies (Feng et al. 2015a)

Site	Rooting	Duration	Facility	O_3 treatment[a]	Dose-response[b]	Reference
Gucheng (39°08′N, 115°48′E)	Pot	1999	OTC	CF, NF, 50, 100, 200 ppb	RY = 100 – 1.30 AOT40	Feng et al. (2003)
Jiaxing (31°53′N, 121°18′E)	Field	2004–2008	OTC	CF, NF, 75/100, 150/200 ppb	RY = 100 – 2.28 AOT40	Wang et al. (2012)
Jiangdu (32°35′N, 119°42′E)	Field	2007–2011	FACE	Ambient air (AA) and 1.5*AA (E-O_3)	RY = 96.1 – 2.5 AOT40	Feng et al. (2012)
					RY = 100 – 8.2 POD_{12}	
Changping (40°12′N, 116°08′E)	Field	2010	OTC	NF, NF + 30, NF + 60, NF + 90 ppb	RY = 100 – 2.2 AOT40	Tong (2011)
					RY = 100 – 3.7 POD_4	
Synthesis of European studies	Pot and field		OTC		RY = 99 – 1.6 AOT40	Mills et al. (2007)
					RY = 100 – 3.8 POD_6	Mills et al. (2011)

FACE free-air O_3 concentration enrichment, *OTC* open-top chamber

[a] *CF* charcoal-filtered ambient air, *NF* non-charcoal-filtered ambient air, *E-O_3* elevated O_3 concentration

[b] *RY* relative yield (%), AOT40 (ppm·h), accumulated hourly O_3 concentration over a threshold of 40 ppb during daytime; *PODy* (nmol m^{-2}), phytotoxic O_3 dose (POD), is used for explaining what POD is, the accumulated stomatal O_3 flux above a flux threshold of *y* nmol O_3 m^{-2} s^{-1} (LRTAP Convention 2010)

Table 12.3 Summary of O_3 impacts on rice yield reported in studies conducted in China compared with findings in European studies (Feng et al. 2015a)

Site	Rooting	Duration	Facility	O_3 treatment	Dose-response	Reference
Gucheng (39°08′N, 115°48′E)	Pot	2000	OTC	CF, NF, 50, 100, 200 ppb	RY = 100 – 0.53 AOT40	Feng et al. (2003)
Jiaxing (31°53′N, 121°18′E)	Field	2004–2008	OTC	CF, NF, 75/100, 150/200 ppb	RY = 100 – 0.95 AOT40	Wang et al. (2012)
Jiangdu (32°35′N 119°42′E)	Field	2007–2011	FACE	Ambient air (AA) and 1.5*AA (E-O_3)	–	
Dongguan (23°01′N, 113°45′E)	Field	2010	OTC	NF, NF + 40, NF + 80, NF + 120 ppb	RY = 100 – 3.9 AOT40; RY = 100 – 2.3 POD_2	Tong (2011)
Synthesis of European studies	Pot and field		OTC		RY = 94 – 0.39 AOT40	Mills et al. (2007)

The abbreviations are as defined in Table 12.2

lasted for 5 years, aimed to study the impacts of elevated [O_3] on the growth, physiological characteristics, and yield components of rice and winter wheat. Experiments at both sites serve as rich sources of information for the regional assessment of ambient and elevated [O_3] effects on the food crop yield loss in the Yangtze River Delta. The experiment at the Jiangdu site also investigated four cultivars of wheat and rice (including Japonica, Indica, and hybrid cultivars) in response to elevated [O_3].

These experiments elucidated the effects of elevated [O_3] on the following aspects: crop growth processes, crop yield, crop quality, and soil processes. These effects are described in more detail in the subsequent sections of this chapter.

12.4.1 Effects of O_3 on Growth Processes in Crop Cultivars

A meta-analysis showed that elevated [O_3], compared with charcoal-filtered air, decreased the leaf photosynthetic rate by 20 % in wheat and by 28 % in rice, and decreased grain yield by 29 % in wheat and by 14 % in rice plants (Feng et al. 2008; Ainsworth 2008), suggesting that decreased photosynthesis was a key factor driving the yield loss in crops exposed to elevated [O_3]. In China, many crop species have been investigated for the effects of O_3. The species studied were winter wheat (e.g. Feng et al. 2011a; Zheng et al. 2005, 2010; Zhu et al. 2011), rice (e.g. Shi et al 2009; Pang et al. 2009; Shao et al. 2014; Zhang et al. 2008), soybean (e.g. Zhang et al. 2014a, b; Zhao et al. 2012), oil rape (Feng et al. 2006; Zheng et al. 2006), maize (Sun et al. 2008), and spinach (Yao et al.2007). Most studies focused on winter wheat and rice. Similar to the results of other experiments outside of China, chronic exposure to elevated [O_3] caused a range of adverse effects on plants, including enhanced lipid peroxidation, reduced or enhanced antioxidant system activity, reduced photosynthetic activity, altered carbon allocation, diminished biomass accumulation, reduced yield, and accelerated senescence, with or without visible injury. Ozone-induced grain yield loss in winter wheat was mainly caused by a reduction in grain mass (Wang et al. 2012; Zhu et al. 2011). On the other hand, a reduction in panicle numbers contributed to the yield loss in rice when exposed to elevated [O_3] (Shi et al. 2009; Pang et al. 2009).

It has also been reported that the effects of O_3 vary by cultivars. Modern wheat cultivars are reported to be more sensitive to O_3 than older accessions, and this was largely attributed to higher stomatal conductance (g_s) in modern cultivars allowing for greater O_3 uptake (Barnes et al. 1990; Pleijel et al. 2006; Biswas et al. 2008). The FACE-O_3 experiment in China was therefore conducted with multiple cultivars to determine their responses to elevated [O_3].

For wheat, two modern cultivars [Yangmai16 (Y16) and Yangfumai 2 (Y2)] of winter wheat with almost identical phenology were investigated. In cultivar Y2, elevated [O_3] significantly accelerated leaf senescence, as indicated by increased lipid oxidation, as well as by faster declines in pigment amounts and photosynthetic rates. The lower photosynthetic rates were mainly due to non-stomatal factors, e.g., lower maximum carboxylation capacity, electron transport rates, and light energy distribution. In cultivar Y16, by contrast, the effects of elevated [O_3] were observed only at the very last stage of flag leaf aging (Feng et al. 2011a). The less sensitive variety Y16 had 33.5 % and 12.0 % higher concentrations of reduced ascorbate in the apoplast and leaf tissue, respectively, than those in the more sensitive variety Y2, whereas no varietal difference was detected in the decline of reduced ascorbate concentration in response to elevated [O_3]. No effects of O_3 or variety were detected in either oxidized ascorbate or the redox state of ascorbate in the apoplast and leaf tissue (Feng et al. 2010b). Since the two cultivars had almost identical phenology and very similar leaf stomatal conductance before senescence, the greater impacts of elevated [O_3] on cultivar Y2 than on cultivar Y16 cannot be explained by differential O_3 uptake, but by apoplast ascorbate contents. Our findings will be useful for scientists to select wheat cultivars that will be tolerant to rising surface [O_3] in East and South Asia.

In rice, varietal difference in the effects of elevated [O_3] was clearer than in wheat. Shi et al. (2009) studied four Chinese rice cultivars: Wujing 15 (WJ15, inbred Japonica cultivar), Yangdao 6 (YD6, inbred Indica cultivar), Shanyou 63 (SY63, three-line hybrid rice cultivar), and Liangyoupeijiu (LYPJ, two-line hybrid rice cultivar) at the FACE-O_3 site. The elevated [O_3] (a mean 25 % enhancement above the ambient [O_3]) strongly accelerated the phenological development of WJ15 and SY63, with maturity being reached by 4 and 8 days earlier, respectively, compared with findings in ambient [O_3], with maturity being reached only 1 day earlier for YD6 and LYPJ. SY63 and LYPJ exhibited significant yield losses by exposure to elevated [O_3] (17.5 % and 15 %, respectively), while WJ15 and YD6 showed no responses to the elevated [O_3]. For all cultivars, no O_3 effect was observed on panicle number per unit area, as a result of there being no changes in either maximum tiller number or productive tiller ratio. However, the number of spikelets per panicle in SY63 and LYPJ showed a significant reduction due to O_3 exposure, while these numbers remained unaffected in WJ15 and YD6. O_3 exposure also caused minor reductions in both filled spikelet percentage and individual grain mass in all tested cultivars. The results of these experiments indicated that yield loss due to O_3 exposure differs among rice cultivars, with the hybrid cultivars (i.e., SY63 and LYPJ) exhibiting greater yield loss than the inbred cultivars (i.e., WJ15 and YD6), a finding which could be attributed to the suppression of spikelet formation in the hybrid cultivars under O_3 stress (Shi et al. 2009). In regard to photosynthetic characteristics, flag leaves in SY63 exhibited an earlier and stronger response to elevated [O_3] than flag leaves in WYJ3. Depression of the light-saturated photosynthetic rate (A_{sat}) was first observed at 237 day of year (DOY) in SY63, and the seasonal mean A_{sat} was reduced by 23.1 %. In contrast, in WYJ3, a conventional inbred cultivar, the impact of elevated [O_3] on A_{sat} was negligible until 266 DOY, and the seasonal mean decrease of A_{sat} was only 9.4 %. The same trend was found in chlorophyll *a* fluorescence parameters. In SY63, the actual quantum yields of photosynthesis system II (PSII) and photochemical quenching (qP) were significantly decreased at 248 DOY, and the downward trends persisted throughout the rest of the life span of flag leaves. No such changes were observed in WYJ3. In the ozonated flag leaves of SY63, necrotic damage occurred and chlorophyll contents declined significantly, by 12.6–43.6 %, throughout the entire functional duration of the leaf. As for WYJ3, chlorophyll remained unaffected until a 25.7 % decrease appeared at 278 DOY under elevated [O_3] (Pang et al. 2009).

12.4.2 Effects of O_3 on Crop Yield

In China, the effects of elevated [O_3] on crops were studied with plants rooted in the field at only two experimental sites. One is Jiaxing, where wheat and rice were investigated for four and five seasons, respectively, in OTCs for the period from 2004 to 2008. The other site is Jiangdu, where four cultivars of winter wheat and five rice cultivars were investigated for five growing seasons (2007–2012) in the FACE-O_3 facility.

At the Jiaxing site, four different levels of [O_3] were set up to build the O_3 dose-response relationship. Results indicated that elevated [O_3] significantly reduced winter wheat yield, by 8.5–58 % and 40–73 % compared with the charcoal-filtered air (CF) control for O_3-1 (AOT40 of 14.3–22.6 ppm·h) and O_3-2 (AOT40 of 24.2–61.9 ppm·h) treatments, respectively. As compared with the subambient [O_3] control, the mean yield losses in rice were 10–34 % and 16–43 %, respectively, when plants were exposed to O_3-1 (AOT40 of 13.8–29.5 ppm·h) and O_3-2 (AOT40 of 32.1–82.6 ppm·h) treatments, respectively. Winter wheat appeared to be more sensitive to O_3 than rice. The O_3-induced yield declines for winter wheat were attributed primarily to the 1,000-grain weight and the harvest index, and the declines for rice were attributed primarily to grain number per panicle and the harvest index (Wang et al. 2012).

At the Jiangdu FACE-O_3 site, a mean 25 % enhancement above the ambient [O_3] (45.7 ppb) significantly reduced the grain yield, by 20 %, with significant variation, in the range from 10 to 35 %, among the combinations of four cultivars and three growth seasons. The reduction of individual grain mass mostly accounted for the yield loss induced by O_3, and showed significant difference between the cultivars (Zhu et al. 2011). The response of relative yield to elevated [O_3] was not significantly different from those reported in China, Europe, and India on the basis of experiments in OTCs.

Elevated [O_3] significantly reduced the grain yield by 12 %, when averaged across all the tested cultivars (hybrid rice cultivars SY63 and LYPJ, and inbred cultivars WJ15 and YD6). However, the hybrid rice cultivars SY63 and LYPJ exhibited greater yield losses, by 17.5 % and 15.0 %, respectively, than the inbred cultivars WJ15 and YD6, which showed no significant yield loss. The different responses between cultivars were attributed to the suppression of spikelet formation in the hybrid cultivars under O_3 stress (Shi et al. 2009).

Our results thus confirmed the rising threat of surface [O_3] on wheat production in China and, indeed, in other parts of the developing world in the near future. Various countermeasures are urgently needed against the crop losses due to O_3, such as mitigation of the increase in surface [O_3] with stricter pollution controls, and enhancement of wheat tolerance to O_3 by breeding and management.

12.4.3 Effects of O_3 on Grain Quality

Elevated [O_3] also significantly changed grain quality. The OTC experiments in 2007 and 2008 indicated that N and S concentrations were increased by exposure to elevated [O_3] (Zheng et al. 2013a). The experiments also showed increases in the concentrations of K, Ca, Mg, P, Mn, Cu, and Zn for winter wheat and Mg, K, Mn, and Cu for rice. The concentrations of protein, amino acid, and lysine in winter wheat and rice were increased and the concentration of amylose was decreased. The increase in the nutrient concentrations was less than the reduction of the grain yield in both winter wheat and rice, and, hence, the absolute amount of the nutrients was reduced by elevated [O_3] (Zheng et al. 2013a).

In the FACE-O_3 study on winter wheat, elevated [O_3] decreased the accumulation rates of amylose, amylopectin, and starch amylase, reduced the accumulation amounts of amylopectin and starch, and decreased the contents of amylopectin and starch, but increased the content of amylose. With the elevation of [O_3], the enzyme activity of grain granule-bound starch synthase (GBSS), soluble starch synthase (SSS), and starch branching enzyme (SBE) decreased after anthesis. The activities of GBSS and SSS had highly significant correlations with amylose, amylopectin, and starch accumulation rates, and the activity of SBE had significant correlations with these rates (Zhang et al. 2013).

Elevated [O_3] significantly increased grain chalkiness and the concentrations of essential nutrients, which was particularly significant for Zn and Cu. The O_3-induced changes in starch pasting properties (e.g., amylose concentration decreased by 15.1 %) indicated a trend of deterioration in the cooking and eating quality of the grain (Wang et al. 2014). The contents of protein, total amino acids (TAAs), total essential AAs (TEAAs) and total non-essential AAs (TNEAAs) in rice grain were increased by 12–14 % with elevated [O_3] (Wang et al. 2014; Zhou et al. 2015). A similar significant response to O_3 was observed for concentrations of the seven essential and eight non-essential AAs. In contrast, elevated [O_3] caused a small but significant decrease in the percentage of TEAAs within TAAs (Zhou et al. 2015).

12.4.4 Effects of O_3 on CH_4 Emission in Rice Paddies

Rice paddies are an important CH_4 source, accounting for nearly 20 % of global anthropogenic CH_4 emission (Intergovernmental Panel on Climate Change; IPCC 2007) and they are also a major source of N_2O, accounting for 22 % of the total emission from croplands in China (Xing 1998). Using OTCs in situ with different [O_3] treatments, CH_4 emissions were measured in a rice paddy in the Yangtze River Delta, China, in 2007 and 2008. The diurnal patterns of CH_4 emission varied temporally with treatments and there was inconsistence in the diurnal variations of CH_4 emissions from the paddy field. CH_4 emissions from the paddy field throughout the growing season were reduced by 46.5 % and 50.6 % in elevated [O_3] of 69.6 ppb and 118.6 ppb, respectively, in 2007, and by 38.3 % and 46.8 % in elevated [O_3] of 82.2 ppb and 138.3 ppb, respectively, in 2008, as compared with the subambient [O_3] level. The seasonal mean CH_4 emissions were negatively correlated with AOT40 ($P<0.01$ in both years), but were positively correlated with the relative rice yield, as well as with the above- and below-ground biomass (Zheng et al. 2011).

Tang et al. (2015) reported, in a FACE-O_3 experiment, that a mean 26.7 % enhancement of [O_3] above ambient [O_3] significantly reduced CH_4 emission at the tillering and flowering stages, leading to a reduction of seasonal integral CH_4 emission by 29.6 % on average across two cultivars: an inbred Indica cultivar, Yangdao 6 (YD6), and a hybrid one, II-Y084. Also, the reduced CH_4 emission was associated with O_3-induced reductions in whole-plant biomass (−13.2 %), root biomass (−34.7 %), and maximum tiller number (−10.3 %). Furthermore, a larger decrease in CH_4 emission with II-Y084 (−33.2 %) than that with YD6 (−7.0 %) was observed at the tillering stage, which may have been due to the larger reduction in tiller number in II-Y084 in elevated [O_3]. Additionally, elevated [O_3] reduced the seasonal mean nitrogen oxides (NOx) flux by 5.7 % and 11.8 % with II-Y084 and YD6, respectively, but these effects were not statistically significant. The relative response of CH_4 emission to elevated [O_3] in the FACE-O_3 experiment was not significantly different from those reported in OTC experiments. The two studies, i.e., the study by Tang et al. (2015) and the study by Zheng et al. (2011) have thus confirmed that increasing [O_3] could mitigate the global warming potential of CH_4 and the findings suggest that the feedback mechanism between O_3 and its precursor emission should be considered in the projection of future O_3 effects on terrestrial ecosystems.

12.4.5 Effects of O_3 on Soil Microbiology

Knowledge is limited regarding the impact of elevated [O_3] on below-ground processes in agro-ecosystems. There are many reports, e.g. Andersen (2003), Grantz et al. (2006), and Feng et al. (2008), that exposure of plants to elevated [O_3] reduces carbon allocation to roots, reduces the root/shoot biomass ratio, and reduces root exudates. As plants serve as the main source of carbon and energy inputs to the plant-soil microbe web, a decrease in the carbon flux from plants due to elevated [O_3] could adversely influence the diversity of soil microbes.

Across the winter wheat growth period, elevated [O_3] significantly reduced soil microbial carbon and changed microbial community-level physiological profiles in rhizosphere soil, but not in non-rhizosphere soil. The relative abundances of fungal and actinomycetous indicator phospholipid fatty acids (PLFAs) were decreased in both rhizosphere and non-rhizosphere soils, while those of bacterial PLFAs were increased by elevated [O_3] (Chen et al. 2009, 2010, 2015).

However, the responses of the soil biota to O_3 pollution are different between cultivars. In a FACE-O_3 experiment, Feng et al. (2015b) found that elevated [O_3] negatively influenced the bacterial community in an O_3-tolerant cultivar of rice, YD6, by decreasing bacterial phylogenetic diversities. In contrast, in the O_3-sensitive rice cultivar IIY-084, the bacterial community responded positively to elevated [O_3] at the tillering stage. However, several keystone bacterial guilds were consistently negatively affected by elevated [O_3] in both cultivars. These findings indicate that elevated [O_3] could negatively influence rice agro-ecosystems and that the crop cultivar is an important determinant of the soil biota responses to elevated [O_3].

In our FACE-O_3 experiment, elevated [O_3] significantly reduced the abundance and percentage of anoxygenic phototrophic purple bacteria (AnPPB) in the total bacterial community in flooded rice soil, via decreasing their genotypic diversity and metabolic versatility (Feng et al. 2011b). Concomitantly, under elevated [O_3], the community composition changed after the rice anthesis stage. These AnPPB responses imply that continuously elevated [O_3] in the future could eventually harm the health of paddy ecosystems through its negative effects on soil microorganisms (Feng et al. 2015b). In a parallel study, elevated [O_3] inhibited methanogenic activity and influenced the composition of paddy methanogenic communities, reducing the abundance and diversity of paddy methanogens by adversely affecting dominant groups, such as acetoclastic Methanosaeta, especially at the rice tillering stage (Feng et al. 2013). These results indicate that elevated [O_3] could negatively influence paddy methanogenic archaeal communities and their critical ecological function.

Both OTC and FACE-O_3 experiments have thus proven that elevated [O_3] significantly changes soil microbial community function and composition, which could influence soil nutrient supply and soil carbon metabolism.

12.5 Estimating the Impacts of Rising O_3 Concentrations on Crops in China

Both weighted-mean concentrations and flux-based approaches are widely used to assess O_3 impacts on crops on regional, national, and global scales (Wang and Mauzerall 2004; Mills et al. 2011; Avnery et al. 2011). However, more and more studies have indicated that O_3 flux is superior to concentration exposure indexes such as AOT40 for assessing O_3 effects (Mills et al. 2011; Büker et al. 2015), because the flux-based approach considers biological and climatic factors that influence daytime stomatal O_3 uptake. To obtain O_3 flux-based assessments, multiplicative stomatal conductance models have been developed and parameterized to estimate O_3 flux through stomata.

12.5.1 Stomatal O_3 Flux Models

Emberson et al. (2000) and Pleijel et al. (2002) developed and parameterized Jarvis-type (Jarvis 1976) multiplicative stomatal conductance models for spring wheat and potato, respectively, and used them to derive relationships between yield loss and stomatal O_3 flux for both species (Danielsson et al. 2003; Pleijel et al. 2002, 2004). These models predict stomatal conductance as a function of phenology and the short-term effects of environmental conditions such as radiation, temperature, vapor pressure deficit (VPD), and [O_3], based on European field observations and conditions (Feng et al. 2015a).

In China, stomatal O_3 flux and flux-response relationships were first derived for winter wheat grown in FACE-O_3 (Feng et al. 2012). A stomatal conductance (g_{sto}) model developed for wheat in Europe was re-parameterized for the Chinese varieties.

Compared with European model parameterizations, the main changes were that the VPD and radiation response functions were made less and more restrictive, respectively, and the temperature function was omitted. The re-parameterized g_{sto} model performed well, with an r^2 value of 0.76. The slope and intercept of the regression between observed and predicted g_{sto} were not significantly different from 1 and 0, respectively.

Tang et al. (2014) parameterized a multiplicative model of g_{sto} for O_3 uptake by rice leaves with the field measurements in a FACE-O_3 experiment. In the g_{sto} model for rice, the entire accumulation period was 1,350 °C days, in contrast to 800–970 °C days for wheat (Pleijel et al. 2007; LRTAP 2010; Feng et al. 2012). The minimum fraction of VPD limitation for gs (f_{VPD}) was set at VPD of 2.7 kPa and the maximum value of fraction of temperature limitation for gs (f_{temp}) occurred at 28 °C. The estimated g_{sto} determined by the parameterized rice g_{sto} model compared well with the observed value ($r^2=0.79$). The regression line was close to and not significantly different from the line of equality, with the slope being 0.932 (no dimension) with a 95 % confidence interval (CI) of 0.823–1.039 and the intercept being 21.66 (mmol m^{-2} s^{-1}), with a 95 % CI of −2.74–46.06 (mmol m^{-2} s^{-1}) (Tang et al. 2014).

12.5.2 O_3 Dose-Response Relationships for Crop Yield in China

Tables 12.2 and 12.3 show the dose-response relationships for O_3. It seems that a similar O_3 dose based on the AOT40 induces larger yield loss in winter wheat with FACE-O_3 than with OTC. Such a comparison is not feasible for the flux-based O_3 dose, since an O_3 flux model has not been established for the OTC experiments in Jiaxing, due to the limited number of observations. An O_3 uptake threshold of 12 nmol m^{-2} s^{-1} was judged most reasonable for the wheat flux-response relationship in subtropical China (Feng et al. 2012). Judging from both flux- and concentration-based relationships, the cultivars investigated in China seem to be more sensitive to O_3 than European cultivars (Feng et al. 2012). The new flux-response relationship can be applied to O_3 risk assessment in wheat in subtropical regions.

For rice, the O_3 dose-response relationship was established based on the AOT40, but not on the stomatal O_3 flux dose. From the results to date, the O_3 impact is estimated to double, based on O_3 flux, or triple, based on O_3 exposure, across the majority of rice-producing areas in the middle and lower reaches of the Yangtze River and in South China between the years 2000 and 2020 (Tang et al. 2014).

12.5.3 Regional Estimates of O_3 Impacts on Crop Production

Past studies (Aunan et al. 2000; Wang and Mauzerall 2004; Tang et al. 2013) have estimated relative yield losses of food crops due to current and projected [O_3] across China, with wide margins of uncertainties in the estimates (Feng et al. 2015a).

Taking winter wheat as an example, [O_3] in 2020 is projected to induce a yield loss in the range of 2.9 % to 7 %, based on the mean [O_3]; a yield loss of 2.3 % to 63 %, based on the seasonal sum of hourly [O_3] exceeding 60 ppb (SUM06); a yield loss of 13.4 % to 16.6 %, based on the AOT40; and a yield loss of 19.2 % to 23.0 %, based on stomatal O_3 flux (Aunan et al. 2000; Wang and Mauzerall 2004; Tang et al. *2013*). The wide margins can be attributed to differences between and within the studies in the estimation of [O_3] at the plant canopy height, O_3 dose metrics, and the O_3 dose-yield loss relationships based on different sets of artificial exposure experiments.

Among the studies mentioned above, only Tang et al. (2013) used the O_3 dose-yield loss relationships established with experiments conducted in China, while the others used the relationships established for Europe and North America. The estimates of the crop losses varied by the dose-response relationships with different O_3 dose metrics. The relative yield loss (RYL) of wheat for the whole of China in 2000, for example, was estimated to be in the range of 6.4 % to 14.9 % (Tang et al. 2013). The POD with the stomatal O_3 flux threshold over 6 nmol m^{-2} s^{-1} (POD6) predicted greater RYL, whereas the 90-day AOT40 gave the lowest estimates. It is noteworthy, however, that the *increase* of RYL from 2000 to 2020 for China was estimated in a much narrower range, of 8.1–9.4 %.

Tang et al. (2013) also showed that the O_3 flux-based estimates of RYL were highly sensitive to perturbations in the meteorological inputs, but that the estimate of the *increase* in RYL from 2000 to 2020 was much more robust than the estimates of RYL per se. Thus, the projected *increase* in wheat production loss in China in the near future is substantial, beyond the uncertainties pointing to the obvious need to curb the rapid increase in surface [O_3] in China.

It must be noted, however, that the O_3 impact estimated by Tang et al. (2013) is based on an experiment at a single site in the Yangze River Delta, whereas there is a vast range of plant varieties, agronomic practices, soil, and climate across the major crop-growing regions of China. Thus, the uncertainties due to the scaling-up from a single site to the entire country of China are yet to be quantified.

12.6 Combined Effects of O_3 and Other Environmental Changes on Chinese Crops

O_3 is unlikely to be the only stressor that plants are subjected to during their growth and development. Previous experiments have demonstrated that plant response to O_3 is altered under other environmental factors that stress crop systems, including atmospheric CO_2 concentration ([CO_2]), temperature, solar radiation, soil moisture, and nitrogen availability. Changes in agricultural productivity can be the result of direct effects of these factors at the plant level, such as alterations in leaf stomatal O_3 uptake, or indirect effects at the system level; for instance, through shifts in crop phenology, nutrient cycling, pest occurrence, and plant diseases. Furthermore, the combined effects of O_3 and other environmental factors on crop systems and alterations to natural emissions of O_3 precursors have the potential to feedback on tropospheric [O_3], with implications for climate change. Nevertheless, in comparison with the studies on crop responses to each individual stressor, much less is known about the interactions of O_3 and these other stressors on plant performance. Below, we outline the limited reports on the combined effects of O_3 and other environmental factors on Chinese crops.

12.6.1 Interaction Between O_3 and CO_2

Generally, O_3 damages photosynthetic tissues and accelerates leaf senescence, whereas increased [CO_2] stimulates photosynthesis and biomass accumulation. The interactive effects of these two gases on plants have received much attention. By conducting a closed-top chamber experiment, Zhao et al. (2015) showed that the combined exposure to elevated [CO_2] (200 ppm above ambient) and [O_3] (60 % higher than ambient) had no significant effects on dry-matter production or nitrogen uptake in a hybrid rice cultivar, Shanyou 63. Similarly, in the same experiment, the combined effects of O_3 and CO_2 on leaf photosynthesis were not significant (Shao et al. 2014). As to the rice grain quality, O_3 affected many quality traits (chalkiness, protein nitrogen, Zn, Cu, hot viscosity, setback) significantly at ambient [CO_2], but such effects became non-significant at elevated [CO_2] (Wang et al. 2014). These results indicate that elevated [CO_2] can modify the O_3 impacts on rice growth, leaf photosynthesis, and grain quality. The mechanisms underlying the amelioration of O_3-induced damage by elevated [CO_2] are not well understood, but reduction in O_3 uptake through stomatal closure induced by elevated [CO_2] has been considered as the major factor responsible for the protection against O_3 aggression. It can be inferred that decreased stomatal conductance due to elevated [CO_2] decreases O_3 uptake; less O_3 uptake will lead to less O_3 damage to leaves, thus leading to more healthy green leaves for carbon assimilation, biomass growth, and grain filling.

12.6.2 O_3 and Aerosol Loading

O_3 and other atmospheric pollutants, e.g., O_3 precursor gases and aerosol particles, usually have a common source from fossil fuel combustion and biomass burning; these species increase atmospheric aerosol loadings (Simpson et al. 2014), which could diminish solar irradiance at regional scales. Therefore, increased tropospheric [O_3] and reduced solar irradiance may occur concomitantly and have combined effects on crop growth. A recent OTC experiment concluded that a 40 % shading of natural solar irradiance significantly exacerbated O_3-induced yield loss in winter wheat, but the extent of the yield loss was less than the additive effects of the individual O_3 and shading effects (Zheng et al. 2013b). These results emphasize that, although reduction in solar irradiance could decrease stomatal conductance and lead to less O_3 uptake by plants, this mitigation may not be able to compensate for the negative effect of this interaction on leaf photosynthesis. Moreover, the interaction of O_3 and solar irradiance could affect soil microbial functional diversity in the wheat rhizosphere, with implications for carbon cycling and sequestration below ground (Wu et al. 2015).

12.6.3 O_3 and Nitrogen Supply

O_3 interactions with nitrogen may also occur with the increased use of nitrogen fertilizer, because O_3-induced acceleration in foliar senescence and reduced translocation of nitrogen from aged leaves may depend on nitrogen availability. In a FACE-O_3 experiment, yields of wheat with high nitrogen fertilizer input were increased significantly compared with yields of wheat with standard nitrogen input, both under elevated [O_3] (Chen et al. 2011). Similar to findings in wheat, an increasing nitrogen supply also mitigated the O_3-induced yield loss in rice, and the application of additional nitrogen at the tillering stage was better than that at the panicle initiation stage (Luo et al. 2013) in terms of mitigating this O_3-induced yield loss. These mitigations of O_3 stress were attributed to significant increases in the net photosynthetic rate and increases in the content of chlorophyll a and chlorophyll b in wheat and rice leaves with high nitrogen availability (Chen et al. 2011; Luo et al. 2012).

12.6.4 O_3-Climate Interactions

Among weather variables, soil moisture is the one most often studied as a factor that interacts with the effects of O_3. This focuses on the soil moisture-O_3 interaction may have occurred because of concerns about the possible overestimation of O_3 impacts on crops. O_3 impacts were estimated with the dose-response relationships based on OTC experiments, where supplementary irrigation was given to compensate for the partial exclusion of rain water and the higher evaporative demand due to the chamber effect. Crop yield losses induced by O_3 were sometimes less in drought-stressed plants than in the well-watered plants. Stomatal closure due to water stress and, hence, reduced O_3 influx has been assumed to be the main cause of the reduced O_3 impacts. There is, however, another interpretation of the reduced O_3 impacts under water stress (Kobayashi et al. 1993), where the drought stress was ameliorated by reduced water use in O_3-stressed plants. In any case, the drought stress caused by O_3 interaction is important, since amelioration of water constraints by irrigation or increased rainfall due to climate change could increase the negative effects of O_3 on the crop. The net effect would be a shift of the major environmental stressor from soil moisture to surface O_3, resulting in a less-than-expected crop yield increase with the increased water availability.

Higher temperatures and altered precipitation can also affect O_3 formation through alterations to natural emissions of O_3 precursors. Finally, understanding how O_3 acts in combination with other stressors, e.g., heat stress, excessive nitrogen deposition, and high atmospheric aerosol loading, will also be important to fill the gaps in our knowledge how to target control efforts. As such, efforts to control O_3 may benefit from coordinated hemispheric- or global-scale action that is closely integrated with efforts on regional and local scales.

12.7 Future Research Needs

12.7.1 Establishing an O_3 Dose Model and Model Parameterization

It is difficult to assess the impact of ambient O_3 on crop productivity over the vast territory of China using a unified dose-response model, since the crop varieties and climates differ greatly by region. In China, past research on wheat and rice has been limited to the Yangtze River Delta, the Pearl River Delta, and the regions of Beijing, Tianjin, and Hebei. A more comprehensive study covering the major agricultural regions and staple food crops is crucial for estimating the surface O_3 impact on food production in China. A robust model of the relationship between crop productivity and O_3 exposure under different conditions should be established and validated against the local field investigations.

12.7.2 Developing Air Quality Standards for Food Security in China

Currently there are no standards in China to protect crops from O_3. To develop a standard, we need to do more monitoring and carry out measurements in rural areas and major crop production regions, including Hebei, Inner Mongolia, Jilin, Heilongjiang, Jiangsu, Anhui, Shandong, Henan, Hunan, Sichuan, Liaoning, Jiangxi, and Hubei Provinces, all of which have a sowing area of more than 3 Mha and in combination account for approximately 76 % of national food production. Combined with the controlled O_3 experiments at different sites mentioned above, O_3 standards for crop production should be developed for China.

Acknowledgments This study was funded by the Hundred Talents Program, Chinese Academy of Sciences, and supported by the State Key Laboratory of Urban and Regional Ecology.

References

Ainsworth EA (2008) Rice production in a changing climate: a meta-analysis of responses to elevated carbon dioxide and elevated ozone concentration. Glob Chang Biol 14:1642–1650

Andersen CP (2003) Source-sink balance and carbon allocation below ground in plants exposed to ozone. New Phytol 157:213–228

Aunan K et al (2000) Surface ozone in China and its possible impact on agricultural crop yields. AMBIO 29:294–301

Avnery S et al (2011) Global crop yield reductions due to surface ozone exposure: 2 year 2030 potential crop production losses and economic damage under two scenarios of O_3 pollution. Atmos Environ 45:2297–2309
Barnes JD et al (1990) Comparative ozone sensitivity of old and modern Greek cultivars of wheat. New Phytol 116:707–714
Biswas DK et al (2008) Genotypic differences in leaf biochemical, physiological and growth responses to ozone in 20 winter wheat cultivars released over the past 60 years. Glob Chang Biol 14:46–59
Büker P et al (2015) New flux based dose-response relationships for ozone for European forest tree species. Environ Pollut 206:163–174
Chen Z et al (2009) Impact of elevated O_3 on soil microbial community function under wheat crop. Water Air Soil Pollut 198:189–198
Chen Z et al (2010) Elevated ozone changed soil microbial community in a rice paddy. Soil Sci Soc Am J 74:829–837
Chen J et al (2011) Nitrogen supply mitigates the effects of elevated [O_3] on photosynthesis and yield in wheat. J Plant Ecol 35:523–530 (in Chinese)
Chen Z et al (2015) Structure and function of rhizosphere and non-rhizosphere soil microbial community respond differently to elevated ozone in field-planted wheat. J Environ Sci 32:126–134
Danielsson H et al (2003) Ozone uptake modelling and flux-response relationships -an assessment of ozone-induced yield loss in spring wheat. Atmos Environ 37:475–485
Emberson LD et al (2000) Modelling stomatal ozone flux across Europe. Environ Pollut 109:403–413
Feng ZW et al (2003) Effects of ground-level ozone (O_3) pollution on the yields of rice and winter wheat in the Yangtze River delta. J Environ Sci 15:360–362
Feng ZZ et al (2006) Response of gas exchange of rape to ozone concentration and exposure regimes. Acta Ecol Sin 26(3):823–829 (in Chinese)
Feng ZZ et al (2008) Impact of elevated ozone concentration on growth, physiology, and yield of wheat (*Triticum aestivum* L.): a meta-analysis. Glob Chang Biol 14:2696–2708
Feng ZZ et al (2010a) Protection of plants from ambient ozone by applications of ethylenediurea (EDU): a meta-analytic review. Environ Pollut 158:3236–3242
Feng ZZ et al (2010b) Apoplastic ascorbate contributes to the differential ozone sensitivity in two varieties of winter wheat under fully open-air field conditions. Environ Pollut 158:3539–3545
Feng ZZ et al (2011a) Differential responses in two varieties of winter wheat to elevated ozone concentration under fully open-air field conditions. Glob Chang Biol 17:580–591
Feng YZ et al (2011b) Elevated ground-level O_3 changes the diversity of an oxygenic purple phototrophic bacteria in paddy field. Microb Ecol 62:789–799
Feng ZZ et al (2012) A stomatal ozone flux-response relationship to assess ozone-induced yield loss of winter wheat in subtropical China. Environ Pollut 164:16–23
Feng YZ et al. (2013) Elevated ground-level O_3 negatively influences paddy methanogenic archaeal community. Sci Rep 3:3193. doi:10.1038/srep03193
Feng ZZ et al (2014) Evidence of widespread ozone-induced visible injury on plants in Beijing, China. Environ Pollut 193:296–301
Feng ZZ et al (2015a) Ground-level O_3 pollution and its impacts on food crops in China: a review. Environ Pollut 199:42–48
Feng YZ et al (2015b) The contrasting responses of soil microorganisms in two rice cultivars to elevated ground-level ozone. Environ Pollut 197:195–202
Grantz DA et al (2006) O_3 impacts on plant development: a meta analysis of root/shoot allocation and growth. Plant Cell Environ 29:1193–1209
IPCC (Intergovernmental Panel on Climate Change) (2007) Climate Change 2007: the physical science basis. contribution of working group I to the fourth assessment report of the intergovernmental panel on climate change. Cambridge University Press, Cambridge, UK
Jarvis PG (1976) The interpretation of the variations in leaf water potential and stomatal conductance found in canopies in the field. Phil Trans R Soc Lond B 273:593–610

Kobayashi K et al (1993) Model analysis of interactive effects of ozone and water stress on the yield of soybean. Environ Pollut 82:39–45

LRTAP Convention (2010) Mapping manual 2004. Manual on methodologies and criteria for modeling and mapping critical loads & levels and air pollution effects, risk and trends. Chapter 3. Mapping critical levels for vegetation, 2010 revision. Available from: http://icpvegetation.ceh.ac.uk

Luo K et al (2012) Effects of elevated ozone on leaf photosynthesis of rice (*Oryza sativa* L.) and mitigation with high nitrogen supply. Ecol Environ Sci 21:481–488 (in Chinese)

Luo K et al (2013) Responses of dry matter production and distribution in rice (*Oryza sativa* L.) to ozone and high nitrogen supply. Chin J Appl Ecol 19:286–292 (in Chinese)

Manning WJ et al (2011) Ethylenediurea (EDU): a research tool for assessment and verification of the effects of ground level ozone on plants under natural conditions. Environ Pollut 159: 3283–3293

Mills G et al (2007) A synthesis of AOT40-based response functions and critical levels of ozone for agricultural and horticultural crops. Atmos Environ 41:2630–2643

Mills G et al (2011) New stomatal flux-based critical levels for ozone effects on vegetation. Atmos Environ 45:5064–5068

Pang J et al (2009) Yield and photosynthetic characteristics of flag leaves in Chinese rice (*Oryza sativa* L.) varieties subjected to free-air release of ozone. Agric Ecosyst Environ 132:203–211

Pleijel H et al (2002) Stomatal conductance and ozone exposure in relation to potato tuber yield results from the European CHIP programme. Eur J Agron 17:303–317

Pleijel H et al (2004) Relationships between ozone exposure and yield loss in European wheat and potato e a comparison of concentration- and flux-based exposure indices. Atmos Environ 38:2259–2269

Pleijel H et al (2006) Differential ozone sensitivity in an old and a modern Swedish wheat cultivar – grain yield and quality, leaf chlorophyll and stomatal conductance. Environ Exp Bot 56:63–71

Pleijel H et al (2007) Ozone risk assessment for agricultural crops in Europe: further development of stomatal flux and flux-response relationships for European wheat and potato. Atmos Environ 41:3022–3040

Shao Z et al (2014) Impact of elevated atmospheric carbon dioxide and ozone concentration on leaf photosynthesis of 'Shanyou 63' hybrid rice. Chin J Eco-Agric 22:422–429 (in Chinese)

Shi GY et al (2009) Impact of elevated ozone concentration on yield of four Chinese rice cultivars under fully open-air field conditions. Agric Ecosyst Environ 131:178–184

Simpson D et al (2014) Ozone-the persistent menace: interactions with the N cycle and climate change. Curr Opin Environ Sustain 9–10:9–19

Sun JW et al (2008) Effects of elevated O_3 concentration on maize active oxygen species metabolism and antioxidative enzymes. J Agric Environ Sci 27(5):1929–1934 (in Chinese)

Tang HY et al (2013) A projection of ozone-induced wheat production loss in China and India for the years 2000 and 2020 with exposure-based and flux-based approaches. Glob Chang Biol 19:2739–2752

Tang HY et al (2014) Mapping ozone risks for rice in China for years 2000 and 2020 with flux-based and exposure-based doses. Atmos Environ 86:74–83

Tang HY et al (2015) Effects of elevated ozone concentration on CH_4 and N_2O emission from paddy soil under fully open-air field conditions. Glob Chang Biol 21:1727–1736

Tong L (2011) O_3 and CO_2 fluxes monitoring and modeling of early rice in southern China and winter wheat in Northern China, Graduate University of Chinese Academy of Sciences, Ph.D. Thesis, pp. 117 (In Chinese)

Wang X, Mauzerall DL (2004) Characterizing distributions of surface ozone and its impact on grain production in China, Japan and South Korea: 1990 and 2020. Atmos Environ 38:4383–4402

Wang XK et al (2007) Assessing the impact of ambient ozone on growth and yield of a rice (*Oryza sativa* L.) and a wheat (*Triticum aestivum* L.) cultivar grown in the Yangtze Delta, China, using three rates of application of ethylenediurea (EDU). Environ Pollut 148:390–395

Wang Y et al (2011) Seasonal and spatial variability of surface ozone over China: contributions from background and domestic pollution. Atmos Chem Phys 11:3511–3525
Wang XK et al (2012) Effects of elevated O_3 concentration on winter wheat and rice yields in the Yangtze River Delta, China. Environ Pollut 171:118–125
Wang YX et al (2014) Effects of elevated ozone, carbon dioxide, and the combination of both on the grain quality of Chinese hybrid rice. Environ Pollut 189:9–17
Wu F et al (2015) Effects of ozone fumigation and depressed solar irradiance on soil microbial functional diversity in winter wheat rhizosphere. Acta Ecol Sin 35:3949–3958 (in Chinese)
Xing G (1998) N_2O emission from cropland in China. Nutr Cycl Agroecosyst 52:249–254
Yao FF et al (2007) Influence of ozone and ethylenediurea (EDU) on physiological characters and foliar symptom of spinach (*Spinacia oleracea* L.) in open-top chambers. Ecol Environ 16(5):1399–1405 (in Chinese)
Yuan XY et al (2015) Assessing the effects of ambient ozone in China on snap bean genotypes by using ethylenediurea (EDU). Environ Pollut 205:199–208
Zhang WW et al (2008) Effects of elevated ozone on rice (*Oryza sativa* L.) leaf lipid peroxidation and antioxidant system. Chin J Appl Ecol 19(11):2485–2489 (in Chinese)
Zhang RB et al (2013) Effects of elevated ozone concentration on starch and starch synthesis enzymes of Yangmai 16 under fully open-air field conditions. J Integ Agric 12(12):2157–2163
Zhang WW et al (2014a) Response of soybean cultivar Dongsheng-1 to different O_3 concentrations in Northeast China. Environ Sci (Chin) 35(4):1473–1478
Zhang WW et al (2014b) Effects of elevated O_3 exposure on seed yield, N concentration and photosynthesis of nine soybean cultivars (*Glycine max* (L.) Merr.) in Northeast China. Plant Sci 226:172–181
Zhao C et al (2009) East China plains: a "Basin" of ozone pollution. Environ Sci Tech 43:1911–1915
Zhao TH et al (2012) Effects of ozone stress on root morphology and reactive oxygen species metabolism in soybean roots. Soybean Sci 31(1):52–57 (in Chinese)
Zhao Y et al (2015) Impact of elevated atmospheric carbon dioxide and ozone concentration on growth dynamic, dry matter production, and nitrogen uptake of hybrid rice Shanyou 63. Acta Ecol Sin 35:1–11 (in Chinese)
Zheng QW et al (2005) Ozone effects on chlorophyll content and lipid peroxidation in the in situ leaves of winter wheat. Acta Bot Boreali-Occidentalia Sin 25(11):2040–2044 (in Chinese)
Zheng QW et al (2006) Impact of different ozone exposure regimes on photosynthetic rate, biomass and yield of field-grown oilseed rape. Asian J Ecotoxicol 1(4):323–329 (in Chinese)
Zheng YF et al (2010) Effects of ozone stress upon winter wheat photosynthesis, lipid peroxidation and antioxidant systems. Environ Sci 31(7):1643–1651 (in Chinese)
Zheng FX et al (2011) Effects of elevated ozone concentration on methane emission from a rice paddy in Yangtze River Delta, China. Glob Chang Biol 17:898–910
Zheng FX et al (2013a) Effects of elevated O_3 exposure on nutrient elements and quality of winter wheat and rice grain in Yangtze River Delta, China. Environ Pollut 179:19–26
Zheng YF et al (2013b) Combined effects of elevated O_3 and reduced solar irradiance on growth and yield of field-grown winter wheat. Acta Ecol Sin 33:532–541 (in Chinese)
Zhou XD et al (2015) Elevated tropospheric ozone increased grain protein and amino acid content of a hybrid rice without manipulation by planting density. J Sci Food Agric 95:72–78
Zhu X et al (2011) Effects of elevated ozone concentration on yield of four Chinese cultivars of winter wheat under fully open-air field conditions. Glob Chang Biol 17:2697–2706

Chapter 13
Effects of Ozone on Chinese Trees

Zhaozhong Feng and Pin Li

Abstract This chapter reviews the effects of elevated ozone on tree species in China, based on the results of studies in the past two decades. The high ozone concentration in summer in most parts of China has induced typical ozone symptoms in urban and mountain forest tree species. In experiments using open-top chambers, elevated ozone affected the growth, gas-exchange rate, foliar microscopy, antioxidant systems, and biogenic volatile organic compound (BVOC) emissions in trees. The effects of ozone on biomass accumulation depended on the ozone concentration, tree species sensitivity, and exposure duration. The ozone uptake of individual tree species was also investigated by the sap flow technique. Further studies were conducted on the interactions between O_3 and other environmental change factors, such as increasing CO_2 concentrations, increased nitrogen deposition, and drought. Future needs for research include the development of an O_3 flux model for the most widely used tree species and the assessment of ozone removal by urban forests on a regional and a national scale.

Keywords Biomass • Elevated CO_2 • Forest tree species • Gas exchange N deposition • ozone

Z. Feng (✉) • P. Li
State Key Laboratory of Urban and Regional Ecology,
Research Center for Eco-Environmental Sciences, Chinese Academy of Sciences,
Shuangqing Road 18,, Beijing 100085, China
e-mail: zhzhfeng201@hotmail.com; fzz@rcees.ac.cn; pinli@rcees.ac.cn

T. Izuta (ed.), *Air Pollution Impacts on Plants in East Asia*,
DOI 10.1007/978-4-431-56438-6_13

13.1 Description of Ozone Injury and Accumulated Amount of Ozone over the Threshold Value of 40 ppb (AOT40) Values at Injury Onset

Among the different effects of ozone on trees, visible injury in leaves is considered to be a valuable tool for the assessment of ozone impacts in the field and for the detection of high potential risk (Schaub et al. 2010). Although visible injury in leaves is not as biologically significant as effects on growth and biomass, it is indicative of ozone stress, and its presence and extent can be used as a surrogate of biological damage to trees (Feng et al. 2014). In China, ozone injury has been reported in some tree seedlings in controlled field experiments with open-top-chambers (OTCs) and in some mature trees species in the field (see Table 13.1). Usually, typical ozone-like visible symptoms, specifically including a characteristic foliar interveinal stipple, reddish stipple, necrotic spots, chlorotic mottling, brown spots, and distinctive leaf yellowing associated with accelerated senescence, are observed on the upper leaf surfaces of older and middle-aged leaves in ozone-sensitive tree species when ozone levels are high enough, exposure is long enough, and environmental conditions are appropriate for gas exchange (Fig. 13.1 and Table 13.1). The Accumulated Ozone over Threshold of 40 ppb (AOT40) at injury onset (AOT40-injury) is used for assessing the potential negative O_3 impacts to trees (Table 13.1). This index has been chosen for its simplicity and flexibility in handling different species or distinct genotypes within a species, as well as for its linear relationships with biomass loss or the reduction in photosynthesis (Li et al. 2016).

13.2 Effects of Ozone on Ecophysiology of Tree Species

13.2.1 Leaf Traits

Ozone sensitivity is species-specific; when comparing the responses of the two main species life types – evergreen and deciduous – differences in ozone sensitivity are evident in a wide range of species: most of the evergreen species are more ozone-tolerant than the deciduous ones. These results put in a wider context the observations by Zhang et al. (2012), who reported a relatively high ozone tolerance in subtropical evergreen species. Among deciduous species, deciduous shrubs (e.g. *Sorbaria sorbifolia*, *Hibiscus syriacus*) and deciduous trees (e.g. *Fraxinus chinensis*, *Liriodendron chinense*) are the species most sensitive to ozone, displaying visible injury with AOT40 values below 10 ppm·h. These species could have high potential as bio-indicators under field conditions in the Beijing area (Feng et al. 2014).

Evergreen species are frequently characterized by their dark green, leathery, and thicker leaves (compared with deciduous species) and, in particular, a higher leaf mass per area (LMA) than that in deciduous species. The higher LMA could be a consequence of the greater leaf thickness, and also of a higher density of mesophyll tissues (palisade and spongy parenchyma layers) (Bussotti 2008). Leaves with high LMA increase the amount of chlorophyll and nitrogen content per unit of leaf surface, supporting more efficient photosynthesis activity, which can better feed

Table 13.1 Ozone injury description and AOT40 (the accumulated hourly mean O_3 concentration over a cutoff threshold of 40 p.p.b. during daylight hours) at injury onset (AOT40-injury, unit: p.p.m. h) in field and OTC (open-top chamber) experiments

Species	AOT40-injury	OTC/field	Description of symptoms	Data source
Sorbaria sorbifolia	6.47	OTC	Yellow-reddish interveinal stippling on leaf upper surface	Li et al. (2016)
Hibiscus syriacus	7.77	OTC/field	Fine light-brown stippling on interveinal parts of the upper surfaces of older leaves. In severely affected leaves, stippling becoming brown. Chlorosis and increased leaf senescence	Li et al. (2016) Feng et al. (2014)
Fraxinus chinensis	8.43	OTC/field	Fine light-brown stippling on interveinal parts of the upper surfaces of older leaves	Li et al. (2016) Feng et al. (2014)
Liriodendron chinense	9.4	OTC	Necrotic spots appeared at the edges of adaxial surfaces of fully expanded leaves, and expanding leaves turned light brown. With the increase in AOT40, the light brown coloration turned dark brown and necrotic spots formed patches. Finally, pre-senescence and earlier defoliation occurred in developed leaves	Zhang et al. (2012)
Ailanthus altissima	12.82	OTC/field	Fine light brown stippling on the upper surface of the leaf. Stippling more abundant in older leaves and leaflets	Li et al. (2016) Feng et al. (2014)
Platanus orientalis	13.37	OTC	Brown stippling on the upper surface of the leaf, finally coalescing to form large brown patches. Stippling more abundant in older leaves	Li et al. (2016)
Liquidambar formosana	13.52	OTC	Light brown foliar stippling and necrotic spots interspersed on the adaxial surfaces of fully developed leaves. Expanding leaves turned light green and/or yellow. Light brown necrotic spots turned into dark brown necrotic patches with increasing AOT40. Finally, early leaf senescence with pronounced leaf abscission occurred after reddening	Zhang et al. (2012)

(continued)

Table 13.1 (continued)

Species	AOT40-injury	OTC/field	Description of symptoms	Data source
Robinia pseudoacacia	16.18	OTC/field	Very fine light brown interveinal stippling, finally also with dark purple stippling. Stippling can also be associated with leaf chlorosis and accelerated senescence	Li et al. (2016) Feng et al. (2014)
Cornus alba	18.48	OTC	Yellow to brown interveinal stippling	Li et al. (2016)
Koelreuteria paniculata	19.07	OTC/field	Interveinal chlorosis and purple-brown stippling, especially in older leaves	Li et al. (2016) Feng et al. (2014)
Cinnamomum camphora	20.06	OTC	Chlorosis first appeared in fully developed leaves and then leaves turned yellow with pale white stippling widespread on the surfaces of both developed and expanding leaves. Young leaves were small, yellow, thin, and weak	Zhang et al. (2012)
Ulmus pumila	23.43	OTC	Leaf yellowing associated with accelerated senescence	Li et al. (2016)
Populus lasiocarpa	23.66	OTC	Dark brown stippling on interveinal parts of older leaves and accelerated senescence	Li et al. (2016)
Cyclobalanopsis glauca	27.85	OTC	Chlorotic mottling, light green and/or yellow coloration and waterlogged spots were widespread on the adaxial surfaces of both developed and expanding leaves. Young leaves turned light green or yellow	Zhang et al. (2012)
Sophora japonica	28.74	OTC	Yellow stippling and yellowing associated with accelerated senescence	Li et al. (2016)
Schima superba	29.35	OTC	Brown spots were interspersed among the abaxial surfaces of leaves in some individuals with no symptoms on the adaxial surfaces of developed leaves. Expanding leaves turned yellow, thin, and easy to curl as well	Zhang et al. (2012)
Lonicera maackii	30.42	OTC	Conspicuous dark brown interveinal stippling. Stipples finally coalescing and forming large brown patches	Li et al. (2016)
Cinnamomum camphora	30.81	OTC	Brown stippling and bronzing on the upper surfaces of younger leaves, as well as additional chlorosis in older leaves	Niu et al. (2011)

Ilex integra	33.36	OTC	Chlorosis first appeared in the middle of adaxial surfaces of fully developed leaves along with red-brown and/or dark brown spots. Dark brown spots formed necrotic patches interspersed on the adaxial surface	Zhang et al. (2012)
Zelkova schneideriana	36.17	OTC	Leaf yellowing associated with accelerated senescence	Li et al. (2016)
Photinia × fraseri	38.24	OTC	No visible symptoms occurred in developed leaves, but the expanding leaves showed curling	Zhang et al. (2012)
Acer truncatum	39.9	OTC	Upper leaf surfaces with chlorotic stippling, reddening, and premature leaf senescence	Li et al. (2015)
Ginkgo biloba		OTC	Light brown flecks	He et al. (2007)
Rhus typhina		Field	Yellow/brown to black stipple	Wan et al. (2014)
Juglans regina		Field	Yellow/brown to black stipple	Wan et al. (2014)
Sophora aureus		Field	Yellow/brown to black stipple	Wan et al. (2014)
Amygdalus triloba		Field	White stipple	Wan et al. (2014)
Salix leucopithecia		Field	White stipple	Wan et al. (2014)
Populus tomentosa		Field	Brown to dark mottle/blotch	Wan et al. (2014)
Morus alba		Field	Dark brown leaves	Wan et al. (2014)
Kerria japonica var. *plenifolia*		Field	Interveinal chlorosis	Wan et al. (2014)
Broussonetia papyrifera		Field	Very fine interveinal light brown stippling	Feng et al. (2014)
Cotinus coggygria var. pubescens		Field	Diffuse, very fine purple-brown coloration of interveinal areas	Feng et al. (2014)
Fraxinus rhynchophylla		Field	Fine light brown stippling on interveinal parts of the upper surfaces of older leaves	Feng et al. (2014)
Pinus bungeana		Field	Chlorotic mottling and accelerated senescence affecting older leaves. Current-year needles are not affected	Feng et al. (2014)
Sambucus williamsii		Field	Dark brown stippling on interveinal parts of older leaves. Stippling evolves to form very distinctive brown necrotic areas, sometimes with associated white flecking. Chlorosis is frequently observed together with stippling	Feng et al. (2014)

Fig. 13.1 Pictures of O_3-like symptoms in the field survey. The species are *1 Ailanthus altissima*, *2 Ampelopsis humulifolia*, *3 Fraxinus rhynchopylla*, *4 Pinus bungeana*, *5 Robinia pseudoacacia*, *6 Hibiscus syriacus*, *7 Humulus scandens*, *8 Koelreuteria paniculata*, *9 Malvaviscus arboreus*, *10 Prunus persica*, *11 Sambucus williamsii*, and *12 Broussonetia papyrifera*.

detoxification processes (Bussotti et al. 2007). Generally, plants with higher LMA are adapted to greater solar radiation, higher temperature, and less rainfall (Bussotti 2008). The LMA reflects the degree of the plant's acclimatization under different ecological conditions: low LMA makes species highly reactive to environmental stimuli through a rapid modulation of photosynthesis, increasing sensitivity to stress; conversely, high LMA increases the resilience and ecological stability of the species (Bussotti 2008). But this evolutionary adaptation to cope with environmental stress by increasing LMA may also confer cross-tolerance to ozone, in part due to a higher detoxification capacity. These results clearly support the hypothesis that higher LMA is related to lower ozone sensitivity in a wide range of species, and they also show that higher LMA values are associated with higher antioxidant content per leaf area (Li et al. 2016). LMA has been considered a relevant factor for explaining the higher tolerance to O_3 in some Mediterranean evergreen shrubs and trees in comparison with related deciduous species (Calatayud et al. 2010, 2011a), and for explaining the relative O_3 tolerance of subtropical species (Zhang et al. 2012), or differences between species of the same genus (Calatayud et al. 2007) or differences between trees of different ages (Wieser et al. 2002).

13.2.2 Microscopy

Microscopy studies can be used for properly discriminating ozone symptoms from others produced by biotic or abiotic factors (Günthardt-Goerg et al. 2007). Light microscopy (LM) combined with transmission electron microscopy (TEM) can show the effects of ozone on leaves whether injured or not. Based on studies in four ozone-sensitive urban tree species in Beijing, *Ailanthus altissima*, *Fraxinus chinensis*, *Platanus orientalis*, and *Robinia pseudoacacia*, a common reaction was found to be a hypersensitive response, followed by programmed cell death (Gao et al. 2016). Specifically speaking, palisade parenchyma cells were, in general, the part of the mesophyll most affected by ozone (see Fig. 13.2). The middle lamella was degraded and thus small pertinacious and polysaccharide extrusions outwards (Figs. 13.3a and 13.4a, e, f) were produced (Figs. 13.3a and 13.4b). Cell walls were progressively altered (Figs. 13.3b, c, 13.4b, c, and 13.5c, e), and fluorescence LM (Fig. 13.2a) and TEM (Fig. 13.4a, b) showed that callose was accumulated between the membrane and the cell wall. Inside the cells, the vacuolar content was altered and the vacuolar content of the affected cells became denser, with the accumulation of tannins (Figs. 13.2b, h and 13.4e, f); as well, there were large accumulations of crystals in the cytoplasm of the affected cells (Fig. 13.4d). Finally, tonoplasts broke and the cells lost turgor (Fig. 13.5b). At a later stage, cells collapsed, leading to an increase of intercellular spaces (Figs. 13.2, 13.3a, c, 13.4b, e, f, and 13.5e). Chloroplasts were also strongly affected in all the species examined, showing increasing electron density and changing their shape. Much bigger starch grains (Figs. 13.3, 13.4b, c, and 13.5a), plastoglobuli (Figs. 13.3a and 13.5e), and lipid-protein bodies (Fig. 13.3) were observed in ozone-treated plants than in control plants. Finally, thylakoid membranes in ozone-treated plants were partly or totally disaggregated (Fig. 13.5b). Mitochondria also experienced degradation processes. In some cases, the accumulation of lipid droplets similar to plastoglobuli was also observed (Fig. 13.3d). Similar but much less conspicuous changes were observed in the spongy parenchyma. Chloroplast stomata were also affected, with large starch accumulations in some cases (Fig. 13.5a). In the vascular bundles, xylem was never affected, but callose deposition appeared to be increased in the sieve tubes of phloem (Figs. 13.2h and 13.5d).

Callose accumulation is a defensive response to abiotic and biotic factors such as wounding, desiccation, metal toxicity, or insect attacks, and it has been postulated that this accumulation might act as a physical barrier to microbial and fungal attack, isolating affected tissues from healthy cells. Callose accumulation after ozone exposure is a common phenomenon in different species (Bussotti et al. 2005; Calatayud et al. 2011b; Gravano et al. 2003).

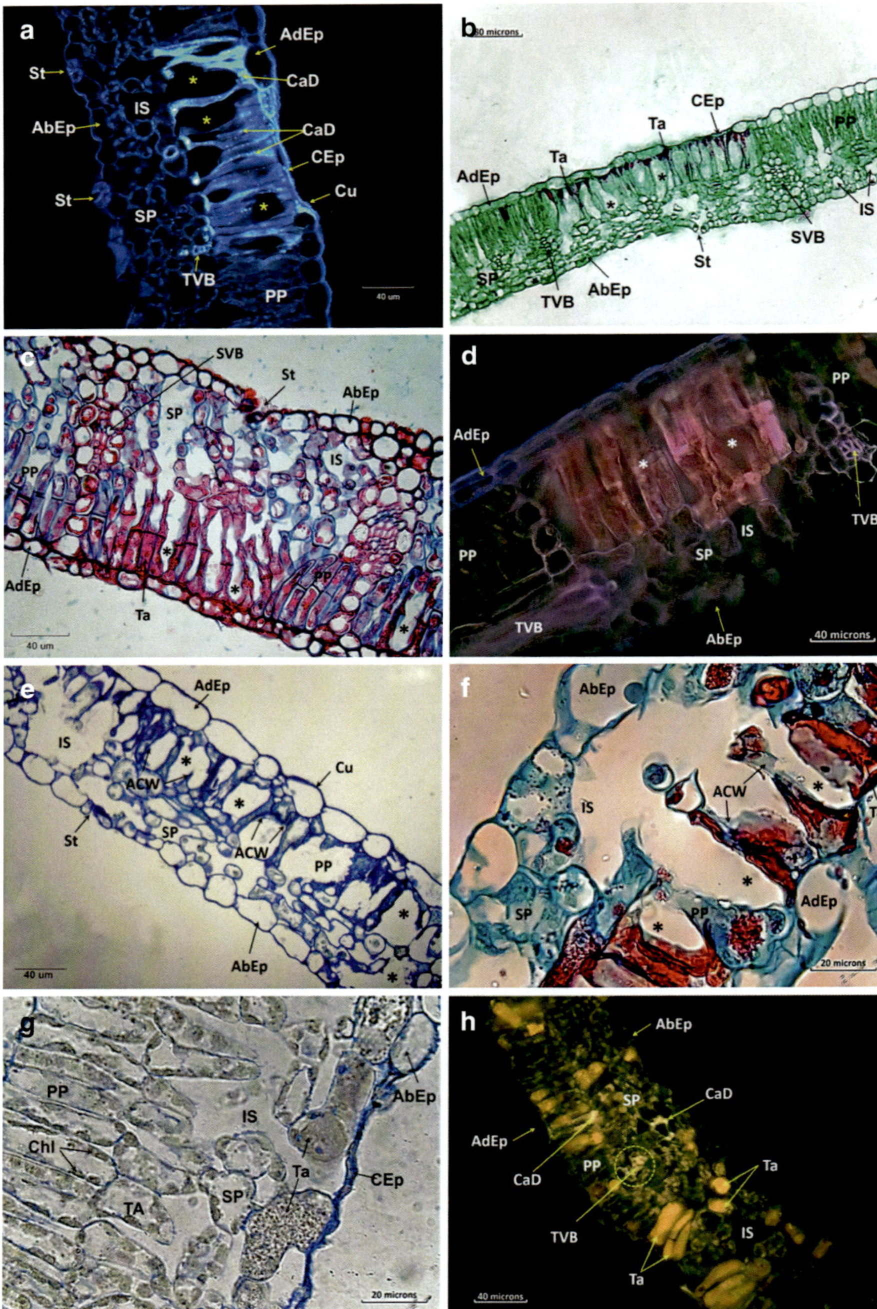
a
AdEp
St
CaD
IS
AbEp
CaD
CEp
St
Cu
SP
TVB
PP
40 um
b
CEp
Ta
PP
Ta
AdEp
IS
SVB
St
SP
TVB
AbEp
c
SVB
St
AbEp
SP
IS
PP
AdEp
Ta
PP
40 um
d
PP
AdEp
TVB
IS
PP
SP
TVB
AbEp
40 microns
e
AdEp
IS
Cu
ACW
St
SP
ACW
PP
AbEp
40 um
f
AbEp
ACW
IS
Ta
AdEp
SP
PP
20 microns
g
PP
AbEp
IS
Chl
Ta
CEp
SP
TA
20 microns
h
AbEp
CaD
SP
AdEp
CaD
PP
Ta
TVB
IS
Ta
40 microns

At the ultrastructural level, after ozone exposure, changes that may be considered as an acceleration of the natural foliar senescence process were seen in organelles (chloroplast-to-gerontoplast transition). Gerontoplast development is seen primarily as a progressive unstacking of grana, a loss of thylakoid membranes, and a massive accumulation of plastoglobuli and other lipid-protein inclusions. Ultrastructural changes in the chloroplast included an increase in electron density; the accumulation of starch, plastoglobuli, and lipid-protein bodies; thylakoid degradation; membrane disruption; and changes in shape. Starch accumulation may be related to difficulties of sucrose transport outside the chloroplasts and to other tissues (Calatayud et al. 2011b). O_3-exposed birch plants were observed to have an enhanced content of soluble sugars in their leaves, as well as the accumulation of starch along veins, which suggested a reduction of carbon export from source leaves. This could be related to the effects of O_3 in altering cell membranes and impairing phloem loading (Grantz 2003). In O_3-injured leaves, phloem cells can experience evident changes in shape, which may obviously impair their functionality and affect sucrose translocation (Calatayud et al. 2011b). On the other hand, the increase in plastoglobuli could originate from the lipid-soluble products of thylakoid membrane degradation (Kivimäenpää et al. 2010).

13.2.3 Gas Exchange in Response to Ozone

Plants take up O_3 mainly through open stomata. O_3-induced decreases in the net photosynthetic rate (A_{sat}), stomatal conductance (g_s), transpiration rate, and quantum yield of photosystem II (PSII) electron transport have been well documented in many tree species in China, including temperate deciduous broadleaf species (e.g. *Ginkgo*

Fig. 13.2 (**a**) Fluorescence micrograph of a cross section of *Ailanthus altissima* leaf stained with aniline blue showing an affected area (*asterisks*) with numerous collapsed cells and with callose depositions (*CaD*) on its cell walls. (**b**) Micrograph of a cross section of an affected *Ailanthus altissima* leaf stained with safranin-fast green, showing numerous collapsed cells (*asterisks*) with tannin (*Ta*) content inside. (**c**) Micrograph of a cross section of an affected *Fraxinus chinensis* leaf stained with trichrome, showing numerous collapsed cells (*asterisks*) with tannin (*Ta*) content inside. (**d**) Autofluorescence micrograph of a cross section of *Fraxinus chinensis* leaf, showing an affected area (*asterisks*) of palisade parenchyma (*PP*) without chlorophyll. (**e**) Micrograph of a semithin cross section of an affected *Platanus orientalis* leaf stained with toluidine blue. Numerous collapsed cells (*asterisks*) leaving between them large intercellular spaces (*IS*) are observed. (**f**) Micrograph of a cross section of an affected *Platanus orientalis* leaf stained with trichrome, showing collapsed cells (*asterisks*) with affected cell wall (*ACW*) and tannin (*Ta*) content inside. (**g**) Micrograph of a semithin cross section of an affected *Robinia pseudoacacia* leaf stained with toluidine blue. Collapsed epidermis (*CEp*) is observed. (**h**) Autofluorescence micrograph of a cross section of an affected *Robinia pseudoacacia* leaf. Numerous cells filled with tannin (*Ta*) are observed. Other abbreviations: *AbEp* abaxial epidermis, *AdEp* adaxial epidermis, *Cu* cuticle, *TVB* tertiary vascular bundle, *SP* spongy parenchyma, *St* stomata, *SVB* secondary vascular bundle

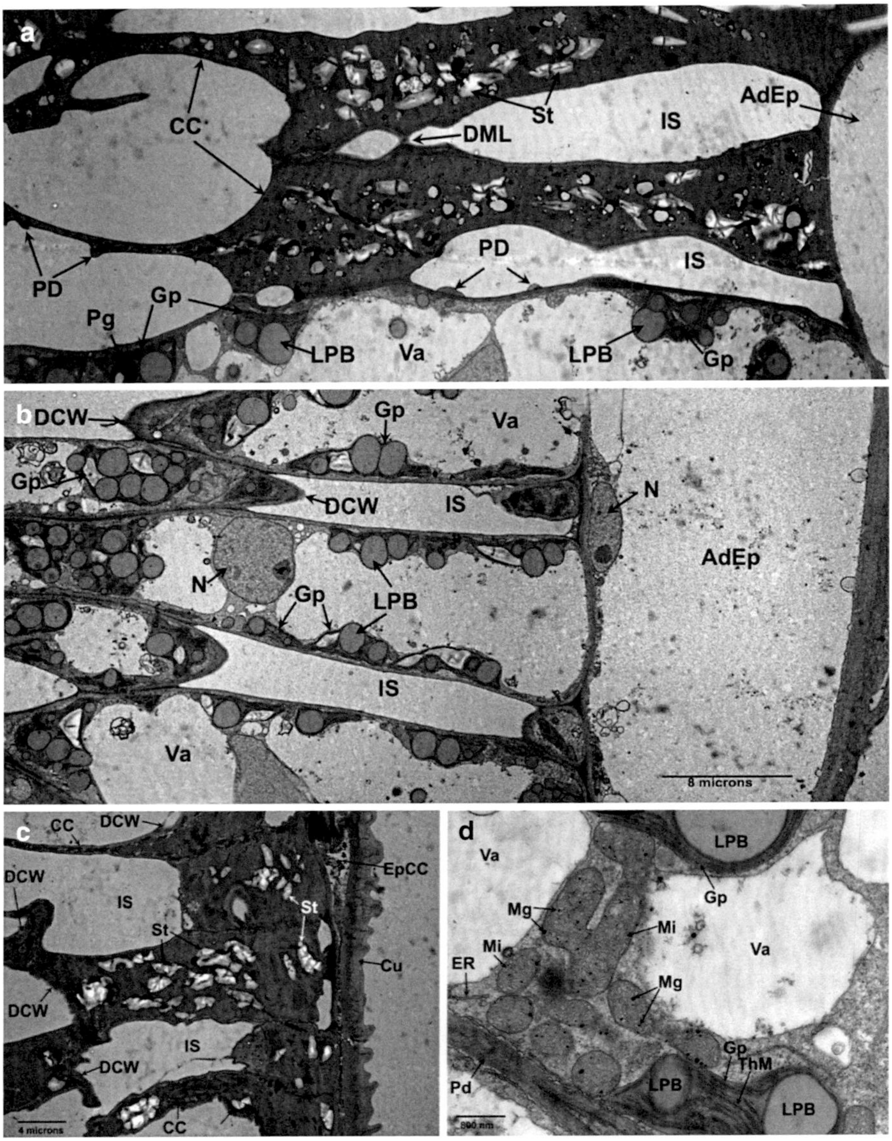

Fig. 13.3 Details of transmission electron microscopy (TEM) micrographs of cross sections of *Ailanthus altissima* leaves. (**a**) Two collapsed cells (*CC*) of palisade parenchyma with great intercellular spaces (*IS*). Pertinacious drops (*PD*) and degraded middle lamella (*DML*) are observed. (**b**) Cells of palisade parenchyma with abundant gerontoplasts (*Gp*) filled with abundant lipid-protein bodies (*LPB*), degraded cell wall (*DCW*), and great intercellular spaces (*IS*). (**c**) Collapsed cells (*CC*) of palisade parenchyma and adaxial epidermis (*EpCC*), showing a degraded cell wall (*DCW*). (**d**) Detail of a cell with several mitochondria (*Mi*) containing some microglobulin (*Mg*). Other abbreviations; *AdEp* adaxial epidermis, *N* nucleus, *Pd* plasmodesma, *Pg* plastoglobuli, *St* starch, *ThM* thylakoid membranes, *Va* vacuole

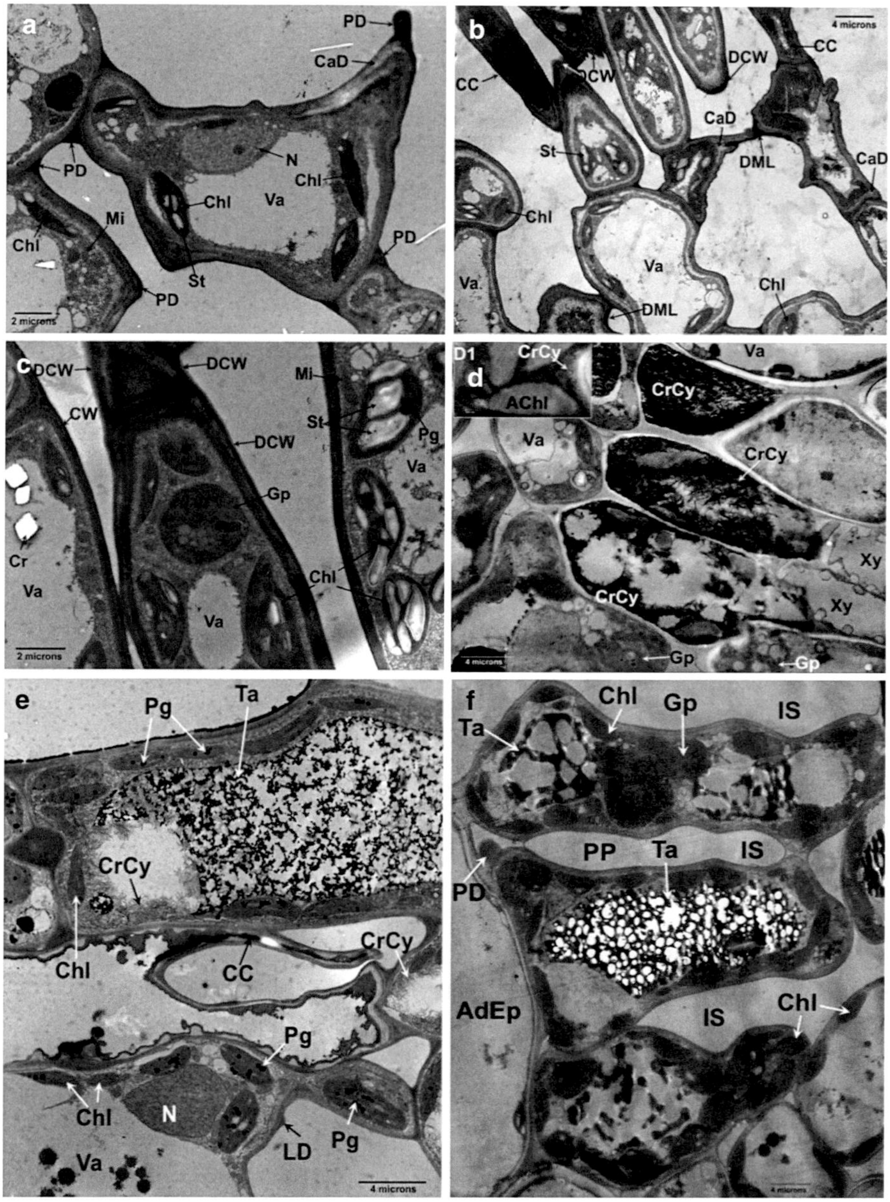

Fig. 13.4 (**a**, **b**) Details of transmission electron microscopy (TEM) micrographs of cross sections of *Fraxinus chinensis* leaves. (**d–f**) Details of TEM micrographs of cross sections of *Platanus orientalis* leaves. (**a**) Spongy parenchyma cell showing an altered medium lamella. Some pertinacious drops (*PD*) and callose depositions (*CaD*) between the cell membrane and cell wall are observed. (**b**) Cells of spongy and palisade parenchyma showing different alterations. Some of them are collapsed (*CC*). (**c**) Detail of palisade parenchyma cells. Degraded cell wall (*DCW*) and gerontoplasts (*Gp*) are observed. (**d**) Cells showing crystalized cytoplasm (*CrCy*). In (D1) a detail of this crystallization is observed. (**e**) Affected palisade parenchyma cells. Collapsed cells (*CC*), vacuoles with tannin (*Ta*), crystalized cytoplasm (*CrCy*), and numerous chloroplasts (*Chl*) with plastoglobuli (*Pg*) are observed. (**f**) Cells of palisade parenchyma showing varying degrees of decomposition of tannin (*Ta*) within the vacuoles (*Va*). Other abbreviations: *AChl* Altered chloroplast, *AdEp* adaxial epidermis, *Cr* (calcium oxalate) crystal, *DML* degraded medium lamella, *N* nucleus, *S* starch, *Xy* xylem

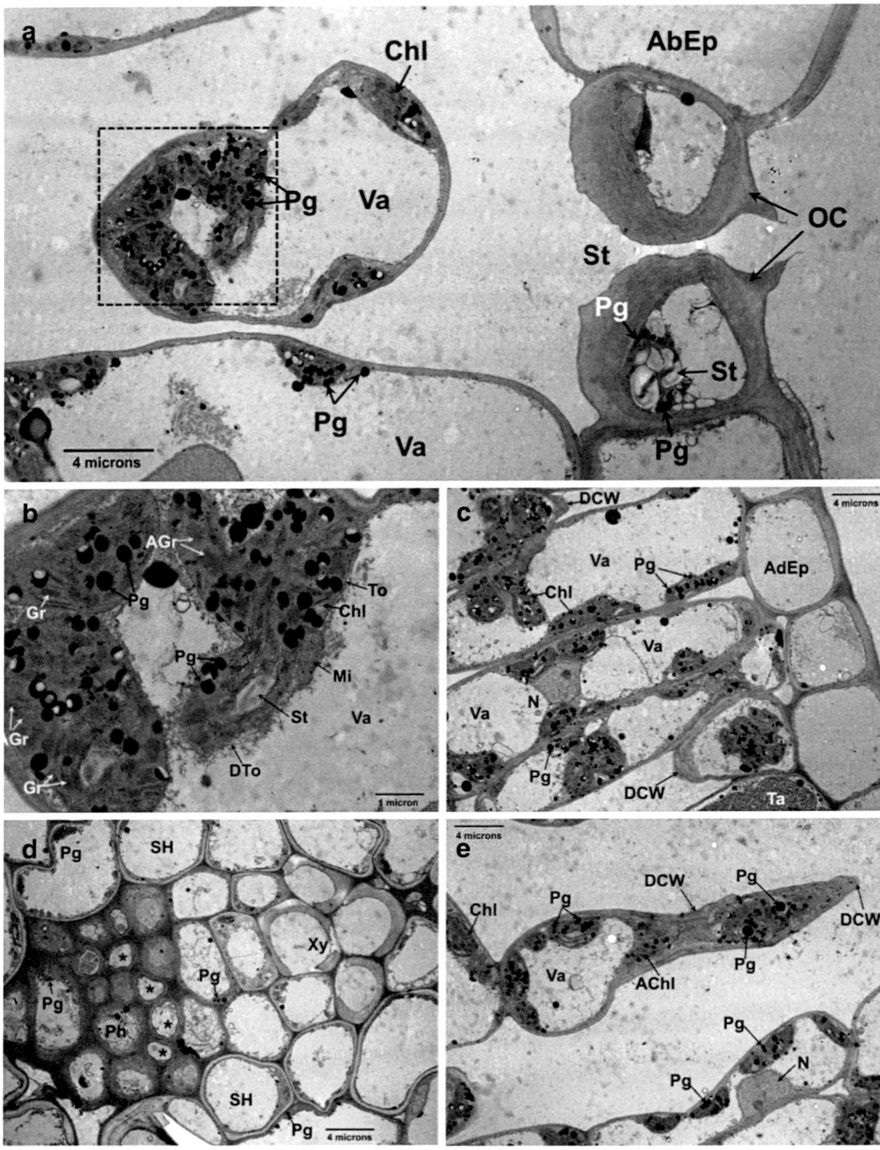

Fig. 13.5 Details of transmission electron microscopy (TEM) micrographs of cross sections of *Robinia pseudoacacia* leaves. (**a**) Detail of a stoma (*ST*) with affected chloroplast and spongy parenchyma. Chloroplasts (*Chl*) in cells of spongy parenchyma are affected and contain numerous plastoglobuli (*Pg*). In the area marked with the *dashed square* shown in **a,** details of affected chloroplasts containing numerous plastoglobuli and many disorganized grana (*AGr*) are shown. (**b**) The tonoplast (*To*) has disintegrated in many areas (*DTo*). (**c**) Palisade parenchyma cells containing chloroplasts with numerous plastoglobuli and degraded cell walls (*DCW*) are seen in some areas. (**d**) Detail of a tertiary vascular bundle with the affected phloem. Callose depositions (*asterisks*) are observed in some phloem (*Ph*) vessels. (**e**) Cells affected in the spongy parenchyma. Their cell walls are degraded (*DCW*) and contain chloroplasts with numerous plastoglobuli (*Pg*). Other abbreviations: *AbEp* abaxial epidermis, *AdEp* adaxial epidermis, *BS* bundle sheath, *Gr* grana, *Mi* mitochondria, *N* nucleus, *OC* occlusive cells, *S* starch, *Ta* tannin, *Va* vacuole, *Xy* xylem

biloba, *Quercus mongolica*) (He et al. 2007; Yan et al. 2010a), subtropical deciduous conifer species (e.g. *Metasequoia glyptostroboides*) (Feng et al. 2008; Zhang et al. 2014), other deciduous broadleaf species (e.g. *Liriodendron chinense*, *Liquidambar formosana*) (Zhang et al. 2011, 2012), and evergreen broadleaf species (e.g. *Cinnamomum camphora*, *Cyclobalanopsis glauca*) (Zhang et al. 2012; Feng et al. 2011; Niu et al. 2014). However, controversy exists concerning the mechanism that underlies the suppressive effects of O_3 on CO_2 assimilation. In several studies, reductions in the in-vivo carboxylation and electron transport parameters have been observed to occur before significant changes in A_{sat}, suggesting that biochemical limitations play an early and primary role in the decline of CO_2 assimilation in O_3-exposed plants (Feng et al. 2011). Furthermore, it was reported that elevated O_3 induced a large degree of A_{sat} reduction in most trees species, with an uncoupling of the relationship between A_{sat} and g_s (Zhang et al. 2012). That is to say, the responses of g_s to elevated ozone were not consistent among species, as shown by an increase in some species and a decrease or lack of change in others. In addition, O_3 was observed to increase stomatal sluggishness (i.e., slowed stomatal response) in two ozone-sensitive urban greening tree species in Beijing, *Fraxinus chinensis* and *Platanus orientalis* (Hoshika et al. 2014). Increased stomatal sluggishness rather than stomatal closure in response to ozone stress suggests that more ozone could enter the intercellular air space and a more severe decrease in photosynthetic capacity could occur in mesophyll cells. Therefore, the observed reduction in A_{sat} may not be attributable to the unavailability of CO_2 due to stomatal limitation (Zhang et al. 2014), but rather to the O_3-induced damage to ribulose-1, 5-bisphosphate carboxylase/oxygenase and the photochemical apparatus (Niu et al. 2014), although the underlying mechanisms are still under investigation.

13.2.4 Antioxidant Response to Ozone

Upon O_3 entering the intercellular air space, antioxidant fluctuations may buffer the formed reactive oxygen species (ROS). Antioxidant systems can be activated to detoxify the accumulation of ROS, including several key enzymatic (e.g. ascorbate peroxidase, glutathione reductase, and superoxide dismutase) and non-enzymatic (e.g. ascorbate, phenolic compounds) antioxidants. Variations of antioxidant systems in some tree species under elevated O_3 in an entire growth season have been reported in China, such as those seen in *Ginkgo biloba* (He et al. 2006), *Pinus armandii* (Yan et al. 2013), and *Quercus mongolica* (Yan et al. 2010b). O_3-induced enhanced activities of antioxidant enzymes (e.g. ascorbate peroxidase, glutathione reductase, monodehydroascorbate reductase, and superoxide dismutase) were observed in the initial stage of ozone exposure; however, subsequently, enzyme activities declined late in the season. It was suggested that the antioxidant enzyme system may respond by acclimating to the early oxidant stress, but the system could not withstand the long-term exposure with the constant higher level of ROS. On the other hand, activation of the synthesis and consumption of the major constituents of non-enzymatic antioxidants, ascorbate and phenolic compounds, is a crucial method of defense and protection against oxidative stress (Zhang et al. 2012). As central metabolites in plant antioxidant systems, ascorbate and phenolic compounds protect plants, acting as chemical scavengers, reducing free radicals, and also acting as substrates of extracellular enzymes, such as ascorbate peroxidase, which detoxify peroxides; thus, the propagation of oxidative signaling is diminished (Dizengremel et al. 2013). However, once the capacity of the antioxidant system is insufficient to offset the high levels of oxidative stress, photosynthesis impairment and visible injury can occur. Of note, based on the constitutive antioxidant levels of tree species in higher areas, such species could be considered to be more tolerant to O_3 stress (Li et al. 2016), but the possible mechanisms of the differential responses to O_3 among species are still under investigation.

13.2.5 Biogenic Volatile Organic Compounds (BVOCs) Response to Ozone

VOCs are the main precursors of tropospheric O_3 and they also contribute to the formation of other secondary pollutants, such as peroxyacetyl nitrates (PANs), aldehydes, and ketones; secondary organic aerosols (SOAs); and particulate matter (PM) (Fehsenfeld et al. 1992; Fuentes et al. 2000; Simpson et al. 2014), which are of concern for plant and human health. Numerous tree species and vegetation emit vast quantities of BVOCs every year. BVOCs account for about 90 % of the global emission of VOCs into the atmosphere (Kesselmeier et al. 1999; Monson et al. 2012). i.e., BVOC emissions are several times higher than the anthropogenic emission of VOCs. Isoprene (2-methyl-1,3-butadiene) (Atkinson et al. 2003) is the most representative BVOC emitted by plants (Guenther et al. 1994), and makes up as much as 80 % of the carbon emitted as VOCs from plants (Calfapietra et al. 2009).

However, variability in BVOC emissions results from complex interactions between plants and their environment (Kesselmeier et al. 1999; Dudareva et al. 2006). External environmental factors have a significant impact on the emission of isoprene by plants (Fuentes et al. 2000; Harley et al. 1999). As a defense response to environmental stress, plants can also increase or decrease BVOC emissions, compared with non-stressed plants. It is important to improve our understanding of the interaction between VOC emission and ozone, because present and predicted climate change is expected to cause concurrent increases in the formation of these gases.

Findings regarding the effects of ozone and other factors on BVOC emissions are sometimes conflicting. Only a few studies have been conducted in China. In short-term ozone fumigation (80 ppb, 30 days, from 8:00 to 17:00 every day) experiments using OTCs in the northeast of China, Li. et al. (2009a) reported that O_3 exposure significantly increased monoterpene emission rates from *Ginkgo biloba* trees in both July and September, and also enhanced the total VOC emission rates. Elevated O_3 may stimulate isoprene emissions by plants, as this molecule plays a role in defense against ozone and other biotic and abiotic stresses. In another study, 4-year-old Chinese pine (*Pinus tabulaeformis*) was exposed to O_3 fumigation with a mean daily O_3 concentration of 66.1 ppb for three growing seasons. Results indicated that elevated O_3 significantly increased the emission rate of VOCs, as compared with non-filtered air (NF), after 30 and 90 days of O_3 exposure (Xu et al. 2012). Trees emit VOCs, which may scavenge O_3 and help provide protection for the photosynthetic apparatus against oxidative stress (Loreto et al. 2001). Isoprene may avoid the cellular damage produced by ROS or may scavenge ROS in the leaf intercellular spaces (Vickers et al. 2009), which may be a helpful response to protect the photosynthetic apparatus from O_3 damage.

However, the results of the above studies are not in agreement with our study, in which elevated O_3 had a significant negative effect on the emission of isoprene in the middle-level leaves of poplar, but had no effects on the upper-level leaves (Yuan et al. 2016). Furthermore, we found, from the results along a vertical leaf profile, that the change in isoprene emission rate induced by O_3 was negatively related to the AOT40 during growth ($P<0.01$), whereas a significant reduction in isoprene emission by poplars was found when a certain ozone threshold (AOT40 >6.6 ppm·h) was exceeded (Yuan et al. 2016). These results suggest that the isoprene emission rate of the hybrid poplar will be reduced significantly with rising ozone concentrations, and this lower isoprene emission rate will reduce O_3 and SOA formation in the air.

13.3 Ozone Uptake of the Urban Tree Canopy Determined by Sap Flow Techniques

Urban forests play an important role in improving urban air quality by taking up and removing ambient O_3, and this depends on the stomatal conductance of the species. The O_3 uptake by urban trees on a regional scale has been examined largely based on modeling approaches that integrate vegetation information with meteorological and ambient O_3 concentration data (Nowak et al. 2000). However, variations in canopy resistance due to trees of different heights and crown configurations, as well as heterogeneous ambient O_3 concentrations, may introduce uncertainties into the estimates (Scott et al. 1998). In recent years, O_3 uptake by urban trees has been measured based on the whole-tree ozone flux, estimated by sap flow measurements, since transpiration and O_3 uptake are coupled through the activity of the stomata (Wieser et al. 2003). The sap flow method has been a popular approach to estimate whole-tree O_3 uptake and has been widely used in heterogeneous and mountainous landscapes (Wieser et al. 2003, 2006; Matyssek et al. 2007, 2008).

In China, a few studies have investigated the O_3 uptake by urban trees at the canopy scale, using the sap flow technique. Wang et al. (2012) evaluated the O_3 uptake in 17 adult trees of six urban species under free atmospheric conditions. During the growing season, the average canopy conductance for O_3 (GO_3) ($P<0.004$), canopy O_3 uptake rate (FO_3) ($P<0.002$), and annual average accumulated stomatal ozone flux over thresholds of 0 (AFst0) ($P<0.013$) differed significantly among the species (Table 13.2). The FO_3 ranged from 0.61 ± 0.07 nmol·m^{-2}·s^{-1} in *Robinia pseudoacacia* to 4.80 ± 1.04 nmol·m^{-2}·s^{-1} in *Magnolia liliiflora* (Table 13.2). However,

Table 13.2 Average canopy conductance for O_3 (GO_3, mmol·m^{-2}·s^{-1}), canopy O_3 uptake rate (FO_3, nmol·m^{-2}·s^{-1}), and accumulated stomatal ozone flux over thresholds of 0 and 1.6 nmol m^{-2} s^{-1} (AFst0, AFst 1.6, mmol·m^{-2}) estimated during the growing season from May 1 to October 31 (Wang et al. 2012)

Species	GO_3	FO_3	AFst0	AFst1.6
A. chinensis	104.2 ± 16.6	4.01 ± 0.61	19.15 ± 3.16	12.38 ± 2.89
C. deodara	62.76 ± 5.21	2.24 ± 0.17	11.29 ± 1.02	5.21 ± 0.81
G. biloba	80.27 ± 0.84	3.09 ± 0.01	15.77 ± 0.02	9.21 ± 0.11
M. liliiflora	129.1 ± 29.2	4.80 ± 1.04	20.97 ± 5.63	14.7 ± 5.13
P. tabulaeformis	58.1 ± 9.16	2.18 ± 0.31	11.12 ± 1.66	5.11 ± 1.28
R. pseudoacacia	17.35 ± 2.14	0.61 ± 0.07	3.43 ± 0.41	0.15 ± 0.08

average FO_3 for deciduous foliage was not significantly higher than that for evergreen foliage (3.13 *vs* 2.21 nmol·m^{-2}·s^{-1}, $P = 0.160$). Species with high GO_3 took up more O_3 than those with low GO_3, but their sensitivities to vapor pressure deficit (D) were also higher, and their FO_3 decreased faster with increasing D, regardless of species. The responses of FO_3 to D and total radiation led to a relatively high flux of O_3 uptake, indicating a high ozone risk for urban tree species.

Wang et al. (2013) found that the diurnal FO_3 for *R. pseudoacacia* showed a single-peak pattern, with the maximum rate occurring at around 15: 00 h. The diurnal FO_3 showed a narrow peak during summer and a wide peak during autumn. The most obvious increase in accumulated AFst occurred around noon time. FO_3 showed a seasonal pattern, with higher values in summer than in autumn. The increase in AFst was more obvious in summer than in autumn. The diurnal and seasonal patterns of O_3 uptake were related to the temporal variations of ambient air O_3 concentrations and GO_3. FO_3 exhibited an asymmetric single-peak pattern: FO_3 increased slightly with increasing total radiation (Rs) when Rs was below 800 W·m^{-2}; however, FO_3 decreased rapidly with increasing Rs when Rs was higher than 800 W·m^{-2}. The annual O_3 uptake by *R. pseudoacacia* trees estimated in their study was 0.16 g·m^{-2}, which was much lower than the values estimated from the Urban Forest Effects Model (UFORE) (Nowak and Crane 2000). This difference suggests the necessity to consider the O_3 uptake flux at the canopy level when evaluating O_3 risks in urban trees.

Poplar is the most afforestation tree species in northern China, it was also investigated for ozone uptake characteristics under natural conditions. The FO_3 and GO_3 of diurnal variation were unimodal. The average FO_3 was 78.22 ± 39.45 nmol·m^{-2}·s^{-1} and the average GO_3 was 127.60 ± 64.35 mmol·m^{-2}·s^{-1}. During the experimental period from May to July (Chen et al. 2015a), the FO_3 was highest in May, then decreased in June and was lowest in July. The diurnal variations of the FO_3 and GO_3 in different tree shape levels were of unimodal type; the changes in FO_3 followed the order of depressed trees (43.24 nmol·m^{-2}·s^{-1}) > middle trees (29.77 nmol·m^{-2}·s^{-1}) > dominant trees (24.56 nmol·m^{-2}·s^{-1}), while the GO_3 value was ranked in the order of dominant trees (101.59 mmol·m^{-2}·s^{-1}) > depressed trees (92.92 mmol·m^{-2}·s^{-1}) > middle trees (81.60 mmol·m^{-2}·s^{-1}). The mean FO_3value in different tree shapes was (32.52 ± 7.87) nmol·m^{-2}·s^{-1}, while that of GO_3 was 92.04 ± 8.18 mmol·m^{-2}·s^{-1} (Chen et al. 2015b).

13.4 Effects of Ozone on Biomass of Tree Species

The experimental effects of ozone on tree biomass depend on the concentration of O_3, tree sensitivity, and the experimental duration. So far, the longest experimental exposure to elevated O_3 in China has been 2 years lower in *Metasequoia glyptostroboides* (Zhang et al. 2014), *Cyclobalanopsis glauca* (Zhang 2011), *Cinnamomum camphora* (Niu et al. 2011), and *Acer truncatum* (Li et al. 2015). Most experiments have lasted for one growing season, and the species included *Ginkgo biloba* (He et al. 2007), poplar (Hu et al. 2015), *Liriodendron Chinense*, *Liquidambar formosana*, *Ilex integra*, *Photinia* × *fraseri*, *Neolitsea sericea*, and *Schima superba* (Zhang et al. 2012). Results showed that exposure to elevated O_3 (~150 ppb) from 6 July to 28 September with an AOT40 of 38.24 ppm·h reduced biomass in two deciduous species, *L. chinense* and *L. formosana*, by 26 % and 21 %, respectively, and in the evergreen species this parameter was reduced by 13 % in *C. camphora*, but little effect was seen in *C. glauca* (−6 %), *N. sericea* (−2 %), *Photinia* × *fraseri* (−3 %), or *I. integra* (−7 %). According to Hu et al. (2015), elevated ozone treatments with an average 8-h concentration of 108.7 ppb reduced total biomass by 29 % for poplar clone '107', by 33 % for clone '84 K' and clone '156', and by 41 % for clone '90' and clone '546'. However, there was no significant difference between biomass production in non-filtered air (NF, an average 8-h O_3 concentration of 53.2 ppb) compared with that in charcoal-filtered air (CF, an average 8-h O_3 concentration of 28.0 ppb) across all poplar clones.

After exposure to elevated O_3 for two growing seasons, NF + 60 ppb treatment (NF + 60) with an AOT40 of 56 ppm·h did not significantly reduce the total biomass of *Metasequoia glyptostroboid*, but NF + 120 ppb treatment (NF + 120) with an AOT40 of 115 ppm·h significantly reduced the biomass, by 11 %, compared with that for ambient O_3 treatment (NF, 5.6 ppm·h) (Zhang 2011). For C. *glauca*, elevated ozone (NF + 60 and NF + 120) did not significantly affect the biomass accumulation or carbon allocation after two growing seasons (Zhang 2011). For C. *camphora*, the stem dry weight was significantly decreased by NF + 120 (−13.7 %), but not by NF + 60. Root dry weight was significantly decreased under both NF + 60 (−9.2 %) and NF + 120 (−18.9 %), with a significant difference between NF + 60 and NF + 120. With regard to total biomass, NF + 120 exerted a significant negative effect (−14.3 %) on total biomass. NF + 60 also decreased the total biomass (−5.7 %), although this decrease was not statistically significant (Niu 2012).

Hu et al. (2015) determined the first stomatal O_3 flux or AOT40 – relative biomass (RB) relationship in poplar and other tree species in China. The R^2 value in the regression of RB and AOT_X increased with an O_3 concentration threshold of up to 60 ppb and then decreased. An O_3 flux threshold of 7 nmol·O_3·m^{-2}·s^{-1} provided the strongest correlation between RB and POD_Y (phytotoxic O_3 dose above a flux threshold of Y nmol·O_3·m^{-2}·s^{-1}), using a step of 1 nmol·O_3·m^{-2}·s^{-1}. The relationships between RB and O_3 dose are listed below.

RB (%) = 99.9–0.41AOT40	$R^2 = 0.88$
RB (%) = 100–0.53AOT60	$R^2 = 0.91$
RB (%) = 99.6–0.75POD$_1$	$R^2 = 0.85$
RB (%) = 99.5–1.2 POD$_7$	$R^2 = 0.91$

Based on these relationships, the critical levels of AOT40, AOT60, POD$_1$, and POD$_7$ for a 5 % reduction of total biomass for poplar were 12.0 ppm·h^{-1}, 9.4 ppm·h^{-1}, 6.1 mmol·m^{-2}, and 3.8 mmol·m^{-2}, respectively.

13.5 Combined Effects of Ozone and Other Environmental Factors on Chinese Trees

13.5.1 The Combined Effects of O_3 and N

N deposition and O_3 exposure generally co-occur, since NO_2 emission and O_3 production are chemically linked (Simpson et al. 2014). N is a limiting factor for tree growth in many forest ecosystems. Adding N moderately via deposition thus has the potential to increase growth, and therefore to sequester CO_2 from the atmosphere (De Vries et al. 2009; Liu et al. 2011). However, the excessive deposition of nitrogen in forest ecosystems can cause soil acidification, induce nutrient imbalances, reduce biodiversity, and increase the sensitivity of trees to other environmental stressors (Lu et al. 2010, 2014).

Anthropogenic reactive N (Nr) emissions into the atmosphere have increased dramatically in China due to rapid agricultural, industrial, and urban development. The average bulk deposition of N increased by approximately 8 kg N ha^{-1} between the 1980s (13.2 kg N·ha^{-1}) and the 2000s (21.1 kg N·ha^{-1}) across China, with an average annual increase of 0.41 kg N·ha^{-1} between 1980 and 2010 (Liu et al. 2013a). The highest bulk N deposition reached 60 kg N·ha^{-1}. The current average N deposition in China is much higher than the N loads needed for the protection of forest ecosystems in Europe and North America (Fowler et al. 2013).

Fewer studies have investigated the interactive effects of O_3 and N on plants in comparison with studies on the effects of these agents as single factors. Some results indicated that the negative effects of O_3 on the photosynthesis, growth, and biomass of trees were ameliorated in *Quercus kelloggii*, exacerbated in *Fagus crenata*, and unaffected in *Quercus serrata* and *Castanopsis sieboldii* by N addition (Watanabe et al. 2007, 2008; Handley et al. 2008; Yamaguchi et al. 2007). In China, we investigated the combined effects of O_3 and N load on the growth and photosynthetic characteristics of *Cinnamomum camphora* seedlings, a dominant evergreen broadleaf tree species in subtropical regions. The seedlings were supplied with N as NH_4NO_3 solution at 0, 30, and 60 kg N ha^{-1}·year^{-1} (simplified as N0, N30, and N60, respectively) and were exposed to NF or elevated O_3 concentration (NF + 60 ppb and NF + 120 ppb) for two growth seasons. In the first growing season, the NF + 60 induced significant negative effects on foliar photosynthesis, including a lower photosynthetic rate, reduced carboxylation efficiency, and reduced quantum yields of PSII and photosynthetic pigment contents, although there was no effect on growth. In contrast, the N load acted as a fertilization effect. A medium N load (N30) increased photosynthetic pigments and stem-base diameter growth relative to N0, whereas a high N load (N60) significantly enhanced the growth, photosynthetic pigments, and dark and light action of photosynthesis in *C. camphora* seedlings. Ozone induced visible foliar symptoms, such as chlorotic bleaching and brown stipples, on the upper surfaces of *C. camphora* leaves, and the leaves became significantly smaller and tended to be thinner under O_3 stress (Feng et al. 2011).

After two growing seasons, the percentages of symptomatic leaves in *C. camphora* seedlings were significantly increased with O_3 concentration (Niu 2012). The elevated O_3 (NF + 120) significantly decreased the mean leaf size and chlorophyll content of both 2009- and 2010-emerged leaves, inhibited the stem height and basal diameter, and reduced biomass accumulation in all plant parts except for the leaves. By comparison, only the chlorophyll content of 2009-emerged leaves and root dry weight were significantly reduced under NF + 60. The specific leaf area, total leaf area, and foliar biomass were not affected, even at a higher O_3 level. On the other hand, the N load, especially N60, exerted significantly positive effects on all variables, except for mean leaf size and the shoot/root ratio. No significant interactions between O_3 and N were detected in regard to growth and biomass accumulation, net photosynthetic rate, and pigment contents of the seedlings, suggesting that the N supply at <60 kg·ha^{-1}·year^{-1} did not significantly modify the response of *C. camphora* to O_3 (Niu 2012).

13.5.2 The Combined Effects of Elevated O_3 and CO_2

Increases in atmospheric concentrations of CO_2 and O_3 can have large but offsetting effects on ecosystem productivity (King et al. 2005), yet their independent and interactive effects on ecosystem processes remain poorly understood. In China, 4-year-old gingko (*Ginkgo biloba*), Chinese pine (*Pinus tabulaeformis*), and *Quercus mongolica* were exposed to ambient and elevated concentrations of CO_2 (700 ppm) and O_3 (80 ppb), using open top chambers, in Shenyang (41°46′N, 123°26′E), China. The research focused on the combined effects of CO_2 and O_3 on growth, gas-exchange, antioxidant systems, endogenous plant hormones, and BVOC emission rates (Yan et al. 2010b; Li et al. 2007, 2009a, b, 2011; Huang et al. 2008; Lu et al. 2009; Xu et al. 2014).

Li et al. (2011) studied *Ginkgo biloba* and found that elevated CO_2 increased leaf area and leaf dry weight, and had no effect on shoot length, increased indole-3-acetic acid (IAA), gibberellin A3 (GA3), zeatin riboside (ZR), dihydrozeatin (DHZR), and isopentenyl adenosine (iPA) content, but decreased abscisic acid (ABA) content. Elevated O_3 significantly decreased leaf area, leaf dry weight, and shoot length, and decreased IAA, GA3, and ZR content, but increased ABA content and had a small effect on iPA and DHZR content. Elevated CO_2 and O_3 decreased IAA, iPA, and DHZR content, while these conditions increased ABA and GA3 content in the early stage of the exposure and then decreased ABA and GA3 content in the late stage (Li et al. 2011). Furthermore, decreases in IAA oxidase (IAAO) and peroxidase activities were observed in plants exposed to elevated CO_2 compared with control. Elevated O_3 had no significant effect on IAAO activity, but it increased peroxidase activity during the early stage of the exposure (Li et al. 2009b). The evidence from this study indicates that elevated CO_2 ameliorated the effects of elevated O_3 on tree growth, and elevated CO_2 may have a largely positive impact on forest tree growth, while elevated O_3 likely has a negative impact.

No significant changes in ROS production and scavenging systems were detected in the seedlings of *Ginkgo biloba* exposed to high CO_2. However, elevated O_3 induced significant increases in H_2O_2 and Malondialdehyde (MDA) content, ascorbate content, and antioxidant enzyme activity. In the combined exposure of elevated CO_2 and O_3, the increase was mitigated by high CO_2, but the effect was not significant (Lu et al. 2009). Elevated CO_2 significantly increased isoprene emission by *Ginkgo biloba* in July and September, while the total monoterpene emission was enhanced in July and decreased in September. Elevated O_3 increased isoprene and monoterpene emissions in July and September. The combination of elevated CO_2 and O_3 did not have any effect on BVOC emissions, except for increases in isoprene and D3-carene in September (Li et al. 2009a). On the other hand, elevated CO_2, alone and in combination with elevated O_3, increased the concentrations of all the determined nonvolatile terpenoids, while elevated O_3 only increased the concentration of bilobalide, demonstrating that terpenoid metabolism in ginkgo leaves was more sensitive to elevated CO_2 than elevated O_3 (Huang et al. 2008).

Li et al. (2007) found that elevated CO_2 increased axial shoot and needle length by 20 % and 10 %, respectively, whereas elevated O_3 reduced these measurements by 43 % and 7 %, respectively, compared with control values in Chinese pine trees (*P. tabulaeformis*). An increase in IAA content and peroxidase activity and a decrease in IAAO activity were observed in the Chinese pine trees exposed to elevated CO_2 concentration compared with control. Elevated O_3 decreased the IAA content and had no significant effect on IAAO activity, but significantly increased peroxidase activity. When trees pre-exposed to elevated CO_2 were transferred to elevated O_3 conditions (EC–EO) or trees pre-exposed to elevated O_3 were transferred to elevated CO_2 (EO–EC), the IAA content was lower than the control (CK), while IAAO activity was higher than that in trees transferred to CK (EC–CK or EO–CK), suggesting that the change in IAA content was also related to IAAO activity (Li et al. 2007). These results indicated that IAAO and peroxidase activities in Chinese pine needles may be affected by changes in the atmospheric environment, resulting in changes in IAA metabolism, which, in turn, may cause changes in growth in Chinese pine trees. Furthermore, elevated CO_2 increased the needle dry weight, net photosynthetic rate, and maximum quantum yield of PSII in previous- and current-year needles. Elevated O_3 significantly decreased dry weight, net photosynthetic rate, and maximum quantum yield of PSII only in previous-year needles. Elevated O_3 increased H_2O_2 and MDA contents in previous-year needles, but did not significantly increase H_2O_2 content in current-year needles, indicating that the oxidative stress induced by elevated O_3 was more severe in previous-year needles than in current-year needles. These results showed that the adverse effect of elevated O_3 in the needles of *P. tabulaeformis* was ameliorated by elevated CO_2 (Xu et al. 2014).

Yan et al. (2010b) investigated the responses of young *Quercus mongolica* trees to combined elevated CO_2 and O_3. Results showed that elevated O_3 increased MDA content and decreased the photosynthetic rate after 45 days of exposure, and prolonged exposure (105 days) induced a significant increase in electrolyte leakage and a reduction of chlorophyll content. All these changes were alleviated by elevated CO_2, indicating that the oxidative stress on cell membranes and photosynthesis was ameliorated. After 45 days of exposure, elevated O_3 stimulated the activities of superoxide dismutase (SOD) and ascorbate peroxidase (APX), but the stimulation was dampened under elevated CO_2 exposure. Furthermore, ascorbate and total phenolic contents were not higher in the trees with combined treatments than in those with elevated O_3 treatment. This indicates that the protective effect of elevated CO_2 against O_3 stress after 45 days of exposure was not achieved by enhancing ROS scavenging ability. After 105 days of exposure, elevated O_3 significantly decreased the activities of SOD, catalase (CAT), and APX, and significantly decreased ascorbate content. Elevated CO_2 suppressed the O_3-induced decrease, which ameliorated the oxidative stress to some extent. In addition, elevated CO_2 increased total phenolic content in the leaves both under both ambient O_3 and elevated O_3 exposure, and this might also have contributed to the protection against O_3-induced oxidative stress (Yan et al. 2010b).

13.5.3 *The Combined Effects of Elevated O_3 and Drought*

Drought coincides with episodes of elevated O_3 concentrations in most regions. Many studies have focused on drought/O_3 interactions to determine possible ways in which drought would affect O_3 injury and plant growth effects (Alonso et al. 2013; Pearson et al. 1994; Matyssek et al. 2006; Pollastrini et al. 2014). Possible interactions have been determined to be antagonistic, neutral (additive), or synergistic.

The results of a meta-analysis indicated that drought appeared to protect plants from O_3 injury and productivity loss, mainly through its influence on reducing stomatal aperture (Wittig et al. 2009). However, individual studies have shown inconsistent interactions between O_3 and drought. Some studies indicated that drought did not change O_3 damage, and noted that the effects of drought and ozone were due to independent mechanisms (Pearson et al. 1994). On the other hand, some studies reported that drought increased plant sensitivity to O_3 (Alonso et al. 2013; Pollastrini et al. 2014) and slowed or even stopped stomatal response to abrupt stresses (Paoletti 2005), a factor which may cause a greater reduction in photosynthesis than in those plants without stomatal sluggishness response. and may actually increase the overall oxidative risk as well, due to limited carbon fixation (Matyssek et al. 2006; Tausz et al. 2007).

In summer, droughts occur frequently in North China, and they will be more severe in future with climate change (Liu et al. 2013b). A 2-year experiment was conducted to investigate the response of Shantung maple (*Acer truncatum*) seedlings to elevated O_3 concentrations and drought stress, using OTCs in a suburb of Beijing, in north China in 2012–2013. The results suggested that both elevated O_3 and drought had significantly reduced leaf mass per area (LMA), g_s, and A_{sat}, as well as above- and below-ground biomass, at the end of the experiment. Drought stress mitigated the expression of O_3 foliar injury, and mitigated O_3-induced reduced LMA, as well as mitigating the effects of O_3 on leaf photosynthetic pigments, height growth, and basal diameter, due to limited carbon fixation. However, elevated O_3 induced more reductions in A_{sat}, g_s, and total biomass, by 23.7 %, 15.5 %, and 8.1 %, respectively, when combined with drought stress. These data suggest that drought did not protect the Shantung maple seedlings from the effects of O_3 when the whole plant was considered (Li et al. 2015). This study also indicated that ozone and drought effects were synergistic but not additive (less than the sum of the factors individually), and this finding will provide an important case for understanding the complex interactive effects of O_3 and drought.

References

Alonso R et al (2013) Drought stress does not protect *Quercus ilex* L. from ozone effects: results from a comparative study of two subspecies differing in ozone sensitivity. Plant Biol 16:375–384

Atkinson R et al (2003) Atmospheric degradation of volatile organic compounds. Chem Rev 103:4605–4638

Bussotti F (2008) Functional leaf traits, plant communities and acclimation processes in relation to oxidative stress in trees: a critical overview. Global Change Biol 14:2727–2739

Bussotti F et al (2005) Ozone foliar symptoms in woody plant species assessed with ultrastructural and fluorescence analysis. New Phytol 166:941–955

Bussotti F et al (2007) Photosynthesis responses to ozone in young trees of three species with different sensitivities, in a 2-year open-top chamber experiment (Curno, Italy). Physiol Plant 130:122–135

Calatayud V et al (2007) Foliar, physiological and growth responses of four maple species exposed to ozone. Water Air Soil Pollut 185:239–254

Calatayud V et al (2010) Contrasting ozone sensitivity in related evergreen and deciduous shrubs. Environ Pollut 158:3580–3587

Calatayud V et al (2011a) Responses of evergreen and deciduous Quercus species to enhanced ozone levels. Environ Pollut 159:55–63

Calatayud V et al (2011b) Physiological, anatomical and biomass partitioning responses to ozone in the Mediterranean endemic plant *Lamottea dianae*. Ecotoxicol Environ Saf 74:1131–1138

Calfapietra C et al (2009) Volatile organic compounds from Italian vegetation and their interaction with ozone. Environ Pollut 157:1478–1486

Chen B et al (2015a) Uptake characteristic of ozone on Poplar canopy by sap flow techniques. J Northeast For Univ 43:72–79 (In Chinese)

Chen B et al (2015b) Ozone uptake characteristics in different dominance hierarchies of poplar plantation. J Beijing For Univ 37:29–36

De Vries W et al (2009) The impact of nitrogen deposition on carbon sequestration by European forests and heathlands. For Ecol Manage 258:1814–1823
Dizengremel et al (2013) Integrative leaf-level ozone phytotoxic ozone dose assessment for forest risk modeling. In: Matyssek R et al (eds) Climate change, air pollution and global challenges, vol 13, Developments in Environmental Sciences. Elsevier, Netherlands, pp 267–288
Dudareva N et al (2006) Plant volatiles: recent advances and future perspectives. Crit Rev Plant Sci 25:417–440
Fehsenfeld et al (1992) Emissions of volatile organic compounds from vegetation and the implications for atmospheric chemistry. Glob Biogeochem Cycles 6:389–430
Feng ZZ et al (2008) Sensitivity of *Metasequoia glyptostroboides* to ozone stress. Photosynthetica 46:463–465
Feng ZZ et al (2011) Effects of ozone exposure on Sub-Tropical evergreen *Cinnamomum Camphora* seedlings grown in different nitrogen loads. Trees 25:617–625
Feng ZZ et al (2014) Evidence of widespread ozone-induced visible injury on plants in Beijing, China. Environ Pollut 193:296–301
Fowler D et al (2013) The global nitrogen cycle in the twenty-first century. Phil Trans R Soc B 368:20130164
Fuentes JD et al (2000) Biogenic hydrocarbons in the atmospheric boundary layer: a review. Bull Am Meteorol Soc 81:1537–1575
Gao F et al (2016) Effects of elevated ozone on physiological, anatomical and ultrastructural characteristics of four common urban tree species in China. Ecol Indic 67:367–379
Grantz D (2003) Ozone impacts on cotton: towards an integrated mechanism. Environ Pollut 126:331–344
Gravano E et al (2003) Foliar response of an *Ailanthus altissima* clone in two sites with different levels of ozone-pollution. Envi ron Pollut 121:137–146
Guenther AB et al (1994) Natural volatile organic compound emission rates for U.S. woodland landscapes. Atmos Environ 28:1197–1210
Günthardt-Goerg MS et al (2007) Linking stress with macroscopic and microscopic leaf response in trees: new diagnostic perspectives. Environ Pollut 147:467–488
Handley T et al (2008) Interactive effects of O_3 exposure on California black oak (*Quercus kelloggii* Newb.) seedlings with and without N amendment. Environ Pollut 156:53–60
Harley PC et al (1999) Ecological and evolutionary aspects of isoprene emission from plants. Oecologia 118:109–123
He XY et al (2006) Responses of the anti-oxidative system in leaves of *Ginkgo biloba* to elevated ozone concentration in an urban area. Bot Stud 47:409–416
He XY et al (2007) Changes in effects of ozone exposure on growth, photosynthesis, and respiration of *Ginkgo biloba* in Shenyang urban area. Photosynthetica 45:555–561
Hoshika Y et al (2014) Determinants of stomatal sluggishness in ozone-exposed deciduous tree species. Sci Total Environ 481:453–458
Hu EZ et al (2015) Concentration- and flux-based ozone dose-response relationships for five poplar clones grown in North China. Environ Pollut 207:21–30
Huang W et al (2008) Influence of elevated carbon dioxide and ozone on the foliar nonvolatile terpenoids in *Ginkgo Biloba*. Bull Environ Contam Toxicol 81:432–435
Kesselmeier J et al (1999) Biogenic volatile organic compounds (VOC): an overview on emission, physiology and ecology. J Atmos Chem 33:23–88
King JS et al (2005) Tropospheric O_3 compromises net primary production in young stands of trembling aspen, paper birch and sugar maple in response to elevated atmospheric CO_2. New Phytol 168:623–635
Kivimäenpää M et al (2010) Visible and microscopic needle alterations of mature Aleppo Pine (*Pinus halepensis*) trees growing on an ozone gradient in eastern Spain. Tree Physiol 30:541–554
Li XM et al (2007) Effects of elevated CO_2 and/or O_3 on hormone IAA in needles of Chinese pine. Plant Growth Regul 53:25–31
Li DW et al (2009a) Impact of elevated CO_2 and O_3 concentrations on biogenic volatile organic compounds emissions from *Ginkgo biloba*. Bull Environ Contam Toxicol 82:473–477

Li XM et al (2009b) Influence of elevated CO_2 and O_3 on IAA, IAA oxidase and peroxidase in the leaves of ginkgo trees. Biol Plant 53(2):339–342

Li XM et al (2011) Effects of elevated carbon dioxide and/or ozone on endogenous plant hormones in the leaves of *Ginkgo biloba*. Acta Physiol Plant 33:129–136

Li L et al (2015) Chronic drought stress reduced but not protected Shantung maple (*Acer truncatum* Bunge) from adverse effects of ozone (O_3) on growth and physiology in the suburb of Beijing, China. Environ Pollut 201:34–41

Li P et al (2016) Differences in ozone sensitivity among woody species are related to leaf morphology and antioxidant levels. Tree Physiology 36:1105–1116.

Liu XJ et al (2011) Nitrogen deposition and its ecological impact in China: an overview. Environ Pollut 159:2251–2264

Liu XJ et al (2013a) Enhanced nitrogen deposition over China. Nature 494:459–462

Liu HY et al (2013b) Drought threatened semi-arid ecosystems in the Inner Asia. Agric For Meteorol 178:1–2

Loreto F et al (2001) Isoprene produced by leaves protects the photosynthetic apparatus, against ozone damage, quenches ozone products, and reduces lipid peroxidation of cellular membranes. Plant Physiol 127:1781–1787

Lu T et al (2009) Effects of elevated O_3 and/or elevated CO_2 on lipid peroxidation and antioxidant systems in *Ginkgo biloba* Leaves. Bull Environ Contam Toxicol 83:92–96

Lu XK et al (2010) Effects of experimental nitrogen additions on plant diversity in an old-growth tropical forest. Global Chang Biol 16(10):2688–2700

Lu XK et al (2014) Nitrogen deposition contributes to soil acidification in tropical ecosystems. Global Chang Biol 20:3790–3801

Matyssek R et al (2006) Interactions between drought and O_3 stress in forest trees. Plant Biol 8:11–17

Matyssek R et al (2007) Promoting the O_3 flux concept for European forest trees. Environ Pollut 146:587–607

Matyssek R et al (2008) The challenge of making ozone risk assessment for forest trees more mechanistic. Environ Pollut 156:567–582

Monson RK et al (2012) Modeling the isoprene emission rate from leaves. New Phytol 195:541–559

Niu JF (2012) Effects of elevated ozone and nitrogen deposition on the growth and physiology of *Cinnamomum camphora* seedlings. Graduate University of Chinese Academy of Sciences, Ph.D. (In Chinese)

Niu JF et al (2011) Impact of elevated O_3 on visible foliar symptom, growth and biomass of *Cinnamomum camphora* seedlings under different nitrogen loads. J Environ Monit 13: 2873–2879

Niu JF et al (2014) Non-stomatal limitation to photosynthesis in *Cinnamomum camphora* seedlings exposed to elevated O_3. PLoS One 9(6):e98572

Nowak DJ and Crane DE (2000) The Urban Forest Effects (UFORE) model: quantifying urban forest structure and functions. In: Mark H, Tom B (eds). Integrated tools for natural resources inventories in the 21st century. Gen. Tech. Rep. NC-212. St. Paul, MN: U.S. Dept. of Agriculture, Forest Service, North Central Forest Experiment Station. 714–720

Nowak DJ et al (2000) A modeling study of the impact of urban trees on ozone. Atmos Environ 34:1601–1613

Paoletti E (2005) Ozone slows stomatal response to light and leaf wounding in a Mediterranean evergreen broadleaf, Arbutus unedo. Environ Pollut 134:439–445

Pearson M et al (1994) Effects of exposure to ozone and water stress on the following season's growth of beech (*Fagus sylvatica* L.). New Phytol 126:511–515

Pollastrini M et al (2014) Severe drought events increase the sensitivity to ozone on poplar clones. Ecotoxicol Environ Saf 100:94–104

Schaub M et al (2010) Assessment of ozone injury. Manual on methods and criteria for harmonized sampling, assessment, monitoring and analysis of the effects of air pollution on forests. UNECE ICP Forests Programme Coordinating Centre, Hamburg, 1–22. Available at: http://icp-forests.net/

Scott KI et al (1998) Air pollutant uptake by Sacramento's urban forest. J Arboric 24:224–234
Simpson D et al (2014) Ozone-the persistent menace: interactions with the N cycle and climate change. Curr Opin Environ Sustain 9–10:9–19
Tausz M et al (2007) Defense and avoidance of ozone under global change. Environ Pollut 147:525–531
Vickers CE et al (2009) A unified mechanism of action for isoprenoids in plant abiotic stress. Nat Chem Biol 5:283–291
Wan WX et al (2014) Ozone and ozone injury on plants in and around Beijing, China. Environ Pollut 191:215–222
Wang H et al (2012) Ozone uptake by adult urban trees based on sap flow measurement. Environ Pollut 162:275–286
Wang H et al (2013) Ozone uptake at the canopy level in *Robinia pseudoacacia* in Beijing based on sap flow measurements. Acta Ecol Sin 33:7323–7331 (In Chinese)
Watanabe M et al (2007) Influences of nitrogen load on the growth and photosynthetic responses of *Quercus serrata* seedling to O_3. Trees 21:421–432
Watanabe M et al (2008) Effects of ozone on the growth and photosynthesis of *Castanopsis sieboldii* seedlings grown under different nitrogen loads. J Agric Meteorol 64:143–155
Wieser G et al (2002) Age effects on Norway spruce (*Picea abies*) susceptibility to ozone uptake: a novel approach relating stress avoidance to defense. Tree Physiol 22:583–590
Wieser G et al (2003) Quantifying ozone uptake at the canopy level of spruce, pine and larch trees at the alpine timberline: an approach based on sap flow measurement. Environ Pollut 126:5–8
Wieser G et al (2006) Quantification of ozone uptake at the stand level in a Pinus canariensis forest in Tenerife, Canary Islands: an approach based on sap flow measurements. Environ Pollut 140:383–386
Wittig V et al (2009) Quantifying the impact of current and future tropospheric ozone on tree biomass, growth, physiology and biochemistry: a quantitative meta-analysis. Global Chang Biol 15:396–424
Xu S et al (2012) Responses of growth, photosynthesis and VOC emissions of *Pinus tabulaeformis* Carr. exposure to elevated CO_2 and/or elevated O_3 in an urban area. Bull Environ Contam Toxicol 88:443–448
Xu S et al (2014) Elevated CO_2 ameliorated the adverse effect of elevated O_3 in previous-year and current-year needles of *Pinus tabulaeformis* in urban area. Bull Environ Contam Toxicol 92:733–737
Yamaguchi M et al (2007) Growth and photosynthetic responses of *Fagus crenata* seedlings to O_3 under different nitrogen loads. Trees 21:707–718
Yan K et al (2010a) Responses of photosynthesis, lipid peroxidation and antioxidant system in leaves of *Quercus mongolica* to elevated O_3. Ecotoxicol Environ Saf 69:198–204
Yan K et al (2010b) Elevated CO_2 ameliorated oxidative stress induced by elevated O_3 in *Quercus mongolica*. Acta Physiol Plant 32:375–385
Yan K et al (2013) Variation of antioxidant system in *Pinus armandii* under elevated O_3 in an entire growth season. Clean-Soil Air Water 41(1):5–10
Yuan XY et al (2016) Interaction of drought and ozone exposure on isoprene emission from extensively cultivated poplar. Plant, Cell & Environment, 39:2276–2287.
Zhang WW (2011) Effects of elevated O_3 level on the native tree species in subtropical China. Graduate University of Chinese Academy of Sciences, Ph.D. (in Chinese)
Zhang WW et al (2011) Effects of ozone exposure on growth and photosynthesis of the seedlings of *Liriodendron chinense* (Hemsl.) Sarg a native tree species of subtropical China. Photosynthetica 49:29–36
Zhang WW et al (2012) Response of native broadleaved woody species to elevated ozone in subtropical China. Environ Pollut 163:149–157
Zhang WW et al (2014) Impacts of elevated ozone on growth and photosynthesis of *Metasequoia glyptostroboides* Hu et Cheng. Plant Sci 226:182–188

Part V
Effects of Acid Deposition on Asian Plants

Chapter 14
Effects of Simulated Acid Rain on Asian Crops and Garden Plants

Yoshihisa Kohno

Abstract Most results with simulated acid-rain exposure experiments have suggested that rain acidity below pH 3.0 could induce direct or indirect deleterious effects on crops and other herbaceous plants, with visible injury development and reductions in growth and yield; however, plants show different sensitivities to acidity below pH 3.0 in terms of visible acute injury or reductions in growth and yield. Current rain acidity, which is regarded as over pH 4.0, may not induce direct adverse effects on field-grown plants. However, acidic precipitation may have a potential threat to induce increasing availability and toxicity of rare earth and heavy metals in plants.

Keywords Simulated acid rain • Rain acidity • Crop response • East Asia

14.1 Introduction

Since the early 1970s, simulated acid rain (SAR) exposure experiments have been performed to assess rain acidification effects on crops in the United States. In 1980, the 10-year National Acid Precipitation Assessment Program (NAPAP) started comprehensive surveys in various related fields in the United States to assess the potential negative impacts of acidic deposition and associated pollutants. The NAPAP concluded that acidic precipitation at ambient levels in the United States was not shown to be responsible for regional crop yield reduction (Shriner et al. 1990).

The Asian region has different natural conditions from those in North America and this is reflected in the development of different agricultural environments. Here the author reviewed East Asian scientific publications describing results of experi-

Y. Kohno
Central Research Institute of Electric Power Industry,
1646 Abiko, Abiko City, Chiba 270-1194, Japan
e-mail: kohno@peach.ocn.ne.jp; yoshihisakohno@gmail.com

T. Izuta (ed.), *Air Pollution Impacts on Plants in East Asia*,
DOI 10.1007/978-4-431-56438-6_14

ments with simulated acid rain (SAR) exposure on crops, including fruit trees, vegetables and other herbaceous plants, and ornamental species.

14.2 Exposure System for Simulated Acid Rain

Different types of exposure systems are used to apply SAR. A primitive exposure tool used in earlier studies was a small hand-held mist sprayer (Fujiwara et al. 1980; Kohno and Fujiwara 1981; Kohno 1994; Taniyama and Saito 1979, 1981). In these studies, a volume of simulated rain was applied to plant surfaces until the leaves were thoroughly wet. Later, a mist sprayer system with an electric pump was used for SAR exposure experiments. This system was operated intermittently to provide certain amounts of SAR during the experimental period (Hosono and Nouchi 1992, 1994; Satoh 1996).

A rain generator system providing 2.5 mm per hour of simulated precipitation, in the form of rain droplets with about 1-mm diameter, for the test area was developed (Kohno and Kobayashi 1989a; Kohno et al. 1994, 1995). Plants were exposed to SAR at a precipitation amount of 10–30 mm for 4–12 h a day. Exposures were carried out in the evening, except for cloudy and rainy days, to prevent the photobleaching effects of exposure under solar radiation.

In the early studies, a single-acid solution, such as sulfuric acid, was diluted with deionized or tap water to prepare the SAR for designated pH levels. Other acids used were a mixture of sulfuric, nitric, and hydrochloric acids at varying concentrations, with a mean major anion composition ratio based on precipitation chemistry in monitoring data.

For example, Kohno and Kobayashi (1989a) prepared SAR as follows. Tap water was purified to a conductivity of 0.1–0.2 μS/cm with a reverse-osmotic pressure membrane system and a mixed ion-exchange resin column system used in turn. The SAR contained sulfate, nitrate, and chloride at an equivalent concentration ratio of 5:2:3, which was close to the mean concentration ratio of the precipitation determined by precipitation chemistry in the monitoring data. Stock solution was volumetrically diluted with deionized water to prepare solution of the designated pH for the exposure experiments. Deionized water was used as the control and as the diluent.

A mist-generating system was also developed by modification of the rain-generator system, with fogger nozzles being used instead of the needles used in the rain-generator system, to simulate misty rain with smaller-diameter droplets. The system was operated intermittently to adjust it to the designated precipitation rate (Kobayashi et al. 1994). The distribution accuracies of precipitation generated by the rain and mist systems were below 10 % and 36 %, respectively of the control value (Kohno et al. 1994).

In general, a system should be developed with plastic and/or stainless steel parts, to avoid unnecessary heavy metal contamination that could be caused by the corrosion of metal parts in contact with an acid solution.

14.3 Visible Injury Development

Yoshida (1971) reported petal discoloration in field-grown morning glory just after mist or drizzly rain events near a heavy industry area, Yokkaichi City, in central Japan in the summer of 1970. Okita and Hashizume (1975) also found petal discoloration, in azaleas grown in Tokyo, Japan, the day after an evening misty event. These observation reports triggered the initiation of precipitation monitoring and investigative cause-effect studies in Japan.

Fujiwara et al. (1980) applied a mist solution of sulfuric, nitric, and sulfurous acids to the petals of morning glory, pear, and tulip and the leaves of bush bean and spinach to find the relationship between petal discoloration or leaf injury development and rain acidity. Generally, petals were more sensitive than leaves to the acid solution treatment, but the sensitive pH level was species-dependent. Morning glory developed necrosis in petals at pH 3.0 or 3.5, the pear petals developed necrosis at pH 2.5, while those of the tulip developed necrosis at pH 1.0 or 2.0. Leaves of bush bean, spinach, pear, morning glory and tulip developed necrosis at pH 1.0–2.5. These results also depended on the components of the acid solution. A comparison of necrosis development in flowers caused by solutions of sulfuric, nitric, or hydrochloric acid suggested that sulfuric acid was most important, then nitric, and then hydrochloric acid. Spinach developed necrosis in leaves at pH 2.0 with each of these three acid solutions, whereas sulfurous acid solution induced injuries at pH 1.5.

Taniyama and Saito (1979, 1981) prepared SAR with sulfuric or sulfurous acid at pH 2, 3, 4, and 5; the control was running water at pH 7.0. Wheat (*Triticum aestivum* cv. Omase), barley (*Hordeum vulgare* cv. Hinodehadaka), and rice (*Oryza sativa* cv. Koshihikari) were sprayed with 500 ml of the SAR from a small-type sprayer. Spraying started at the heading stage and continued for 20 days. Acid mist spray caused visible injuries in leaf blades and sheaths in rice at and below pH 3.

Kohno and Fujiwara (1981, 1982) and Kohno (1994) sprayed sulfuric acid solution at pH 2.0–4.0 every morning for 3 or 4 weeks on radish (*Raphanus sativa* cv. Comet) and for 6 or 8 weeks on bush bean (*Phaseolus vulgaris* cv. Wonder Crop 2). Radish developed necrosis all over the leaves when exposed to pH 2.0, but no visible injury developed at pH 3.0. Foliar injury in bean developed at pH 2.6 and below; however, exposure at pH 3.0 did not cause any visible leaf injuries.

Kohno and Kobayashi (1989a, b) and Kobayashi et al. (1991) exposed soybean (*Glycine max* cv. Early Hakucho or Sapporo Midori) to SAR using a rain generator system. Plants were subjected to SAR twice a week during the growing season, for a 1-h period at a rate of 5 mm h^{-1}. When SAR was below pH 3.0, visible symptoms, such as necrotic spots and necrosis, developed in the young trifoliate leaves. However, at a pH above 3.0, there was no evidence of visible leaf injury.

Kobayashi et al. (1994) exposed nine species to SAR or simulated acid mist (SAM) at an average rate of 2.5 mm/h, 9 h/day, 3 days/week. Perennial ryegrass and Kentucky bluegrass had necrotic symptoms with SAR and SAM exposure at pH 2.5. Pea, perilla, turnip, soybean, white clover, lucerne, and mume (*Prunus mume*) showed visible leaf injuries at pH 3.0 and below. Exposure to SAM at pH 2.5

induced earlier visible injury development than that caused by SAR. However, no visible injuries were observed at pH 4.0.

Hosono and Nouchi (1992) sprayed SAR on radish (cv. Comet), spinach (cv. Young Power and Viroflay), and bush bean (cv. Meal) at a rate of 10 mm per h, three times a week during the growing period. Plants exposed to SAR at pH 3.0 or below developed visible leaf injuries, but no injury developed at pH 4.0. Furthermore, eight crops were subjected to SAR throughout their entire growth periods (Hosono and Nouchi 1994). SAR at pH 3.0 or below produced visible foliar injuries in all crops. The degree of foliar injury varied among species and leaf positions. Injury was more severe in lower leaf positions, particularly cotyledons and primary leaves, than in upper ones. In many cases foliar injury was more severe at the early growth stage than at the later stage.

Satoh (1996) sprayed four weed species (*Polygonum lapathifolium*, *Chenopodium album*, *Erigeron philadelphicus*, and *Digitaria abscendens*), four crop species (*Fagopyrum esculentum*, *Raphanus sativus*, *Dolichos lablab*, and *Oryza sativa*) and two forage plants (*Trifolium repens* and *Festuca arundinacea*) with SAR at pH 3.5–2.5 at a rate of 5 mm per event, with a total of six to eight events, for 2 weeks. *C. album*, *D. abscendens*, and *F. arundinacea* did not develop any visible foliar injuries after SAR treatment; however, the other seven species did develop such injuries.

Leaves of *Arabidopsis thaliana* exposed to SAR at pH 3.0 for 15 days showed many white necrotic spots on part of the upper epidermis (Park 2005). A light microscopic examination of the leaves with necrotic spots showed the upper epidermis was severely compressed with a damaged cuticle layer; the mesophyll cells were also damaged and compressed. However, no noticeable structural changes were observed in vascular bundle cells. Lee et al. (2006) also exposed *A. thaliana* to SAR, at pH 4.0, and it showed normal structure; in contrast, leaves treated with SAR at pH 3.0 or pH 2.0 showed severe collapse, and necrosis was observed in all leaves. Exposure of *A. thaliana* to SAR resulted in the upregulation of genes known to be induced by the salicylic acid (SA)-mediated pathogen resistance response. These results suggested that SAR activates the salicylic acid pathway and activation of this pathway is sensitive to sulfuric acid.

One of the acute direct effects of SAR or SAM is visible injury. Typical symptoms are necrotic spots, necrosis or necrotic lesions, marginal and/or tip necrosis, early senescence or discoloration, and defoliation (Kohno et al. 1994; Kobayashi et al. 1994). Exposure to extremely low-pH SAR or SAM at pH 2.0 or 2.5 induced necrotic lesions or spots in the leaves after several events in crops, grasses, and herbs. When plants were exposed to SAR or SAM at the same precipitation rate, the amount of water adhering to the leaf surface with mist exposure was likely to be greater than that with the rain. Exposure to SAM at pH 2.5 for 9 h per event at an intensity of 2.5 mm per h induced earlier visible injury development than exposure to SAR. This may suggest that the direct effect of acid mist on plants would be more severe than that of acid rain (Kobayashi et al. 1994), when the amount of precipitation is the same as that of the rain.

Based on the above information, SAR at and below pH 3.0 induced visible injuries in most of the plants studied; however, no injury development was observed at pH 4.0 in any of the plants. As the lowest annual mean pH of precipitation in the East Asian region is around pH 4.2 (EANET 2011), it is possible that current rain acidity may not induce direct injuries in crops and herbaceous plants grown in the field in this region.

14.4 Effect on Germination of Seeds

Zabawi et al. (2008) sowed rice seeds of MR 84, the most popular cultivar in Peninsular Malaysia, and sprayed SAR at pH 3.5, 4.0, 4.5, 5.0, and 5.6 (control) every day to maintain the sand surface in a wet condition. The percentage of seed germination decreased as the pH level decreased. Compared with the other treatments, pH 3.5 caused a reduction in germination by 3–5 % and caused a relatively slower germination rate. However, there was no significantly different effect on germination between pH 4.0, pH 4.5, and pH 5.6 as the control.

14.5 Acceleration of Leaching from Plant Surface by Acid Rain

As Tukey (1970) and Nouchi (1992) reported, foliar leaching occurs naturally in any plant species. Various substances will be leached out from plant surface tissues through rain contact. Acidic solution will be neutralized by the leaching of base cations from the tissues through the ion-exchange process. This leaching will be accelerated with increasing acidity of the washing solution.

Kohno and Fujiwara (1982) and Kohno (1994), in their observations with a cryo scanning electron microscope with an energy dispersive X-ray analyzer, demonstrated that the spraying of sulfuric acid solution at pH 2.0 for 4–6 weeks caused visible foliar injuries and the accumulation of a crystalline Ca-S compound on the leaf surface. However, other basic elements, such as potassium and magnesium, were not detected. This observation suggested that mainly tissue Ca neutralized sulfate when acid solution is applied to the leaf surface. In addition, Kohno et al. (2001), using pH-imaging microscopy, reported the process of acid neutralization on leaf surfaces. Leaves neutralized the surface acid by ion exchange with cations derived from their surface tissues and recovered from strong temporary acid stress imposed by acid rain or acid fog in a relatively short period of time.

Generally, in most agricultural fields, fertilizers and/or lime, dolomite, or other basic materials are applied to fields before cultivation to maintain the appropriate soil acidity level for better growth performance in crops. Considering this background, the above findings indicate that acidic precipitation at current acidity levels does not pose a direct threat to plants.

14.6 Effect on Photosynthesis

Taniyama and Saito (1979, 1981) reported a significant decrease in apparent photosynthesis in leaves of wheat, rice, and barley after the plants were sprayed with sulfuric acid mist at pH 4, compared with apparent photosynthesis in the control plants at pH 7. Inhibition of apparent photosynthesis was ranked approximately in the following order according to the relative amounts by which plant CO_2 uptake rates were suppressed by sulfuric and sulfurous acid mists: rice>barley, wheat in sulfuric acid mist; and rice>barley>wheat in sulfurous acid mist.

Hosono and Nouchi (1992) reported that photosynthetic rates in radish leaves were inhibited by SAR at pH 2.5, with a damaged area of 30 %, while bush bean primary leaves showed a damaged area of 90 % with this SAR treatment. However, the photosynthetic rates in the leaves of rice plants with less visible injury caused by spraying SAR at pH 2.5 were not apparently inhibited.

Satoh (1996) reported that photosynthetic activity in radish and kidney bean (*Dolichos lablab*) leaves decreased significantly after the spraying of SAR below pH 3.0. Philadelphia fleabane and white clover leaves showed decreased photosynthetic activity after the spraying of SAR at pH 2.5, while *Polygonum lapathifolium*, common lambsquarters, southern crabgrass, buckwheat, rice, and tall fescue leaves did not show any significant reduction in photosynthetic activity. The respiration in the leaves increased in all these species after the spraying of SAR at pH 2.5. Stomatal conductance did not vary appreciably. The chlorophyll content of the leaves sprayed with SAR decreased, and a positive correlation was observed between the ratio of chlorophyll a content in the treated leaves to that in the control leaves and the relative photosynthetic rate.

Photosynthesis was significantly affected by extremely low pH treatment. Since the lowest annual pH in East Asian regional precipitation is pH 4.2, as described previously, it appears that the current acidity of precipitation may not affect photosynthetic activities.

14.7 Effect on Growth

Kohno and Fujiwara (1981) sprayed sulfuric acid solution every morning, with a hand-held sprayer, at an amount to wet the whole plant surface, for 23 or 28 days in radish and 38 or 42 days in bush bean. The dry-weight growth of radish (cv. Comet) was not significantly decreased at pH 3.0, but it was significantly decreased at pH 2.0 compared with that in the plants sprayed with the solution at pH 3.0. Bush bean plants (cv. Wonder Crop 2) showed maximum dry-weight growth at pH 2.3, even though visible foliar injury developed all over the leaves. These results suggested that pH 2.6–3.0 is a boundary level at which foliar injury may or may not develop. The leaf weight ratio (LWR) in bush bean plants decreased with increasing acidity of the sulfuric acid solution; however, the net assimilation rate (NAR, leaf dry-weight basis) increased with increasing acidity in the later period of the experiment.

This could be the reason that bean plants did not show significant dry-weight growth reduction at pH 2.0, although these plants developed visible foliar injuries. However, the pH level at which growth reduction occurred was not consistent for this species.

Kohno and Kobayashi (1989a, b) and Kobayashi et al. (1991) exposed soybean (*Glycine max* cv. Early Hakucho or Sapporo Midori) to SAR using a rain-generator system. The main anion composition of the SAR was based on the monitoring data of precipitation in Japan, and the SAR was applied at a rate of 5 mm/h, twice a week for the cultivation period. When the SAR was below pH 3.0, plants developed visible symptoms in the young trifoliate leaves. However, at a pH above 3.0, there was no evidence of visible leaf injury; also, tissue dry weights and leaf areas were not affected even after 7 weeks of exposure. The number of root nodules in plants exposed to SAR at pH 4.0 tended to be higher than that in control plants maintained at pH 5.6, but the number decreased subsequently with increasing rain acidity.

Different plant species show different patterns of growth performance as affected by SAR or SAM. Turnips showed reduced plant dry weight with SAR or SAM at pH 2.5 (Kobayashi et al. 1994). Although SAM exposure induced visible injuries in plants earlier than SAR exposure did, white clover, alfalfa, Kentucky bluegrass, and perennial ryegrass showed significantly increased dry weight with increasing acidity in SAR or SAM. In contrast, pea exposed to SAR at pH 2.5 showed increased dry weight; however, with SAM at the same pH, no significant changes were shown. *Perilla* did not show reduced plant growth with SAR at pH 2.5, but growth was significantly increased with SAM at pH 2.5.

Grasses and crops possibly respond to nitrogen input with SAR or SAM that contains nitrate as one of the major anion components. Also, during the acid neutralization process in soil, soil releases available nutrient cations for plant growth. These processes could enhance plant growth (Kobayashi et al. 1991, 1992). Since rain droplets will strike and injure young seedlings in the early emergence stage, strong rain events with large droplet sizes, compared with mist or fog with smaller droplet sizes, may have a significant role in causing growth reduction due to mechanical damage.

The amount of precipitation with mist or fog is generally smaller than that with rain, and a duration of 9 h of exposure time for mist and rain with an extremely high acidity at pH 2.5 is quite unusual. In agricultural crop production, materials such as lime will generally be supplied to adjust soil pH to an appropriate range before plant cultivation. Therefore, the soil acidification stress associated with acidic precipitation at the current acidity level will not induce significant effects on crops with a relatively short rotation.

Hosono and Nouchi (1992) exposed radish, spinach, and bush bean to SAR throughout the entire growing periods. Plants were subjected to SAR events three times a week, at 10 mm/h per event. Plants exposed to SAR at pH 3.0 or below developed visible foliar injuries. The cotyledons of radish and spinach and the primary leaves of bush bean exposed to SAR at pH 2.7 and 2.5 showed severe injuries, which resulted in growth reduction in the early growth stage. However, in these three species, SAR at pH 3.0 or higher did not significantly affect the leaf area or whole dry weight of the plants. Hosono and Nouchi (1994) also exposed eight crop species to SAR throughout their entire growing periods. Plants were subjected to SAR events three times a week, at 7–13 mm/h per event. In the early growth stage, the dry weight of rice plants (cv. Nipponbare, *japonica*) exposed to SAR at pH below 3.0 was reduced compared with that in the control at pH 5.6. In the middle to the late growth stage, however, SAR treatment did not affect the rice plant growth.

In contrast, Zabawi et al. (2008) exposed the rice cultivar MR84 (*indica*) to SAR at a pH of 3.5 (or pH 5.6 as a control), once every 2 days throughout the growing period to assess the effects of SAR on growth. SAR had no adverse effect on plant height. However, the number of tillers produced per plant was significantly lower at pH 3.5 and this resulted in decreasing leaf numbers.

14.8 Effect on Yield

Taniyama and Saito (1979, 1981) reported that grain yield decreased with increasing acidity. Rice (cv. Koshihikari, *japonica*) had a larger yield reduction than barley, but there was no reduction in wheat yield at pH 2. Yield component analysis suggested that the yield reduction in rice was due to a decrease in the percentage of ripened grain.

Kohno and Fujiwara (1981) sprayed sulfuric acid solution every morning, with a hand-held sprayer, at an amount to wet the whole plant surface, for 38 or 42 days, in the bush bean. Pod yield was higher in plants exposed to pH 2.3 than in those exposed to pH 5.6, even though visible foliar injury developed all over the leaves at pH 2.3.

Kohno and Kobayashi (1989b) exposed soybean (*Glycine max* cv. Early Hakucho or Sapporo Midori) to SAR, using a rain-generator system. Dry seed yield in plants treated with SAR at pH 2.0, in three of the four experiments conducted over 3 years, was not significantly affected compared with yield at pH 5.6; however, plants exposed to SAR at pH 4.0 tended to yield more than those treated with pH 5.6 rain. Yield component analysis in the case that significant yield reduction at pH 2.0 was found, suggested that the significant reduction of seed dry weight at pH 2.0 was attributable to a significant decrease in the number of seeds per pod. Such a reduction in seed number at pH 2.0 resulted from an increasing number of empty seeds (Kobayashi et al. 1991). The authors concluded that rain acidity at current levels in Japan may not directly affect soybean yield.

14.9 Other Effects of Acid Rain

Kobayashi et al. (1992) analyzed changes in the elemental condition of cultivated soil and soybean plants exposed to SAR. The pH of the cultivated soil in the upper surface layer was significantly decreased with increasing acidity of SAR. Soluble basic cations, Mn and Zn, in the soil water were significantly increased after exposure to

SAR at pH 2.0. As a result of these changes, the uptake of the basic cations, Cu, Mn, and Zn, by the soybean plants increased. These results suggested that soil chemistry in the upper layer of cultivated soil may be changed significantly by SAR at pH 3.0 and below, but the effects of such changes on the growth and yield of soybean plants would not be important under ordinary manuring practices in agricultural fields.

Zhang et al. (2004) sprayed SAR onto the carmine spider mite, *Tetranychus cinnabarinus*, and its host plant, eggplant (*Solanum melongena*) to examine potential changes in three protective enzymes, peroxidase (POD), superoxide dismutase (SOD), and catalase (CAT) and three hydrolases, acid phosphatase (ACP), alkaline phosphatase (ALP), and carboxylesterase (CarE) in the mite. POD, SOD, and CAT activities in the mite increased significantly with increases in the acidity of SAR, reaching the highest levels at pH 4.0 or 3.0, and then declined. ACP activities were similar to those observed for the three protective enzymes. ALP activities decreased as the pH value declined and CarE decreased at pH 4.0 and 3.0 after 30 days. The authors pointed out that the enhanced levels of the anti-oxidative enzymes (POD, SOD, and CAT) and ACP at pH 4.0 and 3.0 reduced the effects of toxic products on the mites, resulting in a strengthening of their defensive capacity, and increases in their survival and reproductive capacity, thus leading to an increase in the density of mites on the host plant. Wang et al. (2004) reported that the carmine spider mite prefers to aggregate on the leaves of host eggplant treated with SAR at pH 4.0. Development times for the immature stages were shortened at pH 4.0 and survival at immature stages was enhanced. The mites feeding on leaves treated with SAR at pH 4.0 had significantly greater reproductivity and longevity than mites feeding on leaves treated with SAR at a lower pH. These results implied that the population growth of the mite on the host plant was enhanced by SAR at pH 4.0.

Wang et al. (2006) investigated the development, reproduction, and acaricide susceptibility of the carmine spider mite after long-term (about 40 generations) exposure to SAR. Results suggested that the mite suffered reproductive defects after long-term exposure to SAR at pH 2.5 and 3.0. The carmine spider mite is remarkably tolerant to a wide range of acidification conditions, and acid rain, particularly at pH 4.0–5.6, will definitely increase the infestation of the mite. Bioassay results showed that after long-term exposure to acid rain, the susceptibility of the mites to two acaricides, dichlorvos and fenpropathrin, did not change significantly.

Kim (2011) exposed garden balsam (*Impatiens balsamina*) to SAR at pH 2.0, 3.0, 4.0, and 5.6 (control) to investigate the effects of SAR on fatty acid composition and the plant's biochemical defense response. SAR markedly inhibited chlorophyll content. The level of H_2O_2 was significantly increased by SAR. As pH levels decreased from 5.6 to 2.0, the ratio of unsaturated to saturated fatty acids increased. The contents of putrescine and spermidine and the activity of spermidine and catalase in the leaves increased as the pH decreased. These results indicate that SAR generates oxidative stress and retards plant growth. A protective biochemical mechanism might be activated to neutralize the oxidative stresses generated by SAR.

Wang et al. (2012) conducted SAR exposure experiments in the field to test litter decomposition and the allelopathic potential of the invasive plant, *Wedelia trilobata* (creeping daisy). The impact of SAR acidity on the decomposition of creeping daisy litter and the allelopathic potential in the surrounding soil were determined. Litter decomposition was commensurate with increases in the acidity of SAR. Bioassays showed that SAR promoted allelopathic activity in the soil surrounding the litter, as measured by the seedling growth of turnip and radish. The authors concluded that SAR accelerated the decomposition of creeping daisy litter and enhanced the allelopathic potential of its litter in the surrounding soil, suggesting that acid rain may enhance the invasiveness of creeping daisy plants.

14.10 Interactive Effects of Acid Rain and Metals

Acid precipitation is an important factor in increasing the activity and mobility of heavy metals in soils. Liao et al. (2005) sprayed acid rain solution of pH 5.6–3.0 from the tops of transplanted bean plants into potted soil containing Cd^{2+} (0, 0.5–10 mg/kg) or Zn^{2+} (0, 20–3,000 mg/kg). The treatments of Cd or Zn and/or acid rain significantly decreased the fresh weight of the bean plants. Combinations of these three factors resulted in more serious toxic effects than any single factor or combination of any two factors. Hu et al. (2009) conducted field experiments under natural conditions to assess the effects of soil acidification produced by SAR on the availabilities of Cu and Pb. As SAR could decrease soil pH and increase available Cu and Pb, there is a potential threat that acid rain could lead to the accumulation of Cu and Pb in soils and the consequent uptake of these elements by plants. Thus, acid rain should be considered as a factor in food security.

Liu et al. (2011) investigated the combined effect of Cd^{2+} and acid rain on the seed germination of soybean. Single treatment with a low level of Cd^{2+} (0.18, 1.0, 3.0 mg L^{-1}) or acid rain did not affect the seed germination. Single treatment with the high concentration of Cd^{2+} (>6 mg L^{-1}) or acid rain at pH2.5 decreased the seed germination. However, combined treatment of Cd^{2+} and acid rain had a more toxic effect than the single treatment with Cd^{2+} or acid rain. Thus, the combination of Cd^{2+} and acid rain had a more potential threat to seed germination of soybean than the single effect of Cd^{2+} or acid rain. Sun et al. (2012) suggested that the combination of Cd^{2+} and acid rain exacerbated the toxic effects of Cd^{2+} or acid rain on photosynthetic parameters in the soybean, due to serious damage to its chloroplast structure.

Wen et al. (2011) investigated the combined effect of lanthanum ion (La^{3+}) and acid rain on soybean seedlings, as rare earth elements have been accumulated in the agricultural environment. Although the growth and photosynthesis of soybean seedlings treated with a low concentration of La^{3+} were improved, the growth and photosynthesis of soybean seedlings were inhibited in the combined treatment with the low concentration of La^{3+} and acid rain. The combination of La^{3+} and acid rain destroyed the chloroplast ultrastructure of cells and exacerbated the harmful effect of the single La^{3+} and acid rain treatments on soybean seedlings.

Liang and Wang (2013) found that a low concentration of La^{3+} (0.06 mmol L^{-1}) did not affect the activity of the antioxidant enzymes CAT and POD, whereas a high concentration of La^{3+} (0.18 mmol L^{-1}) did affect their activity. Compared with treatment with acid rain (pH 4.5 and 3.0) or La^{3+} alone, the joint stress of La^{3+} and acid rain affected the activity of CAT and POD more severely, and induced more H_2O_2 accumulation and lipid peroxidation. Further, Wang et al. (2014) investigated the combined effects of La^{3+} and acid rain on photosynthesis in rice. Combined treatment of 81.6 μM La^{3+} and acid rain at pH4.5 increased the net photosynthetic rate (*Pn*), stomatic conductance (*Gs*), intercellular CO_2 concentration (*Ci*), Hill reaction activity (HRA), apparent quantum yield (AQY), and carboxylation efficiency (CE) in rice. The combined treatment of 81.6 μM La^{3+} and acid rain at pH 3.5 decreased *Pn*, *Gs*, HRA, AQY, and CE and increased *Ci*.

These reports suggested that soil acidification due to acid precipitation at low pH may induce rare earth element and heavy metal toxicity in polluted agricultural areas.

14.11 Conclusion

Shriner et al. (1990) and Haines and Carlson (1989) summarized the types of visible injuries attributed to SAR in many plants and found that threshold pH values depended on the species, tissue age, and exposure conditions. Experiments with plant exposure to SAR presented both the acute and chronic effects of rain acidity, showing the acute effects of rain acidity on plant tops and the chronic effects of acid input. In general, extremely strong acid precipitation, such as that at pH 3.0 or below, may induce direct harmful effects on plants.

Assuming that the current acidity of rain in the East Asian area is pH 4.0, plants receiving SAR at pH 3.0 or pH 2.0 show acute responses to an acidity stress that are ten or a hundred times greater than those of acid rain. Surface soil receives an input of acid that is ten or a hundred times higher than the acidity of SAR used experimentally. However, acid rain at pH 3.0 or lower, as a strong acid, will be neutralized mainly by ion exchange processes. In addition, fertilizer, including lime, will be supplied to obtain better plant performance in the agricultural field.

As observations of rain acidity below pH 4.0 in the Asian area are rare, the exposure experiments with SAR or SAM suggest that the current acidity of precipitation may not cause direct adverse effects on plant growth and yield.

However, acidic precipitation may have a potential threat to induce increasing availability and toxicity of rare earth and heavy metals in plants.

References

EANET (2011) Second periodic report on the state of acid deposition in East Asia. Part I Regional assessment. Acid Deposition Monitoring Network in East Asia (EANET)

Fujiwara T, Umezawa T, Kohno Y (1980) Effects of acid rain on plants – literature review and results of preliminary experiments. CRIEPI Report 479015, Central Research Institute of Electric Power Industry. (In Japanese with English summary)

Haines BL, Carlson CL (1989) Effects of acidic precipitation on trees. P.1-27, In: Adriano DC, Johnson AH (eds) Acidic precipitation, Volume 2: Biological and ecological effects. Springer-Verlag, New York.
Hosono T, Nouchi I (1992) Effects of simulated acid rain on the growth of radish, spinach and bush bean plants. J Jpn Soc Air Pollut 27(3):111–121 (In Japanese with English summary)
Hosono T, Nouchi I (1994) Effects of simulated acid rain on growth, yield and net-photosynthesis of several crops. J Agr Met 50(2):121–127 (In Japanese with English summary)
Hu X, Hu C, Sun X, Lu M, Su B, Cao A (2009) Effects of simulated acid rain on soil acidification, availabilities and temporal and spatial variations of Cu and Pb in a vegetable field under natural condition. J Food Agric Environ 7(1):92–96
Kim HY (2011) Effect of simulated acid rain on fatty acid composition and antioxidant system in garden balsam (*Impatiens balsamina* L.). Kor J Weed Sci 31(2):152–159
Kobayashi T, Kohno Y, Nakayama K (1991) The effects of simulated acid rain on the growth and yield of soybean. J Agr Met 47(2):83–90 (In Japanese with English summary)
Kobayashi T, Kohno Y, Nakayama K (1992) The effects of simulated acid rain on the uptake of mineral elements in soybean plants. J Agr Met 48(1):11–18 (In Japanese with English summary)
Kobayashi T, Matsumura H, Kohno Y (1994) Effects of simulated acid rain and acid mist on development of visible injury and growth in plants. CRIEPI Report: U94025. (In Japanese with English summary)
Kohno Y (1994) Leaf surface observations of radish and bush bean plants exposed to sulfuric acid by cryo-scanning electron microscope. J Jpn Soc Air Pollut 29(2):71–79, In Japanese with English summary
Kohno Y, Fujiwara T (1981) Effects of acid rain on the growth and yield of radish and kidney bean. CRIEPI Report 481008, Central Research Institute of Electric Power Industry. (In Japanese with English summary)
Kohno Y, Fujiwara T (1982) Cryo-scanning electron microscopic observations and concentrations of inorganic elements in radish and kidney bean exposed to acid rain. CRIEPI Report 481014, Central Research Institute of Electric Power Industry. (In Japanese with English summary)
Kohno Y, Kobayashi T (1989a) Effect of simulated acid rain on the growth of soybean. Water Air Soil Pollut 43:11–19
Kohno Y, Kobayashi T (1989b) Effect of simulated acid rain on the yield of soybean. Water Air Soil Pollut 45:173–181
Kohno Y, Matsumura H, Kobayashi T (1994) Effect of simulated acid rain on the development of leaf injury in tree seedlings. J Jpn Soc Air Pollut 29(4):206–219 (In Japanese with English summary)
Kohno Y, Matsumura H, Kobayashi T (1995) Effect of simulated acid rain on the growth of Japanese conifers grown with or without fertilizer. Water Air Soil Pollut 85:1305–1310
Kohno Y, Matsuki R, Nomura S, Mitsunari K, Nakao M (2001) Neutralization of acid droplets on plant leaf surfaces. Water Air Soil Pollut 130:977–982
Lee Y, Park J, Im K, Kim K, Lee J, Lee K, Park J-A, Lee T-K, Park D-S, Yang J-S, Kim D, Lee S (2006) *Arabidopsis* leaf necrosis caused by simulated acid rain is related to the salicylic acid signaling pathway. Plant Physiol Biochem 44:38–42
Liang C, Wang W (2013) Antioxidant response of soybean seedlings to joint stress of lanthanum and acid rain. Environ Sci Pollut Res 20:8182–8191
Liao B, Liu H, Zeng Q, Yu P, Probst A, Probst J (2005) Complex toxic effects of Cd^{2+}, Zn^{2+}, and acid rain on growth of kidney bean (*Phaseolus vulgaris* L). Environ Int 31:891–895
Liu TT, Wu P, Wang LH, Zhou Q (2011) Response of soybean seed germination to cadmium and acid rain. Biol Trace Elem Res 144:1186–1196
Nouchi I (1992) Acid precipitation in Japan and its impact on plants. 1. Acid precipitation and foliar injury. JARQ 26:171–177
Okita T, Hashizume K (1975) Necrotic spots in petals of *Rhododendron* × *obtusum* by misty rain. Air Pollut News Rep Jpn Soc Air Pollut 89:1 (In Japanese)

Park J-B (2005) Effects of simulated acid rain on the shoot growth and internal tissue of *Arabidopsis thaliana*. J Life Sci 15(6):889–894 (In Korean with English summary)

Satoh M (1996) Effect of simulated acid rain on the physiological activities and chlorophyll content of leaves in some weeds and crops. Weed Res Jpn 41(4):310–314 (In Japanese with English summary)

Shriner DS, Heck WW, McLaughlin SB, Johnson DW, Irving PM, Joslin JD, Peterson CE (1990) Response of vegetation to atmospheric deposition and air pollution. NAPAP SOS/T Report 18. In: Acidic deposition: state of science and technology, vol. III. National Acid Precipitation Assessment Program, Washington, DC

Sun Z, Wang L, Chen M, Wang L, Liang C, Zhou Q, Huang X (2012) Interactive effects of cadmium and acid rain on photosynthetic light reaction in soybean seedlings. Ecotoxicol Environ Saf 79:62–68

Taniyama T, Saito H (1979) Effects of acid rain on apparent photosynthesis and grain yield of rice, barley and wheat plants. Rep Tokai Br Crop Sci Soc Jpn 84:61–64 (In Japanese with English summary)

Taniyama T, Saito H (1981) Studies on injurious effects of air pollutants on crop plants. XVI Effects of acid rain on apparent photosynthesis and grain yield of wheat, barley and rice plants. Rep Environ Sci Mie Univ 6:87–101

Tukey HB Jr (1970) The leaching of substances from plants. Ann Rev Plant Physiol 21:305–324

Wang J-J, Zhao Z-M, Zhang J-P (2004) The host plant-mediated impact of simulated acid rain on the development and reproduction of *Terranychus cinnabarinus* (Acari, Tetranychidae). JEN 128(6) doi:10.1111/j.1439-0418.2004.00856.397-402

Wang J-J, Zhang J-P, He L, Zhao Z-M (2006) Influence of long-term exposure to simulated acid rain on development, reproduction and acaricide susceptibility of the carmine spider mite, *Tetranychus cinnabarinus*. J Insect Sci 6:19, www.insectscience.org

Wang RL, Staehelin C, Dayan FE, Song YY, Su YJ, Zeng RS (2012) Simulated acid rain accelerates litter decomposition and enhances the allelopathic potential of the invasive plant *Wedelia trilobata* (Creeping Daisy). Weed Sci 60:462–467

Wang L, Wang W, Zhou Q, Huang X (2014) Combined effects of lanthanum (III) chloride and acid rain on photosynthetic parameters in rice. Chemosphere 112:355–361

Wen K, Liang C, Wang L, Hu G, Zhou Q (2011) Combined effects of lanthanum ion and acid rain on growth, photosynthesis and chloroplast ultrastructure in soybean seedlings. Chemosphere 84:601–608

Yoshida K (1971) Acid rain and morning glory. Air Pollut News Rep Jpn Soc Air Pollut 66:1 (In Japanese)

Zabawi AGM, Esa SM, Leong CP (2008) Effects of simulated acid rain on germination and growth of rice plant. J Trop Agric Fd Sc 36(2). http://ejtafs.mardi.gov.my/jtafs/36-2/Simulated%20acid%20rain.pdf

Zhang J-P, Wang J-J, Shao Z-M, Dou W, Chen Y (2004) Effects of simulated acid rain on the physiology of carmine spider mite, *Tetranychus cinnabrinus* (Boisduvals) (Acari: Tetranichidae). JEN 128(5) doi:10.1111/j.1439-0418.2004.00853.342-347

Chapter 15
Effects of Simulated Acid Rain on Asian Trees

Hideyuki Matsumura and Takeshi Izuta

Abstract We summarized the results obtained from experimental studies on the growth and physiological responses of Japanese and Chinese tree species to simulated acid rain, mist, or fog. Based on the studies conducted in Japan and China, exposure to simulated acid rain, mist, or fog at pH 4.0 or above for one to three growing seasons did not induce any adverse effects on the growth and physiological functions of Japanese and Chinese tree species. Current acidic rain monitored in Japan and China, where the lowest acidity of ambient precipitation was around pH 4.0, was not found to induce any direct significant effects in the tree species examined.

Keywords Acid rain • Acid mist • Acid fog • Japanese tree species • Chinese tree species • Growth

15.1 Introduction

Petal discoloration of morning glory plants in the Kanto and Kinki districts of Japan was observed after precipitation in 1971, and this prompted the start of Japanese research activity. Because the dieback of trees grown in Japan or in China might

---

H. Matsumura (✉)
Central Research Institute of Electric Power Industry,
1646 Abiko, Abiko City, Chiba 270-1194, Japan
e-mail: matz@criepi.denken.or.jp

T. Izuta
Institute of Agriculture, Tokyo University of Agriculture and Technology,
Fuchu, Tokyo 183-8509, Japan
e-mail: izuta@cc.tuat.ac.jp

T. Izuta (ed.), *Air Pollution Impacts on Plants in East Asia*,
DOI 10.1007/978-4-431-56438-6_15

have a possible link with acid precipitation, field survey and experimental studies were proposed. In this chapter, we summarize recent experimental studies on the effects of simulated acid rain (SAR), mist, or fog on the development of visible foliar injury, and on the growth and physiology (factors such as gas exchange rates) of tree species grown in Japan and China.

15.2 Visible Foliar Injury

The direct effects of SAR on trees are visible foliar injuries, such as necrotic spots, marginal/tip necrosis, and defoliation (Haines and Carlson 1989; Shriner et al. 1990). But sensitivities to SAR based on visible foliar injury are different among tree species. Kohno et al. (1994) investigated the development of visible foliar injuries induced by SAR in a single-season experiment (Table 15.1). Seedlings of 46 tree species (11 conifers: *Pinus densiflora*, *P. thunbergii*, *P. strobus*, *Larix kaempferi*, *Picea abies*, *Abies firma*, *A. homolepis*, *Cryptomeria japonica*, *Chamaecyparis obtusa*, *Ch. pisifera*, and *Juniperus chinensis*; 21 deciduous broad-leaved tree species: *Populus suaveolens*, *Alnus firma*, *Betula platyphylla* var. *japonica*, *Fagus crenata*, *Quercus serrata*, *Q. mongolica* var. *crispula*, *Zelkova serrata*, *Liriodendron tulipifera*, *Hydrangea macrophylla*, *Prunua jamasakura*, *Prunus* x *yedoensis*, *Armeniaca mume*, *Armeniaca vulgaris*, *Spiraea cantoniensis*, *Cytisus scoparius*, *Cornus florida*, *Lespedea thunbergii*, *Acer buergerianum*, *Enkianthus perulatus*, *Syringa vulgaris*, and *Fraxinus japonica*; and 14 evergreen broad-leaved tree species: *Myrica rubra*, *Q. phillyraeoides*, *Castanopsis sieboldii*, *Lithocarpus edulis*, *Machilus thunbergii*, *Camellia japonica*, *Pittosporum tobira*, *Pyracantha coccinea*, *Rhaphiolepis umbellata*, *Photinia glabra*, *Euonymus japonicus*, *Rhododendron indicum*, *R. oomurasaki, and Ligustrum japonicum*) were exposed to SAR (5:2:3

Table 15.1 Visible foliar injuries in 46 tree species exposed to simulated acid rain

			pH				
		Scientific name	5.6	4.0	3.0	2.5	2.0
Conifers	Evergreen	*Pinus densiflora*	(2)	–	–		+
		Pinus thunbergii	(1)	–	–		+
		Pinus strobus	(3)	–	–	–	
		Picea abies	(2)	–	–	–	+
		Abies firma	(2)	–	–		+
		Abies homolepis	(3)	–	–	–	+
		Cryptomeria japonica	(3)	–	–	–	+
		Chamaecyparis obtusa	(1)	–	–		+
		Chamaecyparis pisifera	(1)	–	–		+
		Juniperus chinensis 'Kaizuka'	(1)	–	–		+
	Deciduous	*Larix kaempferi*	(3)	–	–	+	Defoliated

		Scientific name	pH 5.6	4.0	3.0	2.5	2.0
Broad-leaves	Evergreen	*Myrica rubra*	(1)	–	–		+
		Quercus phillyreoides	(1)	–	–		+
		Castanopsis sieboldii	(1)	–	–		+
		Lithocarpus edulis	(1)	–	–		+
		Machilus thunbergii	(1)	–	–		+
		Camellia japonica	(1)	–	+		+
		Pittosporum tobira	(2)	–	+		+
		Pyracantha coccinea	(1)	–	+		+
		Rhaphiolepis indica var. *umbellata*	(1)	–	+		+
		Photinia glabra	(1)	–	+		+
		Euonymus japonicus	(1)	–	+		+
		Rhododendron indicum	(1)	–	–		+
		Rhododendron oomurasaki	(1)	–	–		+
		Ligustrum japonicum	(1)	–	+		+
	Deciduous	*Populus suaveolens*	(3)	–	–	+	
		Alnus firma	(1)	–	–		+
		Betula platyphylla var. *japonica*	(3)	–	–	+	Defoliated
		Quercus serrata	(3)	–	+	+	+
		Quercus mongolica var. *crispula*	(3)	–	+	+	
		Fagus crenata	(3)	–	+	+	+
		Zelkova serrata	(3)	–	–	+	
		Liriodendron tulipifera	(3)	–	+	+	
		Hydrangea macrophylla f. *macrophylla*	(1)	–	+		Defoliated
		Prunus jamasakura	(4)	–	+	+	
		Prunus x *yedoensis*	(1)	–	+		Defoliated
		Armeniaca mume	(4)	–	+	+	
		Armeniaca vulgaris	(1)	–	+		Defoliated
		Spiraea cantoniensis	(1)	–	–		+
		Lespedeza thunbergii	(1)	–	+		Dead
		Cytisus scoparius	(1)	–	–		Dead
		Acer buergerianum	(3)	–	–	+	Defoliated
		Cornus florida	(2)	–	+		Defoliated
		Enkianthus perulatus	(2)	–	+		Dead
		Syringa vulgaris	(1)	–	+		Defoliated
		Fraxinus japonica	(3)	–	+	+	

After Kohno et al. (1994)
–, no visible injury developed. +, visible injury developed, *blank space* not examined
Precipitation: 2.5 mm/h × 8 h/day × three times/week
(1) From July to October 1991, *(2)* from April to June 1992, *(3)* from April to June 1993, *(4)* from July to September 1993

equivalent mixture of sulfate, nitrate, and chloride) of pH 2.0, 2.5, 3.0, or 4.0, or deionized water (control, pH 5.6), 20 mm/event, three events/week for 3–4 months in a greenhouse. SAR at pH 2.0 induced visible necrotic foliar injuries or defoliation in all species examined. Japanese larch (*Larix kaempferi*), which is the only deciduous species in the 11 conifers examined, was classified as the most sensitive conifer species to SAR because it defoliated at pH 2.0 and developed leaf tip discoloration at pH 2.5. The other conifers did not show any visible symptoms at pH 3.0 or higher. Seven of the 14 evergreen broad-leaved tree species and 14 of the 21 deciduous broad-leaved tree species showed injuries at pH 3.0. Of the deciduous broad-leaved tree species exposed to SAR at pH 2.0, 10 species were almost completely defoliated or had died. However, evergreen broad-leaved trees did not show such severity of the symptoms at pH 2.0. No species showed any visible injury symptoms after exposure to SAR of pH 4.0.

In a multiple-season exposure experiment, Kobayashi et al. (1997) reported the development of visible foliar injuries in 16 tree species (8 conifers: *Pinus densiflora*, *P. thunbergii*, *P. strobus*, *Larix kaempferi*, *Picea abies*, *Abies homolepis*, *Cryptomeria japonica*, *Chamaecyparis obtusa*, and 8 deciduous broad-leaved tree species: *Populus suaveolens*, *Betula platyphylla* var. *japonica*, *Fagus crenata*, *Quercus serrata*, *Q. mongolica* var. *crispula*, *Liriodendron tulipifera*, *Acer buergerianum*, and *Fraxinus japonica*). Seedlings were exposed to SAR (5:2:3 equivalent mixture of sulfate, nitrate, and chloride) of pH 2.5, 3.0, or 4.0, or deionized water (control, pH 5.6) for three growing seasons (27 months). Visible foliar injuries were observed in all 16 tree species exposed to SAR at pH 2.5. All 8 deciduous broad-leaved trees and only one deciduous conifer (*Laris kaempferi*) developed foliar injuries at pH 3.0. SAR at pH 4.0 caused slight visible injuries on the leaves of Japanese oak (*Quercus mongolica* var. *crispula*). None of the 7 evergreen conifers showed any visible foliar injury symptoms at pH 3.0 or higher. This result showed almost the same tendency as that shown in the single-season exposure experiment (Kohno et al. 1994).

Based on the results of the experimental studies described above, it appears that deciduous broad-leaved trees and deciduous conifers are relatively sensitive to acid rain, with a threshold pH between 3.0 and 4.0 that induces visible foliar injury development, as compared with evergreen broad-leaved trees and conifers, whose threshold pH for this development was between 2.5 and 3.0. Current acidic rain monitored in Japan, where the lowest acidity of ambient precipitation is around pH 4.0, did not induce any direct significant effects on the tree species examined.

15.3 Growth Response

Several experimental studies have reported the effects of SAR on the growth of Japanese and Chinese tree species. Izuta et al. (1990, 1993), Matsumura et al. (1995), and Watanabe et al. (1999) reported the growth responses of Japanese trees to SAR in single-season exposure experiments. Izuta et al. (1990) reported that exposure to SAR at pH 2.0 (15 mm/event, two events/week) for 2 months induced a reduction in the whole-plant dry mass per seedling of Japanese cedar (*Cryptomeria japonica*), but there were no significant effects on the growth of the seedlings exposed to pH 3.0. Izuta et al. (1993) reported that Japanese fir (*Abies firma*) was more sensitive to SAR than Japanese cedar, as shown by their findings that the dry mass of Japanese fir seedlings was reduced by exposure to SAR at pH 3.0 or above (14.5 mm/event, two events/week) for 30 weeks (Fig. 15.1).

Matsumura et al. (1995) investigated the effects of SAR of pH 2.0, 3.0, or 4.0 (5:2:3 equivalent mixture of sulfate, nitrate, and chloride, 18–20 mm/event, three events/week) or deionized water (control, pH 5.6) for 20 weeks on the dry mass of Japanese cedar, Nikko fir (*Abies homolepis*), and Japanese white birch (*Betula platyphylla* var. *japonica*) seedlings (Fig. 15.2). The whole-plant dry mass per seedling of all three tree species examined was significantly reduced by exposure to SAR at pH 2.0, as compared with the control seedlings. Furthermore, the whole-plant dry mass per fir seedling at pH 3.0 was significantly less than that of the control. On the other hand, Watanabe et al. (1999) reported that 20 weeks of exposure to SAR at pH 3.0 and pH 2.5 reduced the needle dry mass and increased the root dry mass per seedling of silver fir (*Abies veitchii*) as compared with the control, and as a result, the whole-plant dry mass per seedling was not significantly different among the treatments. Therefore, it appears that Japanese fir and Nikko fir are more sensitive than other Japanese conifers to acid rain, in terms of their growth inhibition.

Fan and Wang (2000), Liang et al. (2013), Shan et al. (1994, 1996), and Tong and Zhang (2014) reported the growth responses of trees grown in China to SAR in single-season exposure experiments. Shan et al. (1996) reported that exposure to SAR at pH 2.3 or 3.0 (8:1 equivalent mixture of sulfate and nitrate, six events/week) for 14 weeks induced no reduction in the whole-plant dry mass per seedling of Chinese white pine (*Pinus armandii*) as compared with findings for the control

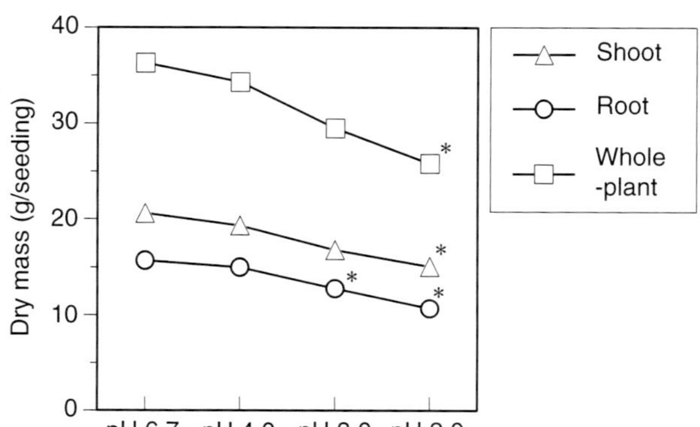

Fig. 15.1 Shoot, root, and whole-plant dry masses per seedling of Japanese fir after 30 weeks of exposure to simulated acid rain. *Asterisks* indicate that the means ($n=8$) are significantly different, at the 0.05 level, from the rain treatment at pH 6.7 (Data from Izuta et al. 1993)

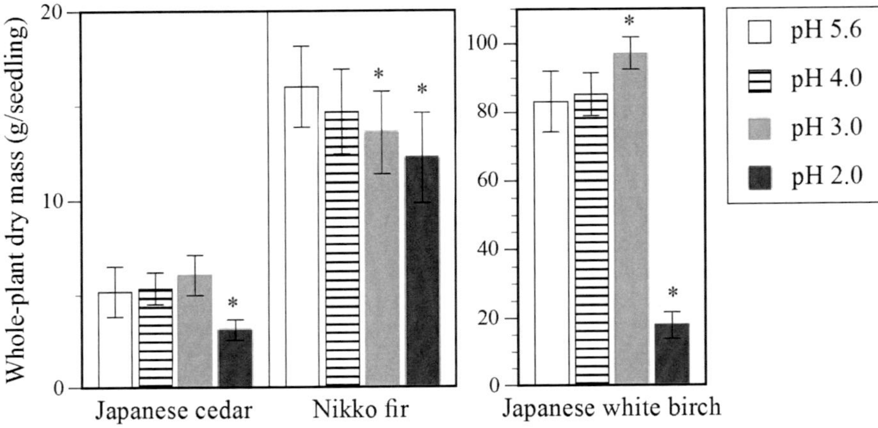

Fig. 15.2 Whole-plant dry masses per seedling of Japanese cedar, Nikko fir, and Japanese white birch after 20 weeks exposure to simulated acid rain. *Asterisks* indicate that the means ($n = 14$ for cedar and fir, $n = 10$ for birch) are significantly different, at the 0.05 level, from the rain treatment of pH 5.6 (Data from Matsumura et al. 1995)

(rain, pH 6.7). Shan et al. (1994) investigated the effects on the dry mass of seedlings of three tree species (two conifers: *Pinus massoniana* and *Cunninghamia lanceolata*, and one evergreen broad-leaved tree species: *Michelia macclurei*) after 6 months exposure to SAR of pH 2.0, 3.0, or 4.5 (8:1 equivalent mixture of sulfate and nitrate). The leaf and root dry mass per seedling of Chinese red pine (*Pinus massoniana*) and Chinese fir (*Cunninghamia lanceolata*) was significantly reduced by exposure to SAR at pH 2.0 as compared with the control (well water, pH 6.6), but SAR exposure at pH 2.0 did not reduce the dry mass of *Michelia macclurei* seedlings. Liang et al. (2013) reported that 4 months exposure to SAR (12:1 equivalent mixture of sulfate and nitrate) at pH 2.5, 3.5, or 4.5 did not affect the shoot, root, or whole-plant dry mass per seedling of Mongolian oak (*Quercus mongolica*) as compared with the control (pH 5.6).

Tong and Zhang (2014) investigated the effects on seedling height and stem diameter in five tree species (one conifer: *Pinus massoniana* (Chinese red pine), and four evergreen broad-leaved tree species: *Castanopsis sclerophylla*, *Cinnamomum camphora*, *Manglietia fordiana*, and *Elaeocarpus glabripetalus*) after exposure to SAR of pH 2.5 or 4.0 (2:1 equivalent mixture of sulfate and nitrate, 2.2 mm/event, two events/day for 5 days). The seedlings of Chinese red pine exhibited significant reduction in height and stem diameter at pH 2.0 as compared with the control (pH 5.6), but SAR exposure at pH 2.0 did not reduce the growth of the four evergreen broad-leaved tree species. Fan and Wang (2000) reported that 6 months exposure to SAR (10:1 equivalent mixture of sulfate and nitrate) at pH 2.0 reduced the shoot and root dry mass per seedling in five tree species (three evergreen broad-leaved: *Castanopsis fissa*, *Cinnamomum camphora, and Ligustrum japonicum*, and two deciduous broad-leaved: *Melia azedarach* and *Koelreuteria bipinnata*) as compared with the control (distilled water), but no significant reduction by SAR at pH 3.5 or higher was observed.

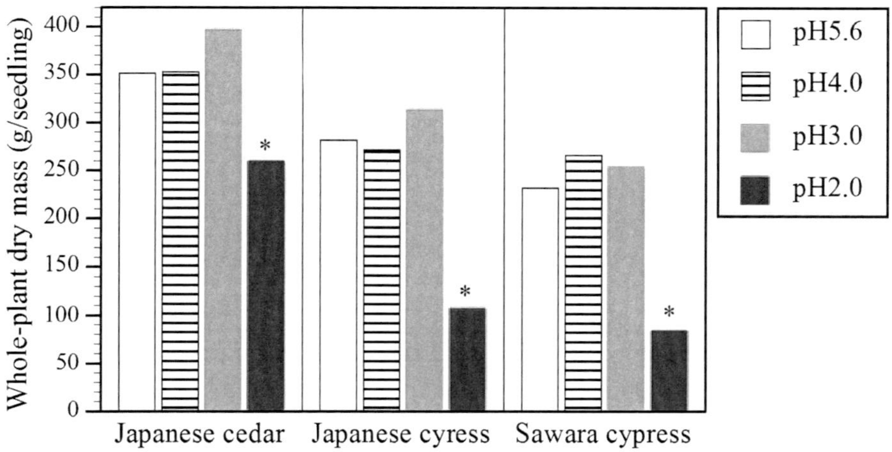

Fig. 15.3 Whole-plant dry masses per seedling of Japanese cedar, Japanese cypress, and Sawara cypress, grown in volcanic ash soil (andosol) with added fertilizer, after 23 months exposure to simulated acid rain of pH 4.0, 3.0, or 2.0, or deionized water (pH 5.6). Precipitation per season, from April to September, was 1440 mm. *Asterisks* indicate that the means ($n = 12$) are significantly different, at the 0.05 level, from the rain treatment at pH 5.6 (Data from Kohno et al. 1995)

Very limited information is available on the growth responses of Japanese trees to SAR in multiple-season exposure experiments. Kohno et al. (1995) investigated the growth responses of Japanese conifers to SAR. Seedlings of Japanese cedar, Japanese cypress (*Chamaecyparis obtusa*), and Sawara cypress (*Chamaecyparis pisifera*) were exposed to SAR (5:2:3 equivalent mixture of sulfate, nitrate, and chloride) of pH 2.0, 3.0, or 4.0, or deionized water (control, pH 5.6), 10–30 mm/event, three times per week in the warm season (April–September) and 5 mm/event, two times/week in the cold season (October–March) for two seasons (23 months) in greenhouses. The seedlings of all conifers showed reduced whole-plant dry mass per seedling after exposure to SAR at pH 2.0, but no growth reduction occurred in any conifer seedlings after 23 months exposure to SAR at pH 3.0 or 4.0 (Fig. 15.3).

Kobayashi et al. (1997) reported the growth responses in eight conifers (*Pinus densiflora*, *P. thunbergii*, *P. strobus*, *Larix kaempferi*, *Picea abies*, *Abies homolepis*, *Cryptomeria japonica*, and *Chamaecyparis obtusa*) and eight deciduous broad-leaved tree species (*Populus suaveolens*, *Betula platyphylla* var. *japonica*, *Fagus crenata*, *Quercus serrata*, *Q. mongolica* var. *crispula*, *Liriodendron tulipifera*, *Acer buergerianum*, and *Fraxinus japonica*) to SAR (5:2:3 equivalent mixture of sulfate, nitrate, and chloride) of pH 2.5, 3.0, or 4.0, or deionized water (control, pH 5.6), 20 mm/event, three events/week in the warm season (April–September) and 5 mm/event, two events/week in the cold season (October–March) for three growing seasons (28 months) (Fig. 15.4). Although one conifer (Japanese black pine [*P. thunbergii*]) showed significant reduction in the whole-plant dry mass per seedling, the other seven conifers did not show a significant reduction by SAR at pH 2.5 as compared with the control (pH 5.6). No growth reduction occurred in the seedlings of

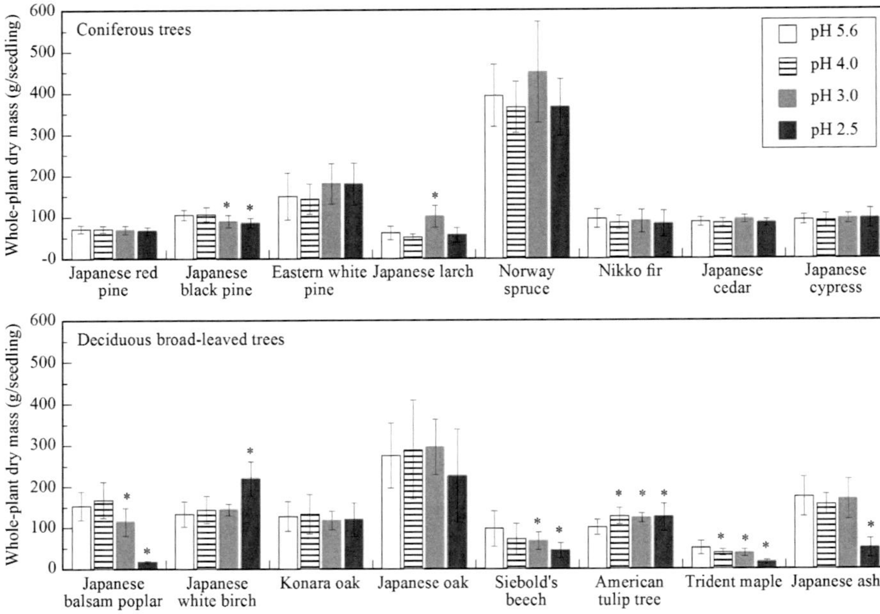

Fig. 15.4 Whole-plant dry masses per seedling of eight coniferous and eight deciduous broad-leaved tree species after three growing seasons (28 months) of exposure to simulated acid rain of pH 4.0, 3.0, or 2.5, or deionized water (pH 5.6). *Asterisks* indicate that the means ($n = 10$) are significantly different, at the 0.05 level, from the rain treatment at pH 5.6 (After Kobayashi et al. 1997)

any of the eight conifers with exposure to SAR at pH 3.0 or higher. In contrast, the deciduous broad-leaved trees had a wide variation in growth response to SAR. Exposure to SAR at pH 2.5 induced significant reductions in the whole-plant dry mass per seedling of four deciduous broad-leaved tree species – Japanese balsam poplar (*Populus suaveolens*), Siebold's beech (*Fagus crenata*), trident maple (*Acer buergerianum*), and Japanese ash (*Fraxinus japonica*). Conversely, exposure to SAR at pH 2.5 significantly increased the whole-plant dry mass per seedling of Japanese white birch (*Betula platyphylla* var. *japonica*) as compared with the control.

There are a few reports concerning the effects of simulated acid mist or fog on Japanese trees. Matsumura (2001) reported no growth reduction in any of the 14 tree species they examined (9 conifers: *Pinus densiflora*, *P. thunbergii*, *Larix kaempferi*, *Picea abies*, *Abies firma*, *A. homolepis*, *A. veitchii*, *Cryptomeria japonica*, and *Chamaecyparis obtusa*; and 5 deciduous broad-leaved tree species: *Populus suaveolens*, *Betula platyphylla* var. *japonica*, *Fagus crenata*, *Quercus mongolica* var. *crispula*, and *Zelkova serrata*) in response to simulated acid mist (SAM; 1:2:1 equivalent mixture of sulfate, nitrate, and chloride) of pH 3.0, three events/week from June to November for three growing seasons, as compared with SAM at pH 5.0. On the other hand, Igawa et al. (1997) reported needle loss and reduction in seedling height in Japanese fir seedlings exposed to simulated acid fog (1 mM nitric acid, 1 mM sodium chloride, and 1 mM ammonium sulfate) at pH 3.0, as compared with the control (pH 5.0). Sugihara et al. (2008) also reported that exposure to simulated acid fog reduced growth and physiological activity in Siebold's beech seedlings.

15.4 Physiological Responses

Limited information is available on the physiological responses of Japanese or Chinese tree species to SAR. Izuta et al. (1993) investigated the effect of SAR on the dark respiration rate of Japanese fir seedlings. Exposure to SAR at pH 2.0 (2:1 equivalent mixture of sulfate and nitrate, 14.5 mm per event, two events per week) for 30 weeks increased the dark respiration rate of Japanese fir needles as compared with the rate in control rain (deionized water, pH 6.7). Matsumura et al. (1995) reported increased dark respiration rates in the leaves of three species (*Abies homolepis, Cryptomeria japonica, and Betula platyphylla* var. *japonica*) and a decreased net photosynthetic rate in Japanese white birch leaves induced by exposure to SAR at pH 2.0 (5:2:3 equivalent mixture of sulfate, nitrate, and chloride, 18–20 mm per event, three events per week) for 20 weeks, as compared with findings in control rain (deionized water, pH 5.6). Exposure to SAR at pH 3.0 or lower induced an increase of the dark respiration rate in the seedlings of three conifers (Japanese cedar, Japanese cypress, and Sawara cypress) (Kohno et al. 1995), Nikko fir (Matsumura et al. 1998), silver fir (Watanabe et al. 1999), Chinese red pine and Chinese fir (Shan et al. 1994), and Chinese red pine (Tong and Zhang 2014). Therefore, increased dark respiration is considered to be a factor related to SAR-induced growth reduction in these species.

Chen et al. (2013) and Liu et al. (2007) reported the photosynthetic responses of Chinese trees to SAR. Chen et al. (2013) reported decreased net photosynthetic rates in Chinese sweetgum (*Liquidambar formosana*) and Chinese guger tree (*Schima superba*) leaves after exposure to SAR at pH 3.0 (containing sulfate anions only, two events/day for 4 days), as compared with the findings in the control (pH 6.0). Liu et al. (2007) reported decreased net photosynthetic rates in Chinese guger tree and Chinese red pine seedlings induced by exposure to SAR at pH 3.0 and 4.0 (2:1 equivalent mixture of sulfate and nitrate, one event/week) for 1 month, as compared with the findings in the control (tap water, pH 4.3).

15.5 Conclusion

Based on the results of the experimental studies described above, it appears that exposure to simulated acid rain, mist, or fog at pH 4.0 or above for one to three growing seasons does not have any adverse effects on the growth and physiological functions of Japanese and Chinese tree species.

Current acidic rain monitored in Japan and China, where the lowest acidity of ambient precipitation was around pH 4.0, did not have any direct significant effects on the tree species examined.

References

Chen J et al (2013) Photosynthetic and antioxidant responses of *Liquidambar formosana and Schima superba* seedlings to sulfuric-rich and nitric-rich simulated acid rain. Plant Physiol Biochem 64:41–51

Fan HB, Wang YH (2000) Effects of simulated acid rain on germination, foliar damage, chlorophyll contents and seedling growth of five hardwood species growing in China. For Ecol Manage 126:321–329

Haines BL, Carlson CL (1989) Effects of acidic precipitation on trees. In: Adriano DC, Johnson AH (eds) Acidic precipitation Vol. 2. Biological and ecological effects. Springer, New York, pp 1–27

Igawa M et al (1997) Effect of simulated acid fog on needles of fir seedlings. Environ Exp Bot 38:155–163

Izuta T et al (1990) Growth responses of *Cryptomeria* seedlings to simulated acid rain. Man Environ 16:44–53 (in Japanese with English abstract, figures and tables)

Izuta T et al (1993) Effects of simulated acid rain on the growth of fir seedlings. J Jpn Soc Air Pollut 28:29–37 (in Japanese with English abstract, figures and tables)

Kobayashi T, Kohno Y, Matsumura H (1997) Effects of simulated acid rain on tree species. In: Kohno Y (ed) Proceedings of CRIEPI International Seminar on Transport and Effects of Acidic Substances, November 28–29, 1996. CRIEPI, Tokyo, pp 184–189

Kohno Y, Matsumura H, Kobayashi T (1994) Effect of simulated acid rain on the development of leaf injury in tree seedlings. J Jpn Soc Air Pollut 29:206–219 (in Japanese with English abstract, figures and tables)

Kohno Y, Matsumura H, Kobayashi T (1995) Effect of simulated acid rain on the growth of Japanese cedar, Japanese cypress and Sawara cypress. J Jpn Soc Atmos Environ 30:191–207 (in Japanese with English abstract, figures and tables)

Liang X et al (2013) Effects of simulated acid rain on growth and bleeding sap amount of root in Quercus mongolica. Acta Ecologica Sinica 33:4583–4590 (in Chinese with English abstract, figures and tables)

Liu J et al (2007) Responses of chlorophyll fluorescence and xanthophyll cycle in leaves of *Schima superba* Gardn. & Champ. and *Pinus massoniana* Lamb. to simulated acid rain at Dinghushan Biosphere Reserve, China. Acta Physiol Plant 29:33–38

Matsumura H (2001) Impacts of ambient ozone and/or acid mist on the growth of 14 tree species: an Open-top Chamber study conducted in Japan. Water Air Soil Pollut 130:959–964

Matsumura H et al (1995) Effects of simulated acid rain on dry weight growth and gas exchange rate of Japanese cedar, Nikko fir and Japanese white birch seedlings. J Jpn Soc Atmos Environ 30:180–190 (in Japanese with English abstract, figures and tables)

Matsumura H, Kobayashi T, Kohno Y (1998) Effects of ozone and/or simulated acid rain on dry weight growth and gas exchange rates of Japanese cedar, Nikko fir, Japanese white birch and Japanese zelkova seedlings. J Jpn Soc Atmos Environ 33:16–35 (in Japanese with English abstract, figures and tables)

Shan Y et al (1994) Ecophysiological effects of simulated acid rain on saplings of *Pinus massoniana*, *Cunninghamia lanceolata*, and *Michelia macclurei*. Environ Sci 2:209–215

Shan Y et al (1996) The individual and combined effects of ozone and simulated acid rain on growth, gas exchange rate and water use efficiency of *Pinus armandii* Franch. Environ Pollut 91:355–361

Shriner DS et al (1990) Response of vegetation to atmospheric deposition and air pollution. NAPAP SOS/T report 18. In: Acidic deposition: state of science and technology, vol III. National Acid Precipitation Assessment Program, Washington, DC

Sugihara A et al (2008) Growth and physiological responses of beech seedlings to long-term exposure of acid fog. Sci Total Environ 391:124–131

Tong S, Zhang L (2014) Differential sensitivity of growth and net photosynthetic rates in five tree species seedlings under simulated acid rain stress. Pol J Environ Stud 23:2259–2264

Watanabe T et al (1999) Effects of simulated acid rain on growth, gas exchange rates and nutrient status of *Abies veitchii* seedlings. J Jpn Soc Atmos Environ 34:407–421 (in Japanese with English abstract, figures and tables)

Chapter 16
Combined Effects of Simulated Acid Rain and Other Environmental Factors on Asian Trees

Hideyuki Matsumura and Takeshi Izuta

Abstract We summarized the results obtained from experimental studies on the combined effects of concurrent or sequential exposure to simulated acid rain, mist, or fog, and ozone, on the growth and physiology of Japanese and Chinese tree species. Based on the limited number of studies conducted in Japan and China, we found that the combined effects of acid precipitation and ozone on Japanese and Chinese tree species were additive or greater than additive. Moreover, there is a possibility that acid precipitation with a relatively low acidity increased the extent of ozone-induced inhibition of assimilate translocation from the shoot to roots, and this ozone effect on trees exacerbated by acid precipitation may be associated with nitrate input from precipitation.

Keywords Acid rain • Acid mist • Acid fog • Ozone • Combined effect • Japanese tree species • Chinese tree species • Growth

16.1 Introduction

The decline of trees such as Japanese red pine (*Pinus densiflora*), Japanese larch (*Larix kaempferi*), Japanese fir (*Abies firma*), Japanese cedar (*Cryptomeria japonica*), Japanese white birch (*Betula platyphylla* var. *japonica*), Japanese oak (*Quercus*

H. Matsumura (✉)
Central Research Institute of Electric Power Industry,
1646 Abiko, Abiko City, Chiba 270-1194, Japan
e-mail: matz@criepi.denken.or.jp

T. Izuta
Institute of Agriculture, Tokyo University of Agriculture and Technology,
Fuchu, Tokyo 183-8509, Japan
e-mail: izuta@cc.tuat.ac.jp

T. Izuta (ed.), *Air Pollution Impacts on Plants in East Asia*,
DOI 10.1007/978-4-431-56438-6_16

mongolica var. *crispula*), and Siebold's beech (*Fagus crenata*) in Japan, and the decline of Chinese red pine (*Pinus massoniana*) and Chinese white pine (*Pinus armandii*) in China may possibly be attributable to the co-occurrence of acid precipitation and gaseous air pollutants, such as ozone, in forests. Accordingly, it is necessary to investigate the combined effects of acid rain, mist, or fog, and ozone, on tree species grown in East Asia.

As described in the previous chapter, based on the results of experimental studies, exposure to simulated acid rain, mist, or fog at pH 4.0 or above did not have any adverse effects on the growth and physiological functions of the Japanese and Chinese tree species examined. On the other hand, many studies of the effects of ambient ozone on the juvenile trees of forest species native to Japan, North America, or Europe have demonstrated decreased biomass in sensitive tree species (Wittig et al. 2009; Izuta et al. 2001). However, experimental evidence of significant interaction between acid precipitation and ozone has been contradictory in conifers (pine, spruce) and deciduous broad-leaved tree species (oak, maple, and ash) native to North America or Europe (Temple 1988).

To date, limited information is available on the responses of Asian trees to the combined effects of these pollutants. In studies of multiple-season exposure under field conditions, much less attention has been paid to these combined effects than to their single effects. In this review, we describe experimental studies on the effects of simulated acid rain, mist, or fog, in combination with ozone, on the growth and physiology of Japanese and Chinese tree species.

16.2 Combined Effects of Simulated Acid Rain and Ozone

There are several reports concerning the combined effects of simulated acid rain (SAR) and ozone on Japanese and Chinese tree species. Miwa et al. (1993) investigated the combined effects of SAR and ozone on dry-matter production in Japanese cedar seedlings. The seedlings were exposed to sixteen factorial combinations of four rain pHs (pH 7.0 [deionized water, control], or pH 4.5, 3.5, or 2.5 [2:1 equivalent mixture of sulfate and nitrate], 15 mm/event, two events/week) and four levels of ozone (charcoal-filtered air [at or below one-third of the ambient ozone level], or 0.1, 0.2, or 0.3 ppm, 4 h/day, 3 days/week) in a single growing season (12 weeks) in open-top chambers with an exclusion cover for ambient rain. There were no significant effects of SAR or ozone treatment, or interactions between the two treatments, on the shoot, root, or whole-plant dry masses in the Japanese cedar. The ratio of the dry mass of shoot to roots (shoot/root ratio) in the seedlings exposed to the highest ozone concentration or rain acidity was significantly enhanced compared with the ratio in seedlings exposed to the lowest ozone concentration or rain acidity, but there were no significant interactive effects. Shan et al. (1996) exposed Chinese white pine seedlings to combinations of SAR (pH 6.7 [control] or 3.0 or 2.3 [16:1 equivalent mixture of sulfate and nitrate, six events/week) and ozone (charcoal-filtered air [CF] or 0.3 ppm, 8 h/day, 6 days/week in an indoor exposure chamber) in one growing season (14 weeks), and reported that high ozone treatment significantly reduced the root and whole-plant dry mass and the net needle photosynthetic rate, and enhanced the shoot/root ratio as compared with findings for CF ozone, but no significant effects of interactions between SAR and ozone treatment were observed on these parameters.

On the other hand, Kohno et al. (1995) reported that a significant effect of interaction between SAR and ozone treatment was observed on the shoot/root ratio in Japanese cedar, but not on the shoot/root ratio in Japanese cypress (*Chamaecyparis obtusa*) and Sawara cypress (*Chamaecyparis pisifera*), when the seedlings were exposed to various combinations of SAR (pH 5.6 [deionized water, control] or pH 3.0 [5:2:3 equivalent mixture of sulfate, nitrate, and chloride], 20 mm/event, three events/week from April to September; and 10 mm/event, two events/week from October to March) and ozone (0, 0.06, 0.12, or 0.18 ppm, 6 h/day, daily) for two growing seasons (23 months). The ratio in Japanese cedar increased linearly with increasing levels of ozone, and the ratio was more greatly enhanced by exposure to SAR at pH 3.0 than at pH 5.6. In another study, Matsumura et al. (1998) investigated dry matter production and gas exchange rates in Japanese cedar, Nikko fir (*Abies homolepis*), Japanese white birch, and Japanese zelkova (*Zelkova serrata*) seedlings exposed to various combinations of SAR (pH 5.6 [deionized water, control] or 3.0 [5:2:3 equivalent mixture of sulfate, nitrate, and chloride], 16–25 mm/event, three events/week) and ozone (0.4, 1.0, 2.0, or 3.0 times the average profile of ozone concentration in ambient air, daily) for one growing season (20 weeks). The dry shoot, root, and whole-plant masses per seedling, and the net photosynthetic rates in Japanese cedar and the two broad-leaved tree species (Japanese white birch and Japanese zelkova) were significantly decreased linearly with increasing levels of ozone, but there were no significant effects of interactions between SAR and ozone on the shoot, root, and whole-plant dry mass, net photosynthetic rate, or dark respiration rate in any of the four species. However, significant effects of interactions were observed on the shoot/root ratio in all four species (Fig. 16.1). The shoot/root ratios in the two conifers and the birch exposed to SAR at pH 3.0 were significantly increased linearly with increasing levels of ozone, but the increase was not significant at pH 5.0. The shoot/root ratio in zelkova was significantly increased with increasing levels of ozone, and the increase in the shoot/root ratio on exposure to SAR at pH 3.0 was more greatly enhanced than that at pH 5.6. Similar effects of interactions between acid precipitation and ozone on shoot/root ratios were reported in American conifers (Hogsett et al. 1989; Reich et al. 1987) and broad-leaved tree species (Chappelka et al. 1988). These results may be related to the fact that chronic exposure to ozone inhibits the translocation of assimilates from the shoot to the roots (Spence et al. 1990). The ozone-induced inhibition of assimilate translocation seems to be closely related to the processes associated with phloem loading (Okano et al. 1984; Günthardt-Goerg et al. 1993). Therefore, there is a possibility that, in the four tree species examined by Matsumura et al. (1998), acid rain with a relatively low pH increased the extent of the ozone-induced inhibition of assimilate translocation from the shoot to the roots.

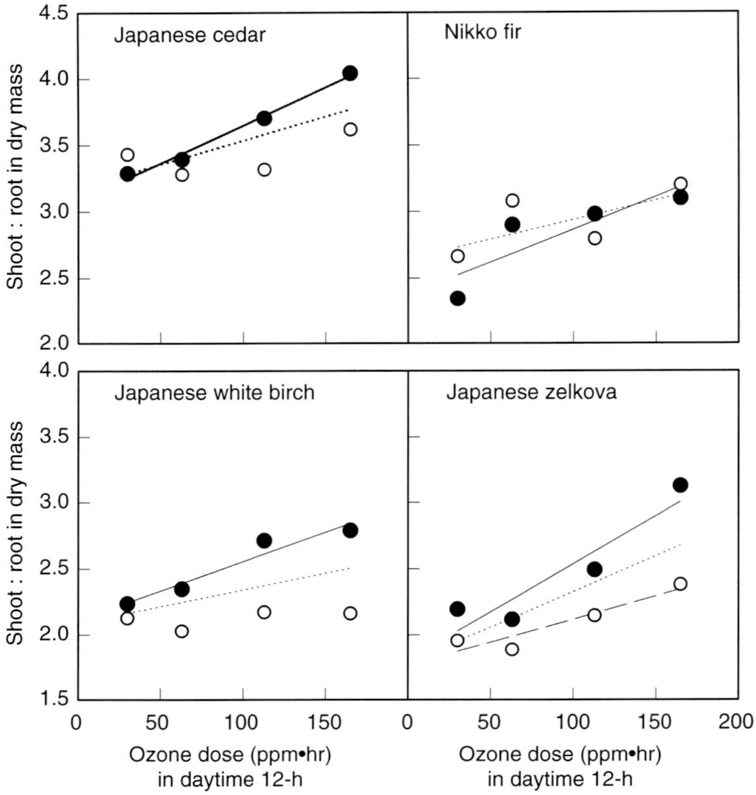

Fig. 16.1 Responses of the shoot (leaf + stem)/root dry-mass ratio in Japanese cedar, Nikko fir, Japanese white birch, and Japanese zelkova seedlings exposed to combinations of simulated acid rain and ozone, as a function of total ozone dose in the daytime 12-h for 20 weeks. *Symbols* (*closed*, pH 5.6; *open*, pH 3.0) represent the mean values of 12 seedlings. Linear regression lines for the rain treatment at pH 5.6 (*broken line*), at pH 3.0 (*solid line*), and for pooled data (*dotted line*) were drawn when linear responses to ozone were significant at the 0.05 level (After Matsumura et al. 1998)

16.3 Combined Effects of Simulated Acid Fog or Mist and Ozone

There are two reports concerning the combined effects of simulated acid mist or fog and ozone on Japanese trees. Sugihara et al. (2009) investigated the combined effects of simulated acid fog (SAF) and ozone on the dry-matter growth of Siebold's beech seedlings. For two growing seasons (14 months), the seedlings were exposed to various combinations of SAF (pH 5.0 or 3.0 [at a mixture of anions and cations that mimicked the highly acidic fog observed in mountainous areas in Japan], 2 h/event, two events/week from May to July and from September to December in the first season and from April to July in the second season) and ozone (ambient air or

0.06 ppm, 7 h/day, 4 days/week from September to December in the first season and from April to July in the second season). SAF at pH 3.0 reduced the number of leaves, whole-plant dry mass, and starch concentration in roots, and enhanced the shoot/root ratio as compared with SAR at pH 5.0, while 0.06 ppm ozone treatment reduced the number of leaves and starch concentration in roots as compared with ambient air treatment. Unfortunately, the interactions between SAF and ozone could not be determined, because only one-way analysis of variance was performed; thus, the combined effect of SAF and ozone was determined only for the starch concentration in roots, and the effect was shown to be additive.

Matsumura (2001) investigated the combined effects of simulated acid mist (SAM) and ozone on the growth of native trees and trees introduced to Japan. The seedlings of 14 tree species (9 conifers: *Pinus densiflora*, *P. thunbergii*, *Larix kaempferi*, *Picea abies*, *Abies firma*, *A. homolepis*, *A. veitchii*, *Cryptomeria japonica*, and *Chamaecyparis obtusa;* and 5 deciduous broad-leaved tree species: *Populus suaveolens*, *Betula platyphylla* var. *japonica*, *Fagus crenata*, *Quercus mongolica* var. *crispula*, and *Zelkova serrata*) were exposed, in open-top chambers with an exclusion top cover for ambient rain, to combinations of SAM at two pH values (pH 5.0 or 3.0 [1:2:1 at an equivalent mixture of sulfate, nitrate, and chloride, mimicking the highly acidic fog observed in mountainous areas in Japan], 0.25 mm × 3 min/event, three or four intermittent events/night, three nights/week from June to November; trees received an annual average SAM precipitation of 230 mm) and two levels of ambient ozone (charcoal-filtered air [CF] or non-filtered air [NF], all day every day; the CF and NF chambers at the experimental sites had 30 % and 90 % of ambient ozone) for three growing seasons (30 months). Significant decreases in shoot, root, and whole-plant biomass in the NF-ozone treatments were observed in seven of the conifers (Japanese red pine [*Pinus densiflora*], Japanese larch [*Larix kaempferi*], Norway spruce [*Picea abies*], Japanese fir [*Abies firma*], Nikko fir [*A. homolepis*], Veitchii's silver fir [*A. veitchii*], and Japanese cedar [*Cryptomeria japonica*]) and four of the deciduous broad-leaved tree species (Japanese balsam poplar [*Populus suaveolens*], Japanese white birch [*Betula platyphylla var. japonica*], Siebold's beech [*Fagus crenata*], and zelkova [*Zelkova serrata*]) (Japanese red pine, Japanese larch, Norway spruce, Japanese fir, Nikko fir, Veitchii's silver fir, and Japanese cedar) and four of the deciduous broad-leaved tree species (Japanese balsam poplar [*Populus suaveolens*], Japanese white birch, Siebold's beech, and zelkova) as compared with the findings for CF-ozone treatment. No visible foliar injury attributable to NF-ozone treatment could be found in any of the tree species examined. These results indicated that current ambient levels of ozone in Japan are high enough to have adverse effects on the biomass of the tree species examined, except for Japanese black pine (*P. thunbergii*), Hinoki cypress (*Chamaecyparis obtusa*), and Japanese oak (*Quercus mongolica* var. *crispula*). On the other hand, the shoot and whole-plant biomasses of all tree species examined, except for 3 species (Norway spruce, Nikko fir, and Japanese oak) were significantly increased in pH 3.0 SAM as compared with findings in pH 5.0,. These stimulations induced by pH 3.0 SAM were attributed to nitrogen fertilization from acid mist (Temple 1988). It was reported that not only did the seasonal mean pH of fog water samples collected in

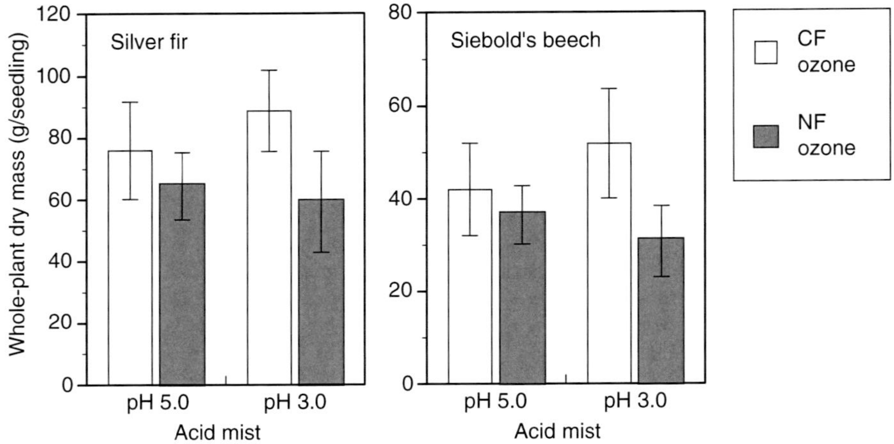

Fig. 16.2 Whole-plant dry mass per seedling in Veitchii's silver fir and Siebold's beech after two seasons of exposure to combinations of ozone (charcoal-filtered air [CF] or non-filtered air [NF]) and simulated acid mist (pH 5.0 or pH 3.0) at the Akagi experimental site. *Columns* represent the mean values (± SD) of 24 seedlings (After Matsumura 2001)

some mountainous areas in the Kanto district of Japan average 3.5 or higher , but that a small percentage of the fog water samples had a pH of 3.0 or below (Murano 1993; Takahashi and Fujita 1992). Therefore, it appears that direct adverse effects of the current acidity of fog observed in Japan could be small on tree growth. Synergistic interactions were found on Veitchii's silver fir and Siebold's beech, and the biomass decreases induced by NF-ozone were greater at pH 3.0 mist treatment than at pH 5.0 mist treatment (Fig. 16.2). These results suggested that the ozone effect on trees exacerbated by acid precipitation might be associated with nitrate input from precipitation.

16.4 Conclusion

The results obtained from the limited number of experimental studies on the effects of concurrent or sequential exposure to simulated acid rain, mist, or fog, and ozone, on the growth and physiology of the Japanese and Chinese tree species described above indicate that: (1) the combined effects of acid precipitation and ozone on Japanese and Chinese tree species were additive or greater than additive, (2) there is a possibility that acid precipitation with a relatively low acidity increased the extent of ozone-induced inhibition of assimilate translocation from the shoot to the roots, and (3) the ozone effect on trees exacerbated by acid precipitation may be associated with nitrate input from precipitation.

References

Chappelka AH, Chevone BI, Burk TE (1988) Growth response of green and white ash seedlings to ozone, sulfur dioxide and simulated acid rain. For Sci 34:1016–1029

Günthardt-Georg M et al (1993) Differentiation and structural decline in the leaves and bark of birch (*Betula pendula*) under low ozone concentrations. Trees 7:104–114

Hogsett WE et al (1989) Sensitivity of western conifers to SO_2 and seasonal interaction of acid fog and ozone. In: Olson RK, Lefohn AS (eds) Effects of air pollutants on western forests, APCA Transactions Series No.16 (Tr-16). Air & Waste Management Association, Pittsburg, pp 469–491

Izuta et al (2001) Experimental studies on the effects of ozone on forest tree species. J Jpn Soc Stmos Environ 36:60–77, in Japanese with English abstract, figures and tables

Kobayashi T, Kohno Y, Matsumura H (1997) Effects of simulated acid rain on tree species. In: Kohno Y (ed) Proceedings of CRIEPI International Seminar on Transport and Effects of Acidic Substances, November 28–29, 1996. CRIEPI, Tokyo, pp 184–189

Kohno Y, Matsumura H, Kobayashi T (1995) Effect of simulated acid rain on the growth of Japanese cedar, Japanese cypress and Sawara cypress. J Jpn Soc Atmos Environ 30:191–207, in Japanese with English abstract, figures and tables

Matsumura H (2001) Impacts of ambient ozone and/or acid mist on the growth of 14 tree species: an open-top chamber study conducted in Japan. Water Air Soil Pollut 130:959–964

Matsumura H, Kobayashi T, Kohno Y (1998) Effects of ozone and/or simulated acid rain on dry weight growth and gas exchange rates of Japanese cedar, Nikko Fir, Japanese white birch and Japanese zelkova seedlings. J Jpn Soc Atmos Environ 33:16–35, in Japanese with English abstract, figures and tables

Miwa M, Izuta T, Totsuka T (1993) Effects of simulated acid rain and/ or ozone on the growth of Japanese cedar seedlings. J Jpn Soc Air Pollut 28:279–287, in Japanese with English abstract, figures and tables

Murano K (1993) Current status of acid fog research. J Jpn Soc Air Pollut 28:185–199, in Japanese with English abstract, figures and tables

Okano K et al (1984) Alteration of ^{13}C-assimilate partitioning in plants of *Phaseolus vulgaris* exposed to ozone. New Phytol 97:155–163

Reich PB et al (1987) Effects of ozone and acid rain on white pine (*Pinus strobus*) seedlings grown in five soils. I. Net photosynthesis and growth. Can J Bot 65:977–987

Shan Y et al (1996) The individual and combined effects of ozone and simulated acid rain on growth, gas exchange rate and water use efficiency of *Pinus armandii* Franch. Environ Pollut 91:355–361

Spence RD, Rykiel EJ Jr, Sharpe PJH (1990) Ozone alters carbon allocation in loblolly pine: assessment with carbon-11 labeling. Environ Pollut 64:93–106

Sugihara A et al (2009) Effects of acid fog and ozone on the growth and physiological functions of Fagus crenata seedlings. J For Res 14:394–399

Takahashi A, Fujita S (1992) Contribution of nitrogen oxides to the acidification of rain – uptake and reaction of nitric acid/nitrate in the fog droplet –. CRIEPI Research report, T91082. CRIEPI, Tokyo, in Japanese with English abstract, figures and tables

Temple PJ (1988) Injury and growth of Jeffrey pine and giant sequoia in response to ozone and acidic mist. Environ Exp Bot 28:323–333

Wittig VE et al (2009) Quantifying the impact of current and future tropospheric ozone on tree biomass, growth, physiology and biochemistry: a quantitative meta-analysis. Glob Change Biol 15:396–424

Chapter 17
Effects of Soil Acidification on Asian Trees

Takeshi Izuta

Abstract Acid deposition is a serious environmental problem in Asia and may adversely affect Asian tree species due to soil acidification, the leaching of nutrients essential for normal plant growth from rhizospheric soil, and enhancement of the solubility of toxic metals such as aluminum (Al) and manganese (Mn) in the soil solution. The results of limited experimental studies indicate that soil acidification increases the concentrations of Al and Mn in the soil solution and reduces the growth, physiological functions such as net photosynthetic rate, and concentrations of nutrients such as Ca and Mg in several Asian forest tree species such as *Cryptomeria japonica*, *Pinus densiflora*, *Betula platyphylla*, *Fagus crenata*, and *Pinus massoniana*. A relatively low pH of culture solution below 3.5 reduces fresh and dry-weight growth and increases the concentrations of Ca and Mg in the shoots of Asian tree species such as *C. japonica*. Excessive Al or Mn in soil solution or water culture medium reduces fresh and dry-weight growth, shoot water content, and concentrations of Ca and Mg in the shoots and roots of Asian tree species such as *C. japonica* and *P. densiflora*. The extent of the negative effects of Al on growth, physiological functions, and water and nutrient status is greater than that of Mn when the concentrations of both elements in water culture medium are the same. Because the atmospheric deposition of acid substances is increasing in Asian countries, further studies are needed to elucidate the effects of soil acidification on many Asian forest tree species.

Keywords Acid deposition • Soil acidification • Asian forest tree species • Aluminum (Al) • Manganese (Mn)

T. Izuta
Institute of Agriculture, Tokyo University of Agriculture and Technology,
Fuchu, Tokyo 183-8509, Japan
e-mail: izuta@cc.tuat.ac.jp

T. Izuta (ed.), *Air Pollution Impacts on Plants in East Asia*,
DOI 10.1007/978-4-431-56438-6_17

17.1 Introduction

Acid deposition is a serious environmental problem in Asia. Acid deposition may adversely affect Asian forest ecosystems due to soil acidification, and the reduction in soil pH may result in the leaching of nutrients essential for normal plant growth from rhizospheric soil and an enhancement of the solubility of toxic metals such as aluminum (Al) in the soil solution. To prevent the negative impacts of soil acidification due to acid deposition on Asian forest ecosystems, and to develop countermeasures, it is necessary to clarify the effects of soil acidification on Asian forest tree species and its mechanisms. In this chapter, I introduce experimental studies on the effects of soil acidification, Al, and manganese (Mn) on the growth, physiological functions such as photosynthesis, and nutrient status of Asian forest tree species.

17.2 Effects of Soil Acidification on Japanese Forest Tree Species

In Japan, experimental studies on the effects of soil acidification due to acid deposition on Japanese forest tree species started in the late 1980s (Izuta et al. 1990, 1997; Miwa et al. 1994; Ohtagaki et al. 1996; Miwa et al. 1998; Shan et al. 1997, 2000).

Izuta et al. (1990) investigated the effects of artificial soil acidification on the growth and nutrient status of *Cryptomeria japonica* seedlings. Two-year-old seedlings were grown in humic andosols acidified by adding H_2SO_4 solution, once per 2 days during the growth period for 80 days. The whole-plant dry mass and fresh weight of the seedlings grown in soil that received 1.4 (soil pH (H_2O) at the end of the growth period: 4.03) and 2.8 g H^+ m^{-2} soil surface area (soil pH (H_2O): 3.82) during the growth period were not significantly less than those of the seedlings grown in the control soil that received deionized water (soil pH (H_2O): 4.40). However, the whole-plant dry mass and fresh weight of the seedlings grown in soil that received 4.2 g H^+ m^{-2} (soil pH (H_2O): 3.74) during the growth period were significantly less than those of the seedlings grown in the control soil that received deionized water. In the seedlings grown in the soil that received 4.2 g H^+ m^{-2} during the growth period, the Al concentration in the above-ground part of the seedlings was significantly higher, but Ca concentrations in the above- and below-ground parts of the seedlings were significantly lower than those in the seedlings grown in the control soil. The results obtained in this study suggest that soil acidification reduced the growth of *Cryptomeria japonica* seedlings, and this was caused by the toxicity of Al dissolved in the soil solution.

Miwa et al. (1994) reported the effects of artificial soil acidification on the growth of *Cryptomeria japonica* seedlings grown in three typical Japanese soils, andosols, red-yellow soil, and brown forest soil. Two-year-old seedlings were grown in artificially acidified soils at 10, 30, 60, or 100 mg H^+ L^{-1} air-dried soil (obtained by adding H_2SO_4 solution), or control soils (without the addition of H_2SO_4 solution) in a greenhouse for 12 weeks from June to September. The dry-weight growth of the seedlings grown in the three types of soils for 12 weeks was reduced with increasing amounts of H^+ added to the soils. However, the extent of reduction in the whole-plant dry mass was greater in red-yellow soil and brown forest soil than in andosol. The pH (H_2O) of the three soils was reduced and the water-soluble Mn and Al concentrations were increased with increasing amounts of H^+ added to the soils. The concentration of Al in the roots of the seedlings grown in artificially acidified soils was markedly higher than that in the roots of the seedlings grown in the control soils. The authors concluded that the growth reduction of *Cryptomeria japonica* seedlings induced by the addition of H_2SO_4 solution to the soils was greater in red-yellow soil and brown forest soil than that in andosol, and the growth reduction was closely related to the reduction of soil pH and the increase of Al concentration in the soils and in the roots of the seedlings.

Miwa et al. (1998) investigated the dry-weight growth of *Cryptomeria japonica* seedlings grown in artificially acidified brown forest soils with different parent materials (volcanic ash, granite, or sandstone and slate) in relation to the pH (H_2O), water-soluble Al concentration, and the (Ca+Mg+K)/Al molar ratio in the soils. H^+ of 10, 30, 60, and 100 mg was added, as H_2SO_4 solution, to 1 L of the acidified soils, but not to the control soils. Two-year-old seedlings were transplanted in the potted control and acidified soils, and were grown in a greenhouse for 12 weeks from June to September. The dry-weight growth of the seedlings grown in the brown forest soils was reduced with decreasing soil pH (H_2O), and with increasing water-soluble Al concentrations in the soils. The growth of the seedlings grown in the soil originating from volcanic ash was reduced when the concentration of water-soluble Al in the soil increased to more than 10.5 $\mu g\ g^{-1}$ on the basis of air-dried soil, and the growth of the seedlings grown in the other soils was reduced when the Al concentration increased to more than 30 $\mu g\ g^{-1}$. The growth of the seedlings grown in the soils derived from volcanic ash and granite was reduced when the (Ca+Mg+K)/Al molar ratio in the soils decreased to less than 5, but that of the seedlings grown in the soil originating from sandstone and slate was reduced when this molar ratio in the soil was 9.21 or less, and the concentration of water-soluble Al in the soil was 37.8 $\mu g\ g^{-1}$. The results obtained in this study show that the dry-weight growth of *Cryptomeria japonica* seedlings grown in brown forest soils acidified by adding H_2SO_4 solution was reduced, regardless of the concentration of water-soluble base cations in the soils, when the concentration of water-soluble Al in the soils was above 30 $\mu g\ g^{-1}$, and this growth was reduced when the (Ca+Mg+K)/Al molar ratio was below 5, even though the Al concentration was below 30 $\mu g\ g^{-1}$.

Ohtagaki et al. (1996) investigated photosynthetic activity in *Cryptomeria japonica* seedlings grown in potted brown forest soil derived from granite, acidified at 20 or 30 mg $H^+\ L^{-1}$ air-dried soil (soil pH (H_2O): 3.97 and 3.85, respectively) by adding H_2SO_4 solution. Two-year-old seedlings were grown in a greenhouse for 12 weeks from June to August. Control seedlings were grown in the soil without the addition of H_2SO_4 solution (soil pH (H_2O): 4.12) during the same period. Eight or 12 weeks after transplanting, the net photosynthetic rate of the seedlings grown in the acidified soil was significantly reduced as compared with the control value. Kinetic analyses

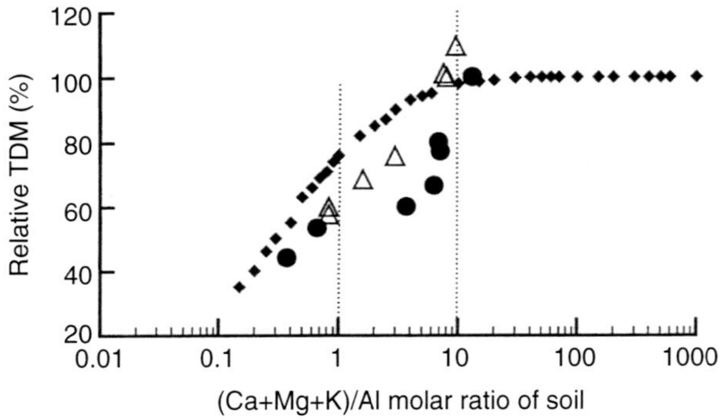

Fig. 17.1 Relationships between relative tree dry mass (TDM) of *Cryptomeria japonica* (●), *Pinus densiflora* (△), and *Picea abies* (◆) seedlings and the (Ca+Mg+K)/Al molar ratio of soil solution. The relative TDM (%) is the ratio of the whole-plant dry mass of seedlings grown in acidified soil to that of seedlings grown in control soil (Data source: Izuta et al. 1997; Lee et al. 1997a; Sverdrup et al. 1994)

of light-response and CO_2-response curves of the net photosynthetic rate showed that the quantum yield did not change, but the carboxylation efficiency was reduced in the seedlings grown in the acidified soil. There was a highly negative correlation between the Al concentration in the above-ground part of the seedlings and the carboxylation efficiency of the seedlings.

Izuta et al. (1997) investigated the growth and nutrient status of 2-year-old seedlings of *Cryptomeria japonica* grown for 100 days in brown forest soil derived from granite, acidified with H_2SO_4 solution with or without the leaching of cations from the soil. The Al concentration in the acidified soil increased with increasing amounts of H^+ added to the soil. When the Al concentration in the soil was below approximately 30 μg g^{-1}, the degree of reduction in the whole-plant dry mass was greater in the seedlings grown in the acidified soil with leaching of cations than in those grown in the acidified soil without leaching. A positive correlation was observed between the whole-plant dry mass of the seedlings and the (Ca+Mg+K)/Al molar ratio of the soil (Fig. 17.1). The concentrations of Ca, Mg, and K in the roots of the seedlings tended to decrease with decreasing soil molar ratios of Ca/Al, Mg/Al, and K/Al, respectively. These results suggest that the (Ca+Mg+K)/Al molar ratio of the soil is a useful indicator for assessing the critical load of acid deposition in forest ecosystems in Japan.

Since the 1990s, several Japanese researchers have conducted experimental studies on the effects of artificial soil acidification on Japanese deciduous broad-leaved tree species such as *Betula platyphylla* and *Fagus crenata*. Izuta et al. (1996a) reported the growth and nutrient status of *Betula platyphylla* seedlings grown in andosol and brown forest soil acidified by adding H_2SO_4 solution. Two-year-old seedlings were grown for 100 days in andosol and brown forest soil derived from granite, artificially acidified by adding H_2SO_4 solution. The amount of H^+ added to the soils was 4.68, 9.36, or 14.04 mg H^+ L^{-1} air-dried soil. There were no adverse effects of H^+ addition to the andosol on the whole-plant dry mass or nutrient status of the seedlings. However, the whole-plant dry mass of the seedlings grown in brown forest soil was reduced with increasing amounts of H^+ added to the soil. Furthermore, the concentrations and uptake rates of Ca and Mg in the seedlings grown in acidified brown forest soil were significantly less than those in the seedlings grown in the control soil not treated with H^+ as H_2SO_4 solution. The concentration of water-soluble Al increased in the brown forest soil, but not in the andosol, with increasing amounts of H^+ added to the soil. Therefore, the large amount of Al dissolved in brown forest soil due to the addition of H^+ was considered to be one of the factors limiting the plant growth and the availability of essential elements in this soil.

Izuta et al. (2001) investigated the dry-matter production, net photosynthetic rate, leaf nutrient status, and trunk anatomical characteristics of 3-year-old seedlings of *Fagus crenata* grown in brown forest soil derived from granite, acidified by adding H_2SO_4 solution. The soil acidification led to a decreased (Ca+Mg+K)/Al molar ratio in the soil solution. Dry mass per plant of the seedlings grown in the soil treated with H^+ at 120 mg L^{-1} air-dried soil was significantly reduced compared with the control value at 0 mg L^{-1}. When the net photosynthetic rate was reduced in the seedlings grown in the soil treated with H^+ at 120 mg L^{-1}, the carboxylation efficiency and maximum net photosynthetic rate at saturated CO_2 concentration were lower than the control values. The addition of H^+ to the soil at 120 mg L^{-1} reduced the concentration of Ca in the leaves. By contrast, Al concentration in the leaves was increased with increasing amounts of H^+ added to the soil. The annual rings formed in the seedlings grown in the soil treated with H^+ at 120 mg L^{-1} were significantly narrower than those in the seedlings grown in the soil treated with H^+ at 0 (control), 10, 30, 60, or 90 mg L^{-1}. Based on the results obtained in this study, the authors concluded that *Fagus crenata* was relatively sensitive to a reduction in the (Ca+Mg+K)/Al molar ratio of soil solution compared with *Picea abies* (Sverdrup et al. 1994). Izuta et al. (2004) also reported the growth, net photosynthesis, and leaf nutrient status of 3-year-old seedlings of *Fagus crenata* grown for two growing seasons in brown forest soil derived from granite, acidified with H_2SO_4 or HNO_3 solution. The whole-plant dry mass of the seedlings grown in the soil acidified by the addition of H_2SO_4 or HNO_3 solution was significantly less than that of the seedlings grown in control soil not supplemented with H^+ as H_2SO_4 or HNO_3 solution. However, the degrees of reduction in the whole-plant dry mass and net photosynthetic rate of the seedlings grown in the soil acidified by the addition of H^+ as H_2SO_4 solution at 100 mg L^{-1}, on the basis of air-dried soil volume (S-100 treatment), were greater than those in the seedlings grown in the soil acidified by the addition of H^+ as HNO_3 solution at 100 mg L^{-1} (N-100 treatment). The concentrations of Al and Mn in the leaves of the seedlings grown in the S-100 treatment were significantly higher than those in the N-100 treatment. A positive correlation was obtained between the molar ratio of (Ca+Mg+K)/(Al+Mn) in the soil solution and the relative whole-plant dry mass of the seedlings grown in the acidified soils to that of the seedlings grown in the control soil (Fig. 17.2). Based on these results, the authors concluded that the negative effects of soil acidification due to sulfate deposition were greater than those due to nitrate deposition on the growth, net photosynthesis,

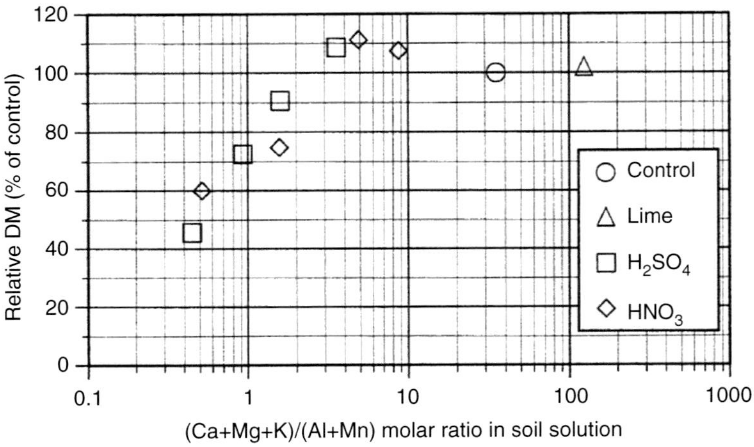

Fig. 17.2 The relationship between the molar ratio of the total concentration of Ca, Mg, and K to the total concentration of Al and Mn in soil solution [(Ca+Mg+K)/(Al+Mn)] and the relative whole-plant dry mass per plant of *Fagus crenata* seedlings grown in acidified soil to that of the seedlings grown in the control soil (relative DM) (After Izuta et al. 2004)

and leaf nutrient status of *Fagus crenata*. They also concluded that the molar ratio of (Ca+Mg+K)/(Al+Mn) in soil solution was a suitable soil parameter for evaluating critical loads of acid deposition in efforts to protect *Fagus crenata* forests in Japan.

17.3 Effects of Soil Acidification on Asian Forest Tree Species

Several researchers have investigated the effects of artificial soil acidification on the growth, physiological functions, and nutrient status of Korean (Lee et al. 1997a, 1998, 2005; Choi et al. 2005, 2008) and Chinese tree species (Yang et al. 1996; Tang et al. 2002).

Lee et al. (1997a) investigated the growth and nutrient status of 2-year-old seedlings of *Pinus densiflora* grown in brown forest soil, acidified by adding H_2SO_4 solution. For investigating soil acidification without leaching of cations from the soil, H^+ was added as H_2SO_4 solution to the soil at 10, 30, 60, or 90 mg H^+ L^{-1} air-dried soil. Soil without the additional supply of H^+ was used as a control. For investigating soil acidification with leaching of cations from the soil, the soil was placed in a container after soil acidification had been carried out by the above-mentioned method. Water at three times the volume of the soil was then poured into the container. After 3 days, water was gradually removed from the bottom of the container. The seedlings were transplanted to the acidified soils or the control soil, and then they were grown in a

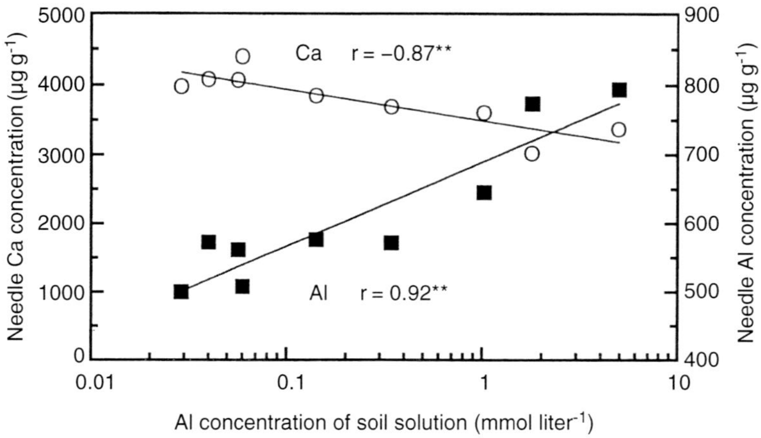

Fig. 17.3 Relationships among the Al concentration in soil solution, needle Ca concentration, and needle Al concentration in *Pinus densiflora* seedlings grown in acidified and non-acidified brown forest soil (After Lee et al. 1997a)

greenhouse for 120 days. The Al concentration in the soil solution of the acidified soil increased with increasing amounts of H^+ added to the soil. The whole-plant dry mass of the seedlings was reduced by the addition of H_2SO_4 solution to the soil. Increased Al concentration in the below-ground part of the seedlings reduced the concentrations of Ca and Mg in the above-ground part. The concentration of Ca in the needles was reduced with increasing concentrations of Al in the soil solution and the needles (Fig. 17.3). The whole-plant growth of the seedlings was dependent not only on the Al concentration but also on the balance of Al and cations in the soil solution. When the (Ca+Mg+K)/Al molar ratio of the soil solution was less than 7.0, the whole-plant dry mass was reduced, and it was reduced to approximately 60 % of the control when the molar ratio was 1.0 (Fig. 17.1).

Lee et al. (1998) reported the growth and photosynthetic responses of 2-year-old seedlings of *Pinus densiflora* grown in brown forest soil originating from granite, acidified by adding H_2SO_4 solution. The seedlings were grown in a greenhouse for 120 days. The pH of the soil solution was reduced, but the Al concentration in the soil solution was increased with increasing amounts of H^+ added as H_2SO_4 solution to the soil. The dry-weight growth of the seedlings was reduced as the amount of H^+ added to the soil was increased. There was a highly positive correlation between the dry-weight growth of the seedlings grown in the acidified soil and the (Ca+Mg+K)/Al molar ratio in the soil solution. The net photosynthetic rate of the seedlings grown in the acidified soil declined as the amount of H^+ in the soil increased. In the seedlings grown in acidified soil, the quantum yield and carboxylation efficiency of photosynthesis decreased, which suggests that soil acidification inhibited the photochemical reactions of photosynthesis and reduced the activity and/or amount of ribulose 1,5-bisphosphate carboxylase/oxygenase (Rubisco).

Yang et al. (1996) investigated the effects of SO_2 and soil acidification, alone and in combination, on the growth and nutrient status of *Pinus massoniana* seedlings (Masson's pine). Before transplanting the seedlings, the pH of red-yellow soil was adjusted to pH 3.7 or 4.5 by adding a solution of H_2SO_4. The seedlings were transplanted in the treated soil, and were then grown for 18 weeks in open-top chambers (OTC). During the growth period, the seedlings were exposed to charcoal-filtered air or SO_2 at 50 or 100 nmol mol^{-1} (ppb) in the OTC for 8 h a day (9: 00–17: 00), 5 days a week, for 17 weeks. Yellowish and whitish visible injuries were observed on the needles of the seedlings grown in the soil adjusted to pH 3.7 and exposed to SO_2 at 100 ppb. The exposure to SO_2 or soil acidification reduced the dry-weight growth of the seedlings. However, there was no significant interactive effect of SO_2 and soil acidification on the dry-weight growth of the seedlings. Although the exposure to SO_2 did not change the nutrient status of the seedlings, the concentrations of Al and Mn in the shoots were increased, while Ca concentration was reduced in the seedlings grown in the soil adjusted to pH 3.7. From the results obtained in this study, the authors concluded that the forest declines of *Pinus massoniana* observed in China may be closely related to relatively high concentrations of atmospheric SO_2, such as those above a daily average value of 100 ppb, and low soil pH below 4.0 induced by the wet or dry deposition of acid substances.

Tang et al. (2002) investigated the effects of acidification of red-yellow soil on the seed germination of *Pinus tabulaeformis* (Chinese red pine). Acid treatment with H_2SO_4 solution significantly changed the physical and chemical properties of red-yellow soil by lowering its pH value and by the leaching of Al ions. The seed germination rate in soil treated with H_2SO_4 solution was less than that in untreated soil, and this was mainly due to low soil pH below 3.5 and large amounts of Al ions in the soil solution. The addition of H_2SO_4 solution to the soil decreased soil aggregation and increased the number of micro-aggregates (under 250 μm in diameter). Such changes in soil chemical and physical properties increased the soil viscosity, and prevented the seedlings from fully developing after germination.

Choi et al. (2005) investigated the effects of soil acidification on 3-year-old seedlings of *Pinus koraiensis* (Korean pine) grown in brown forest soil derived from granite. The soil was acidified with an acid solution (anion molar ratio: SO_4^{2-}:NO_3^-:Cl^- = 5:3:2) and the seedlings were grown in soils with H^+ ion concentrations of 0 (control), 10, 30, 60, and 90 mmol kg^{-1} for 182 days. With increasing amounts of H^+ added to the soil, the concentrations of Ca, Mg, K, Al, and Mn increased, especially below a soil pH of 3.8. The concentrations of Ca, Mg, and K in the needles and stems of the seedlings increased with increasing H^+ added to the soil, whereas their concentrations in the roots decreased. Conversely, the concentrations of N and P in each organ of the seedlings were higher in all soil treatments than in controls. Also, the concentrations of Al and Mn increased significantly in all organs of the seedlings with increasing H^+ added to the soil. A strong positive correlation was found between the total dry mass per plant of the seedlings and the (Ca+Mg+K)/Al molar ratio calculated from the concentrations of water-soluble elements in the soil. When the (Ca+Mg+K)/Al molar ratio reached 1.0, the relative total dry mass per plant had fallen to 40 % of the control value. These results show that the sensitivity to soil acidification of the whole-plant dry mass of *Pinus koraiensis* was less than that of *Pinus densiflora* (Lee et al. 1997a).

Choi et al. (2008) evaluated the effects of ectomycorrhizal colonization on the growth and physiological activity of *Larix kaempferi* (Japanese larch) seedlings grown under soil acidification. They grew the seedlings with three types of ectomycorrhizae for 180 days in acidified brown forest soil derived from granite. The soil had been treated with an acid solution (0 as control, 10, 30, 60, and 90 mmol H^+ kg^{-1}). The water-soluble concentrations of Ca, Mg, K, Al, and Mn increased with increasing amounts of H^+ added to the soil. Ectomycorrhizal development was significantly increased in the soil treated with 10 and 30 mmol H^+ kg^{-1}, but was significantly reduced in the soil treated with 60 and 90 mmol H^+ kg^{-1}. The concentrations of Al and Mn in the needles and roots of the seedlings increased with increasing amounts of H^+ added to the soil. The total N in the seedlings significantly increased with increasing amounts of H^+ in the soil and with colonization with ectomycorrhiza. The maximum net photosynthetic rate at light and CO_2 saturation (P_{max}) was greater in the soil treated with 10 mmol H^+ kg^{-1} than in the control soil, and was less in the soils treated with >30 mmol H^+ kg^{-1}, especially in those treated with 60 and 90 mmol H^+ kg^{-1}. However, colonization with ectomycorrhiza significantly reduced the concentrations of Al and Mn in the needles and roots and increased the values of P_{max} and total dry mass per plant. The relative total dry mass per plant of the seedlings was approximately 40 % at a (Ca+Mg+K)/Al molar ratio of 1.0. However, seedlings with ectomycorrhizal colonization had a 100–120 % greater total dry mass per plant at a (Ca+Mg+K)/Al ratio of 1.0 than non-ectomycorrhizal seedlings, even though the acid treatment reduced their overall growth.

17.4 Effects of Al and/or Mn on Asian Forest Tree Species

Soil acidification due to acid deposition increases the concentrations of Al and Mn in the soil solution, and excessive concentrations of these elements are considered to have detrimental effects on the growth, physiological functions such as photosynthesis, and nutrient status of forest tree species. However, limited information is available on the effects of Al and/or Mn on Asian tree species (Miyake et al. 1991; Izuta et al. 1995, Izuta et al. 1996a, b; Lee et al. 1997b, 1999; Hirano and Hijii 1998; Tahara et al. 2005; Hirano et al. 2007).

Miyake et al. (1991) investigated the effects of Al on the growth of 1-year-old seedlings of *C. japonica* grown in culture solutions containing Al at 0 (control), 30, 60, and 90 mg kg^{-1} (ppm), added as $AlCl_3$, for 60 days. The root dry weight and shoot and root fresh weights were significantly reduced by Al treatments at 60 ppm or higher, as compared with the control values. Furthermore, the shoot fresh weight was reduced by Al treatment at 30 ppm. However, the shoot dry weight was not significantly affected by Al treatment. Aluminum was highly accumulated in the roots of the seedlings by Al treatment, but was transported only in negligible amounts to the needles. The concentrations of Ca and Mg in the plant organs were reduced by Al treatment, while the concentration of P in the roots was somewhat increased by this treatment. The root respiration rate was reduced by Al treatment. The authors concluded that Al inhibited the growth of *Cryptomeria japonica* seedlings by inhibiting the growth and respiratory activity of roots and by inhibiting the uptake and transportation of nutrients such as Ca and Mg.

Izuta et al. (1996b) reported the effects of low pH or excessive Al on the growth, water content, and nutrient status of *Cryptomeria japonica* seedlings. Two-year-old seedlings were grown for 12 weeks in a nutrient solution adjusted to pH 3.0, 3.5, 4.0, or 4.5, or in nutrient solution (pH 4.0) containing no Al (control) or Al at 10, 30, 60, or 90 mg kg^{-1} (ppm). The root growth was not significantly affected by pH treatment, but was significantly reduced by Al treatment at $\geq$10 ppm. The shoot water content was not significantly affected by pH treatment, but was significantly reduced by Al treatment at 90 ppm. The concentrations of Ca and Mg in the shoot were significantly increased by relatively low solution pH values of $\leq$3.5, but the concentrations of these elements in the shoot and roots were significantly decreased by Al treatment at $\geq$10 ppm. The P concentrations in the shoot and roots were not significantly affected by the solution pH, but were significantly reduced in the shoot and increased in the roots by Al treatment at $\geq$10 ppm.

Hirano and Hijii (1998) investigated the effects of low pH or excessive Al on the root morphology and nutritional status of 2-year-old seedlings of *Cryptomeria japonica*, using glass beads as a medium. They compared the effects of pH or Al, determined by the glass-beads method, with those determined by soil acidification treatment using brown forest soil. An increased number of branching roots, reduced root length, and reduced concentrations of Ca and Mg in the white roots were induced by low pH and excessive Al. Browning and reduction of K were specific symptoms in the white roots subjected to low pH treatment. The effects of excessive Al caused an increase of root diameter and increased the concentrations of P and Al in the white roots. Furthermore, the effects of Al were very similar to those resulting from the exposure to pH 2.0 solution in brown forest soil. These results suggest that the root morphology of *C. japonica* seedlings was adversely affected not only by low pH, but also by excessive Al in brown forest soil, indicating that this type of soil may have the potential for producing a decline in the root systems of this tree species.

Tahara et al. (2005) investigated the effects of ectomycorrhizal association on the tolerance to 1 mM Al of *Pinus densiflora* seedlings with or without ectomycorrhizal association with the fungus *Pisolithus tinctorius*. Association with *P. tinctorius* alleviated the Al-induced inhibition of root elongation and biomass growth in the seedlings with ectomycorrhizal association. The secretion of malate and citrate, both low-molecular-weight organic acids that can detoxify Al by the formation of stable complexes, was investigated in *P. tinctorius* mycelia and in the roots of the seedlings with and without *P. tinctorius* association. Citrate secretion from *P. tinctorius* mycelia in vitro was stimulated by Al treatment. Citrate secretion from the roots of the ectomycorrhizal-associated seedlings was also stimulated by Al treatment, but was not detected in the seedlings without ectomycorrhizal association. These results suggest that citrate secreted from the roots of the ectomycorrhizal-associated seedlings was produced in the hyphae of *P. tinctorius* and the citrate secretion may play a role in enhancing the Al tolerance of the host seedlings.

Izuta et al. (1995) reported the effects of excessive Mn on the growth, water content, and nutrient status of *Cryptomeria japonica* seedlings. Two-year-old seedlings were grown in a nutrient solution containing Mn at 1 (control), 30, 60, and 90 mg kg^{-1} (ppm) for 60 days. The relative growth rate (RGR) and net assimilation rate (NAR) of the seedlings during the Mn treatment decreased with increasing Mn concentrations in the nutrient solution, and the fresh and dry weights per plant were significantly reduced by Mn treatment at 90 ppm as compared with the control values. The Mn treatment at 30 ppm or higher reduced the concentrations of Ca and Mg in the needles and reduced the water content of the seedlings. However, Fe concentration in the needles was significantly increased by Mn treatment at 90 ppm. The results obtained in this study suggest that the excessive Mn-induced reduction in the growth of the seedlings was mainly due to that in the efficiency of dry-matter production, which was caused not only by excessive Mn accumulation in the needles, but also by inhibition of the uptake of essential elements, such as Ca and Mg, from the roots and inhibition of their transportation from the roots to the needles.

Lee et al. (1997b) investigated the effects of Al and Mn, singly and in combination, on the dry-weight growth and nutrient status of 2-year-old *Pinus densiflora* seedlings grown in a nutrient culture solution for 90 days. Aluminum was added as $AlCl_3$ at 1, 10, 30, or 60 mg kg^{-1} (ppm), and Mn was added as $MnCl_2$ at 1, 30, or 60 mg kg^{-1} to the nutrient culture solution. The pH of the solution was maintained at 4.0 by adding HC1 or NaOH solution. The interactive effects of Al and Mn on the dry-weight growth of the seedlings were not statistically significant, although Al and Mn had additive effects on this growth parameter. There were single effects of Al or Mn on the dry-weight growth of the seedlings, as well as on the concentrations of Al and Mn in the seedlings. Al treatment at $\geq$10 ppm or Mn treatment at 60 ppm significantly reduced the dry-weight growth of the seedlings, which indicates that the effect of Al on the dry-weight growth was stronger than that of Mn. The chlorophyll content of the needles was not affected by Al treatment, but was significantly reduced by Mn treatment at 60 ppm. Furthermore, Al treatment at 60 ppm or Mn treatment at $\geq$30 ppm caused a significant reduction in the root dark respiration rate of the seedlings. The net photosynthetic rate of the seedlings was reduced with increasing concentrations of Al or Mn in the nutrient culture solution, which suggests that the Al- and Mn-induced reductions in the RGR and NAR of the seedlings were mainly due to the inhibition of net photosynthesis.

17.5 Concluding Summary

Based on the above experimental studies on the effects of soil acidification, Al, and Mn on Asian tree species, it is considered that, in these species, soil acidification due to acid deposition induces low soil pH, and causes the leaching and reduction of base cations such as Ca, Mg, and K, as well as the leaching of Al and excessive Mn in soil. These

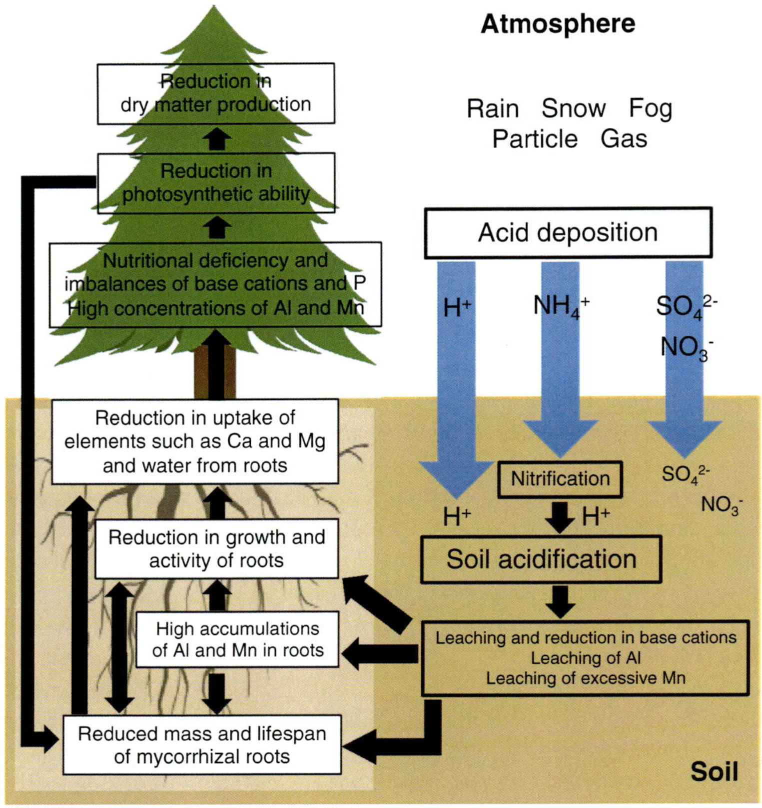

Fig. 17.4 Schematic image of the negative effects of soil acidification due to acid deposition on forest tree species

phenomena lead to the accumulation of Al and Mn in the roots, reductions in the uptake of base cations and water from the roots, reduced mass and lifespan of roots with ecto-mycorrhizal association, and reduction in root growth and activity. High concentrations of Al and Mn occur in the above-ground parts of the Asian tree species examined, with nutritional deficiency and imbalances of base cations and P also occurring in the above-ground parts, along with reductions in photosynthetic capacity and dry-matter production (Fig. 17.4). Therefore, as soon as possible, we should take effective measures to prevent the adverse effects of soil acidification on Asian forest tree species.

References

Choi DS, Jin HO, Lee CH, Kim YC, Kayama M (2005) Effect of soil acidification on the growth of Korean pine (*Pinus koraiensis*) seedlings in a granite-derived forest soil. Environ Sci 12(1):33–47

Choi DS, Jin HO, Chung DJ, Sasa K, Koike T (2008) Growth and physiological activity in *Larix kaempferi* seedlings inoculated with ectomycorrhizae as affected by soil acidification. Trees 22(5):729–735

Hirano Y, Hijii N (1998) Effects of low pH and aluminum on root morphology of Japanese red cedar saplings. Environ Pollut 101(3):339–347

Hirano Y, Mizoguchi T, Brunner I (2007) Root parameters of forest trees as sensitive indicators of acidifying pollutants: a review of research of Japanese forest trees. J For Res 12(2): 134–142

Izuta T, Yokota F, Miyake H, Totsuka T (1990) Effects of soil acidification on the growth of Cryptomeria seedlings. Man Environ 16(4):55–61 (in Japanese with English abstract)

Izuta T, Noguchi K, Aoki M, Totsuka T (1995) Effects of excess manganese on growth, water content and nutrient status of Japanese cedar seedlings. Environ Sci 3(4):209–220

Izuta T, Seki T, Totsuka T (1996a) Growth and nutrient status of *Betula platyphylla* seedlings grown in andosol and brown forest soil acidified by adding H_2SO_4 solution. Environ Sci 4(4):233–247

Izuta T, Yamada A, Miwa M, Aoki M, Totsuka T (1996b) Effects of low pH and excess Al on growth, water content and nutrient status of Japanese cedar seedlings. Environ Sci 4(2): 113–125

Izuta T, Ohtani T, Totsuka T (1997) Growth and nutrient status of *Cryptomeria japonica* seedlings grown in brown forest soil acidified with H_2SO_4 solution. Environ Sci 5(3):177–189

Izuta T, Yamaoka T, Nakaji T, Yonekura T, Yokoyama M, Matsumura H, Ishida S, Yazaki K, Funada R, Koike T (2001) Growth, net photosynthetic rate, nutrient status and secondary xylem anatomical characteristics of *Fagus crenata* seedlings grown in brown forest soil acidified with H_2SO_4 solution. Water Air Soil Pollut 130:1007–1012

Izuta T, Yamaoka T, Nakaji T, Yonekura T, Yokoyama M, Funada R, Koike T, Totsuka T (2004) Growth, net photosynthesis and leaf nutrient status of *Fagus crenata* seedlings grown in brown forest soil acidified with H_2SO_4 or HNO_3 solution. Trees (Struct Funct) 18(6): 677–685

Lee CH, Izuta T, Aoki M, Totsuka T (1997a) Growth and element content of red pine seedlings grown in brown forest soil acidified by adding H_2SO_4 solution. J Jpn Soc Atmos Environ 32(1):46–57 (in Japanese with English abstract)

Lee CH, Izuta T, Aoki M, Totsuka T (1997b) Effects of Al and Mn, alone and in combination, on growth and nutrient status of red pine seedlings hydroponically grown in nutrient culture solution. J Jpn Soc Atmos Environ 32(5):371–382 (in Japanese with English abstract)

Lee CH, Izuta T, Aoki M, Totsuka T, Kato H (1998) Growth and photosynthetic responses of red pine seedlings grown in brown forest soil acidified by adding H_2SO_4 solution. Jpn J Soil Sci Plant Nutr 69(1):54–62 (in Japanese with English abstract)

Lee CH, Jin HO, Izuta T (1999) Growth, nutrient status and net photosynthetic rate of *Pinus densiflora* seedlings in various levels of aluminum concentrations. J Korean For Soc 88(2): 249–254

Lee CH, Lee SW, Kim EY, Kim YK, Byun JK, Won HG, Jin HO (2005) Growth of *Pinus densiflora* seedlings in artificially acidified soils. Korean J Ecol 28(6):389–393

Miwa M, Izuta T, Totsuka T (1994) Effects of soil acidification on the growth of Japanese cedar seedlings grown in three soils from different parent materials. J Jpn Soc Air Pollut 29(5): 254–263 (in Japanese with English abstract)

Miwa M, Izuta T, Totsuka T (1998) Dry weight growth of Japanese cedar seedlings grown in artificially acidified brown forest soil. J Jpn Soc Atmos Environ 33(2):81–92 (in Japanese with English abstract)

Miyake H, Kamei N, Izuta T, Totsuka T (1991) Effects of aluminium on the growth of hydroponically grown seedlings of *Cryptomeria japonica* D. Don. Man Environ 17(1):10–16, in Japanese with English abstract

Ohtagaki T, Miwa M, Izuta T, Totsuka T (1996) Photosynthetic response of Japanese cedar seedlings grown in brown forest soil acidified by adding H_2SO_4 solution. J Jpn Soc Atmos Environ 31(1):11–19 (in Japanese with English abstract)

Shan Y, Izuta T, Aoki M, Totsuka T (1997) Effects of O_3 and soil acidification, alone and in combination, on growth, gas exchange rate and chlorophyll content of red pine seedlings. Water Air Soil Pollut 97(3–4):355–366

Shan Y, Izuta T, Totsuka T (2000) Phenological disorder induced by atmospheric nitrogen deposition: original causes of pine forest decline over Japan. Part I. Phenological disorder, cold death of apical shoots of red pine subjected to combined exposures of simulated acid rain and soil acidification, and implications for forest decline. Water Air Soil Pollut 117:191–203

Sverdrup H, Warfvinge P, Nihlgård B (1994) Assessment of soil acidification effects on forest growth in Sweden. Water Air Soil Pollut 78:1–36

Tahara K, Norisada M, Tange T, Yagi H, Kojima K (2005) Ectomycorrhizal association enhances Al tolerance by inducing citrate secretion in *Pinus densiflora*. Soil Sci Plant Nutr 51(3):397–403

Tang H, Wang R, Izuta T, Aoki M, Totsuka T (2002) Effects of red-yellow soil acidification on seed germination of Chinese pine. J Environ Sci 14(1):115–119

Yang L, Izuta T, Aoki M, Totsuka T (1996) Effects of SO_2 and soil acidification, alone and in combination, on growth of masson pine seedlings. J Jpn Soc Atmos Environ 31(1):1–10 (in Japanese with English abstract)

Chapter 18
Effects of Nitrogen Load on Asian Trees

Tatsuro Nakaji and Takeshi Izuta

Abstract The effects of increasing nitrogen (N) load on Asian forest trees have been studied mainly in Japan and China during the past decade. In this chapter, we summarize the expected mechanisms of the harmful effects of excessive N load on the eco-physiological function of trees, and introduce experimental studies on the growth responses and foliar nutrient status of young Asian tree seedlings to increasing N loads. By comparing the growth responses to various N loads in 12 tree species in Japan and China, we confirmed that: (1) the threshold of the N load that induced growth reduction was quite different among the species; (2) the threshold value ranged between 50 and 100 kg N ha^{-1} $year^{-1}$ in relatively sensitive (low-tolerance) species. Furthermore, a significant relationship between foliar nutrient balance and growth indicated that (3) the threshold N/P ratio for growth reduction was slightly higher than that in European tree species, and (4) an Mn/Mg ratio of over 0.8 was observed, together with growth reduction, in the sensitive tree species.

Keywords Excessive N load • Growth response • Mn/Mg ratio • N/P ratio • Nutrient imbalance • Species difference • Tree

18.1 Nitrogen Saturation in Forest Ecosystems

Generally, nitrogen (N) is a limiting nutrient factor in many temperate forests (Ingestad and Kähr 1985; Crane and Banks 1992). In forest ecosystems where N limits primary production, because most of the N supplied by atmospheric deposition is absorbed and assimilated by microbes and plants, N output as nitrate (NO_3^-) in stream water is lower than N input by atmospheric deposition (Stoddard 1994).

T. Nakaji (✉)
Tomakomai Experimental Forest, Field Science Center for Northern Biosphere, Hokkaido University, Tomakomai 053-0035, Japan
e-mail: nakaji@fsc.hokudai.ac.jp

T. Izuta
Institute of Agriculture, Tokyo University of Agriculture and Technology, Fuchu, Tokyo 183-8509, Japan
e-mail: izuta@cc.tuat.ac.jp

T. Izuta (ed.), *Air Pollution Impacts on Plants in East Asia*,
DOI 10.1007/978-4-431-56438-6_18

However, several researchers have observed relatively high N outputs in stream water, exceeding the N input, and they have suggested that the N status of some forests in Central Europe and North America is changing from 'N-limited' to 'N-saturated' (Skeffington and Wilson 1988; Stoddard 1994; Fenn et al. 1996). In Europe, harmful effects of excessive N load have been expected with increasing N-dominated acid deposition (see Chap. 3). High nitrate deposition is a risk in forest ecosystems, since a high input of ammonium (NH_4^+) to the soil induces soil acidification and nutrient imbalance, and the risk of excessive ammonium load has also been been predicted in some forested areas (Nihlgård 1985).

18.2 Effects of Excessive N Load on Asian Trees

Several researchers have investigated the effects of increasing inputs of nitrate and ammonium on tree health to clarify these phenomena and their mechanisms (e.g. Wilson and Skeffington 1994b; Seith et al. 1996; Izuta and Nakaji 2003; Wang and Liu 2014). Most of these studies dealt with young (few-years-old) tree seedlings, and the N effects on photosynthesis (Nakaji et al. 2002; Azuchi et al. 2004; Guo et al. 2014), growth (Nakaji et al. 2005; Izuta et al. 2005; Mao et al. 2014), mycorrhizal roots, and nutrient status (e.g. Wallenda and Kettle 1998; Seith et al. 1996; George et al. 1999; Sogn and Abrahamsen 1998; Nakaji et al. 2005) were investigated experimentally. Furthermore, based on the results of fertilizer experiments and monitoring studies, mainly in Europe, an empirical critical load of N and a nutrient index were proposed (De Vries et al. 2000a, b; WHO 2000). In European forest ecosystems, the thresholds of the N load for the appearance of N saturation and forest damage were approximately 10 kg N ha^{-1} $year^{-1}$ and 25 kg N ha^{-1} $year^{-1}$, respectively (Wright et al. 1995). Bobbink et al. (2015) have reported that the empirical critical load of N that would not cause nutrient imbalance, reduction in root biomass, or species change showed large variations among ecosystem types, and it ranged from 3 to 20 kg N ha^{-1} $year^{-1}$. In Asian countries such as China, Japan, India, and South Korea, high atmospheric N deposition, similar to or higher than that in Central Europe and North America, has been observed (Dentener et al. 2014; Kulshrestha et al. 2014). For example, in Japan, although the mean wet N deposition by precipitation over Japan has been maintained at 7 to 10 kg N ha^{-1} $year^{-1}$ (Katoh et al. 1990; Hara 1992), the N input by wet deposition in coniferous forests near suburban areas ranged from 10 to 20 kg N ha^{-1} $year^{-1}$ (Ohrui and Mitchell 1997), and sometimes it reached 40 kg N ha^{-1} $year^{-1}$ near urban areas (Okita et al. 1993; Baba and Okazaki 1998). In China, wet N deposition has been observed at a range of 9 to 23 kg N ha^{-1} $year^{-1}$ (Du and Liu 2014), and dry N deposition in North China has been estimated to be about 25 kg N ha^{-1} $year^{-1}$ (Shen et al. 2014). These reports suggested that the N deposition in these countries would already reach the threshold range of N saturation. In Europe, thresholds of element concentration and element balance in tree leaves were proposed in the 1990s by European Commission -United Nations/Economic Commission for Europe EC-UN/ECE et al. (1997) for evaluating the nutrient status of major tree species, such as spruce, pine, oak, and beech

Table 18.1 Criteria used for the assessment of foliar nutrient concentrations in four European tree species

		Spruce			Pine			Oak			Beech		
Class[a]		1	2	3	1	2	3	1	2	3	1	2	3
Nutrient concentration (mg g^{-1})	N	<12	12–17	>17	<12	12–17	>17	<15	15–25	>25	<15	15–25	>17
	P	<1.0	1.0–2.0	>2.0	<1.0	1.0–2.0	>2.0	<1.0	1.0–1.8	>1.8	<1.0	1.0–1.7	>1.7
	K	<3.5	3.5–9.0	>9.0	<3.5	3.5–10.0	>10.0	<5.0	5.0–10.0	>10.0	<5.0	5.0–10.0	>10.0
	Ca	<1.5	1.5–6.0	>6.0	<1.5	1.5–4.0	>4.0	<3.0	3.0–8.0	>8.0	<4.0	4.0–8.0	>8.0
	Mg	<0.6	0.6–1.5	>1.5	<0.6	0.6–1.5	>1.5	<1.0	1.0–2.5	>2.5	<1.0	1.0–1.5	>1.5
	S	<1.1	1.1–1.8	>1.8	<1.1	1.1–1.8	>1.8	–	–	–	<1.3	1.3–2.0	>2.0
Nutrient ratio (g g^{-1})	N/P	<6.0	6.0–17.0	>17.0	<6.0	6.0–17.0	>17.0	<8.3	8.3–25.0	>25.0	<10.6	10.6–25.0	>17.0
	N/K	<1.3	1.3–4.9	>4.9	<1.2	1.2–4.9	>4.9	<1.5	1.5–5.0	>5.0	<1.8	1.8–5.0	>5.0
	N/Ca	<2.0	2.0–11.3	>11.3	<2.0	2.0–11.3	>11.3	<1.9	1.9–8.3	>8.3	<2.3	2.3–6.3	>6.3
	N/Mg	<8.0	8.0–28.3	>28.3	<8.0	8.0–28.3	>28.3	<6.0	6.0–25.0	>25.0	<12.0	12.0–25.0	>25.0

Source: European Commission -United Nations/Economic Commission for Europe EC-UN/ECE et al. (1997)
[a] *1* = low, *2* = normal or adequate, *3* = optimal to high

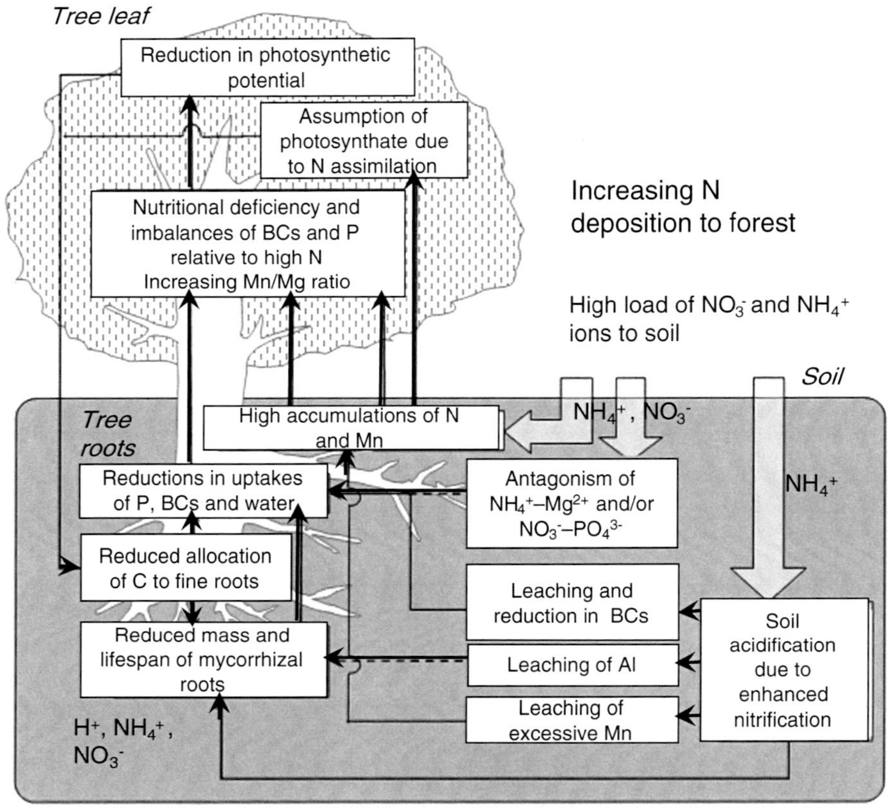

Fig. 18.1 Schematic images of the negative effects of increasing N deposition in sensitive forest tree species

(Table 18.1). Because these nutrient indices have not been extensively examined in Asian countries, studies on the critical N load and nutrient balance are needed for evaluating the current status and sufficient management of Asian forests.

Figure 18.1 shows schematic images of the negative effects of excessive N load to the soil on sensitive (i.e., low-tolerant) tree species. Although NO_3^- and NH_4^+ in soil generally play roles as plant fertilizers, over-nutrition of N by the forms of NO_3^- and NH_4^+ induces soil acidification by H^+ originating from HNO_3 and the nitrification of NH_4^+ (Nilsson 1986). Soil acidification causes leaching of base cations such as Ca and Mg from soil to the watershed and enhances the solubility of Mn and Al (Van Breemen et al. 1982; Ulrich and Sumner 1991). This change can induce a lack of mineral nutrients and cause the excessive accumulation of Mn and toxic Al ions in plants. High concentrations of soil NO_3^- and acidity affect the species component of mycorrhiza and reduce the mycorrhizal infection rate in fine roots (i.e., mycorrhizal roots), as well as reducing mycorrhizal lifespan (Majdi and Nylund 1996; Wallenda and Kottke 1998; Wöllecke et al. 1999). Mycorrhizal roots play important roles in the uptake of water and nutrients such as Mg and phosphorus (P) (Marschner and Dell 1994). Because high concentrations of NO_3^- and NH_4^+ can cause chemical antagonism with Mg^{2+} and PO_3^- at root uptake sites (Boxman and Roelofs 1988; Wilson and Skeffington 1994a), these biotic and chemical changes in the soil environment could cause deteriorated nutrient status in trees via the excessive accumulation of Mn and/or deficiencies in Mg and P. This nutrient imbalance may also increase sensitivity to other environmental stressors such as drought and frost (Nihlgård 1985; Cowling et al. 1988; Schulze 1989; Izuta and Nakaji 2003). In sensitive tree species such as *Pinus densiflora* (Japanese red pine), excessive N-induced low foliar concentrations of Mg and P reduced photosynthetic activity, due to the reduction of foliar chlorophyll and ribulose-1,5-bisphosphate (RuBP) carboxylase/oxygenase (Rubisco), despite a high foliar N concentration (Nakaji et al. 2001, 2002). Furthermore, a high foliar Mn concentration compared with Mg (i.e., a high Mn/Mg ratio) tends to inhibit the activation of Rubisco (Nakaji et al. 2001; Manter et al. 2005). Consequently, in sensitive tree species and acid soil, excessive N will reduce tree growth via an imbalance of nutrients.

The nutrient imbalances and growth reduction of below-ground organs have been reported in many experimental studies of European tree species. For example, Seith et al. (1996) investigated growth and foliar nutrients in *Picea abies* (Norway spruce) seedlings grown in two N treatments (150 mg and 300 mg N kg^{-1} soil), and they reported N-induced reductions in needle concentrations of N, Ca, Mg, and Mn and increases in those of P and K. They also reported that experimental N addition to the soil induced reduction of fine root growth in *Picea abies* seedlings (Seith et al. 1996; George et al. 1999). Sogn and Abrahamsen (1998) reported that a 5-year experiment with varied N supply at 30 kg and 90 kg N ha^{-1} $year^{-1}$ reduced needle concentrations of K, Mg, and P in *Pinus sylvestris* (Scotch pine) seedlings without growth reduction. Figure 18.2 shows comparisons of growth response to increasing N load in 12 tree species in Japan and China (Nakaji et al. 2005; Izuta et al. 2005; Guo et al. 2010, 2014; Liu et al. 2011; Mao et al. 2014; Wang and Liu 2014). These studies conducted manipulation experiments during one to three growing seasons, using natural forest soil and 1- to 2-year-old seedlings. In this comparison, the relative growth of whole-plant dry mass was calculated in each experiment. The results show interesting trends in regard to threshold and species differences in tolerance to excessive N load. None of the 12 tree species showed growth reduction by N addition below 50 kg N ha^{-1} $year^{-1}$, but the threshold for growth reduction was quite different among the tree species. For example, the whole-plant dry mass of three species, *Fagus crenata* (Japanese beech), *Castanopsis sieboldii* (Sudajii, a Japanese evergreen oak), and *Pinus densiflora* (Japanese red pine) tended to be reduced at values between 50 and 100 kg N ha^{-1} $year^{-1}$ (Fig. 18.2). These species can be classified as N-sensitive (low-tolerant) species. On the other hand, *Cryptomeria japonica* (Japanese cedar), *Picea asperata* (dragon spruce), and *Quercus acuta* (Japanese red oak) can be classified as N-tolerant tree species, because they showed enhanced growth rates even with very high N loads, over 200 kg N ha^{-1} $year^{-1}$ (Fig. 18.2). Although the threshold for two tree species was not known, the four remaining species, *Acer truncatum* (purple brow maple), *Pinus tablaeformis* (Chinese pine), *Lithocarpus edulis* (Matebashii, a Japanese evergreen oak species), and *Quercus glauca* (ring-cup oak) seemed to be intermediate in response, with a threshold of

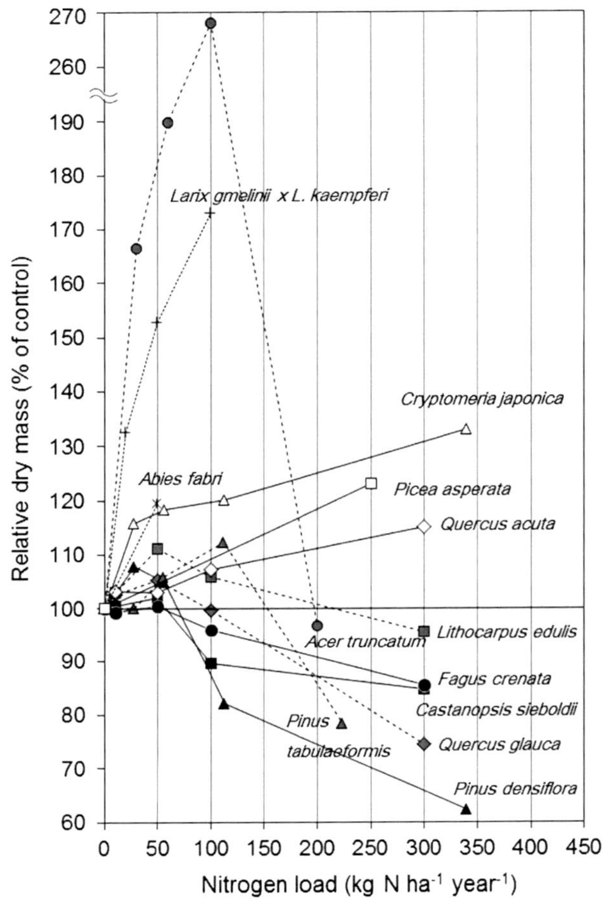

Fig. 18.2 Growth responses of 12 Asian tree species to increasing N load in soil. In each growth experiment, the response was calculated as the relative whole-plant dry mass of seedlings grown under varied N loads compared with that in seedlings grown in non-N-added natural soil (control = 100 %). The lengths of the experimental periods varied: one growing season, *A. truncatum* (Guo et al. 2014), *A. fabri* (Guo et al. 2010), and *L. gmelinii* x *L. kaempferi* (Mao et al. 2014); two growing seasons, *P. asperata* (Liu et al. 2011), *F. crenata* (Izuta et al. 2002, unpublished data), *C. sieboldii*, *L. edulis*, *Q. glauca*, and *Q. acuta* (Izuta et al. 2005); three growing seasons, *P. tabulaeformis* (Wang and Liu 2014), *C. japonica*, and *P. densiflora* (Nakaji et al. 2005)

growth reduction at over 100 kg N ha^{-1} year^{-1}. The responsible mechanism has not yet been clarified; however, this result suggests that we must pay attention to species differences in N sensitivity when estimating the empirical critical load of N in Asian forests for maintaining tree growth and biodiversity.

The growth reduction in sensitive tree species caused by excessive N load has been explained mainly by the depression of photosynthesis due to nutrient imbalances of N/P and Mn/Mg appearing with reduced mycorrhizal infection (Nakaji et al. 2001, 2002; Izuta and Nakaji 2003; Izuta et al. 2005). Figure 18.3 shows the relationship between foliar nutrient status and growth response to an increasing N load in eight Japanese tree species (Izuta et al. 2005; Nakaji et al. 2005; Izuta et al. 2002, unpublished data). When the data were pooled for all tree species, significant correlations were observed in the relationships of dry-mass responses and P ($r=0.41$, $P=0.010$), the N/P ratio ($r=-0.34$, $P=0.035$), Mn ($r=-0.53$, $P<0.001$), and Mn/Mg ($r=-0.53$, $P<0.001$). There was no significant relationship in regard to other foliar elements such as K and Ca. This suggests that P and Mn are important nutrient factors related to the N sensitivity of Japanese tree

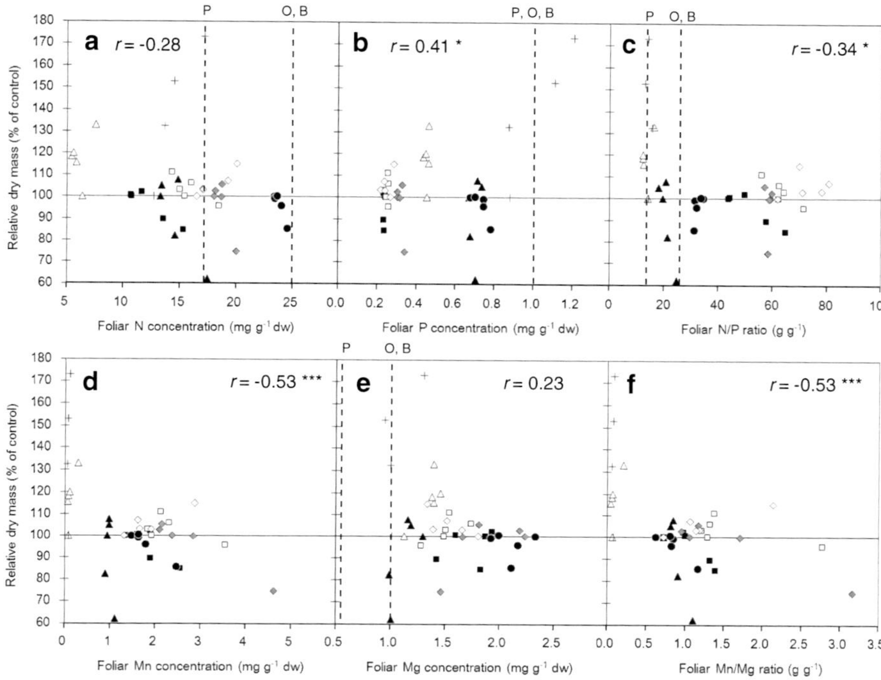

Fig. 18.3 The relationship between foliar nutrient status and growth response to increasing N load in eight Japanese tree species. The nutrient concentration and nutrient balance were investigated in current-year leaves (**a**) N; (**b**) P; (**c**) N/P; (**d**) Mn; (**e**) Mg; (**f**) Mn/Mg. Correlation coefficients for all the plant species and significance levels of the relationship are shown in each panel (* $P<0.05$, *** $P<0.001$). *Vertical dashed lines* indicate threshold values showing optimal-to-high nutrient status for N or low nutrient status for P and Mg in European pine (P), oak (O), and beech (B) species (see Table 18.1) (EC-UN/ECE et al. 1997). Symbols: + *L. gmelinii* × *L. kaempferi* (Mao et al. 2014), ● *F. crenata* (Izuta et al. unpublished data), ■ *C. sieboldii*, □ *L. edulis*, ◆ *Q. glauca*, ◇ *Q. acuta* (Izuta et al. 2005), △ *C. japonica*, ▲ *P. densiflora* (Nakaji et al. 2005)

seedlings. When we compared the criteria for nutrient balance for European tree species and these data, we found that foliar N levels in the Japanese tree species were lower than the European criteria for high nutrient balance (pine = 17 mg N g^{-1}, oak and beech = 25 mg N g^{-1}), and the foliar P concentration was lower than the criteria for P deficit in European tree species (1 mg g^{-1}) (Fig. 18.3a, b). Lower P in Japanese tree species could be related to the fact that the origin of most soil in Japan is volcanic ash. As for the nutrient balances, 10% growth reduction in *P. densiflora* (pine), *F. crenata* (beech), and *C. sieboldii* (evergreen oak) was observed with higher N/P ratios of about 20, 35, and 55, respectively. Mn/Mg ratios over 0.8, 1.0, and 1.3 were observed with growth reduction in *P. densiflora*, *F. crenata*, and *C. sieboldii* (Fig. 18.3). Since the Mn/Mg ratio is related to the activation of Rubisco (Nakaji et al. 2001; Manter et al. 2005), this imbalance is also important. As the species are different in Europe and Japan, European data would not be appropriate, and these comparison results indicate that an original N threshold value is needed for Japanese tree species.

References

Azuchi F, Kinose Y, Matsumura T, Kanomata T, Uehara Y, Kobayashi A, Yamaguchi M, Izuta T (2014) Modeling stomatal conductance and ozone uptake of Fagus crenata grown under different nitrogen loads. Environ Pollut 184:481–487

Baba M, Okazaki M (1998) Acidification in nitrogen-saturated forested catchment. Soil Sci Plant Nutr 44:513–525

Bobbink R, Tomassen H, Weijters M, Van den Berg L, Strengbom J, Braun S, Nordin A, Schütz K, Hettelingh J-P (2015) Effects and empirical critical loads of nitrogen for Europe. In: de Vries W, Hettelingh J-P, Posch M (eds) Critical loads and dynamic risk assessments. Springer, pp 85–128. Dordrecht, Heidelberg, London, New York

Boxman AW, Roelofs JGM (1988) Some effects of nitrate versus ammonium nutrition on the nutrient fluxes in Pinus sylvestris seedlings. Effects of mycorrhizal infection. Can J Bot 66:1091–1097

Cowling E, Krahl-Urban B, Schimansky C (1988) Hypothesis to explain forest decline. In: Krahl-Urban B, Papke HE, Peters K, Schimansky C (eds) Forest decline. U.S. EPA and German Ministry of Research and Technology, Corvallis, pp 120–125

Crane WJB, Banks JCG (1992) Accumulation and retranslocation of foliar nitrogen in fertilized and irrigated Pinus radiata. For Ecol Manag 52:201–223

De Vries W, Reinds GJ, Klap JM, Van Leeuwen EP, Erisman JW (2000a) Effects of environmental stress on forest crown condition in Europe. Part III: estimation of critical deposition and concentration levels and their exceedances. Water Air Soil Pollut 119:363–386

De Vries W, Reinds, GJ, Van Kerkvoorde MS, Hendriks CMA, Leeters EEJM, Gross CP, Voogd JCH, Vel EM (2000b) Intensive monitoring of forest ecosystems in Europe: technical report 2000. Convention on long-range transboundary air pollution international co-operative programme on assessment and monitoring of air pollution effects on forests and European Union Scheme on the protection of forests against atmospheric pollution. EC-UN/ECE, Geneva/Brussels, p 193

Dentener F, Vet R, Dennis RL, Du E, Kulshrestha UC, Galy-Lacaux C (2014) Progress in monitoring and modeling estimates of nitrogen deposition at local, regional and global scales. In: Sutton MA, Mason KE, Sheppard LJ, Sverdrup H, Haeuber R, Hicks WK (eds) Nitrogen deposition, critical loads and biodiversity. Springer, pp 49–56. Dordrecht, Heidelberg, London, New York

Du E, Liu X (2014) High rates of wet nitrogen deposition in China: a synthesis. In: Sutton MA, Mason, KE, Sheppard LJ, Sverdrup H, Haeuber R, Hicks WK (eds) Nitrogen deposition, critical loads and biodiversity. Springer, pp 49–56. Dordrecht, Heidelberg, London, New York

EC-UN/ECE, Stefan KA, Fürst A, Hacker R, Bartels U (1997) Forest foliar condition in Europe, results of large-scale foliar chemistry surveys 1995. EC, UN/ECE, Geneva, p 207

Fenn ME, Poth MA, Johnson DW (1996) Evidence for nitrogen saturation in the San Bernardino mountains in southern California. For Ecol Manag 82:211–230

George E, Kircher S, Schwarz P, Tesar A, Seith B (1999) Effect of varied soil nitrogen supply on growth and nutrient uptake of young Norway spruce plants grown in a shaded environment. J Plant Nutr Soil Sci 162:301–307

Guo J, Yan Y, Wang G, Yan L, Sun X (2010) Ecophysiological responses of Abies fabri seedlings to drought stress and nitrogen supply. Physiol Plant 139:335–347

Guo X, Wang R, Chang R, Liang X, Wang C, Luo Y, Yuan Y, Guo W (2014) Effects of nitrogen addition on growth and photosynthetic characteristics of Acer truncatum seedlings. Dendrobiology 72:151–161

Hara H (1992) Precipitation. In: The Chemical Society of Japan (ed) Chemistry of terrestrial water. Gakkaisyuppan Center, Tokyo, pp 69–78, In Japanese with English summary

Ingestad T, Kähr M (1985) Nutrition and growth of coniferous seedlings at varied relative nitrogen addition rate. Physiol Plant 65:109–116

Izuta T, Tominaga K, Watanabe M, Matsumura H, Kohno Y (2005) Effects of N load on growth and leaf nutrient status of Japanese evergreen broad-leaved tree species. J Agric Meteorol 60:1125–1128

Izuta T, Nakaji T (2003) Effects of high nitrogen load and ozone on forest tree species. Eur J For Res 6:155–170

Katoh T, Konno T, Koyama I, Tsuruta H, Makino H (1990) Acidic precipitation in Japan. In: Bresser AHM, Salomons W (eds) Acidic precipitation, volume 5. International overview and assessment. Springer-Verlag, Berlin, pp 41–102

Kulshrestha UC, Kulshrestha MJ, Satyanarayana J, Reddy LAK (2014) Atmospheric deposition of reactive nitrogen in India. In: Sutton MA, Mason, KE, Sheppard LJ, Sverdrup H, Haeuber R, Hicks WK (eds) Nitrogen deposition, critical loads and biodiversity. Springer, pp 49–56. Dordrecht, Heidelberg, London, New York

Liu Q, Yin H, Chen J, Zhao C, Cheng X, Wei Y, Lin B (2011) Belowground responses of Picea asperata seedlings to warming and nitrogen fertilization in the eastern Tibetan Plateau. Ecol Res 26:637–648

Majdi H, Nylund J-N (1996) Does liquid fertilization affect fine root dynamics and lifespan of mycorrhizal short roots? Plant Soil 185:305–309

Manter D, Kavanagh KL, Rose CL (2005) Growth response of Douglas-fir seedlings to nitrogen fertilization: importance of Rubisco activation state and respiration rates. Tree Physiol 25:1015–1021

Mao Q, Watanabe M, Kobayashi M, Kita K, Koike T (2014) High nitrogen deposition may enhance growth of a new hybrid larch F1 growing at two phosphorus levels. Landsc Ecol Eng 10:1–8

Marschner H, Dell B (1994) Nutrient uptake in mycorrhizal symbiosis. Plant Soil 159:89–102

Nakaji T, Fukami M, Dokiya Y, Izuta T (2001) Effects of high nitrogen load on growth, photosynthesis and nutrient status of *Cryptomeria japonica* and *Pinus densiflora* seedlings. Trees 15:453–461

Nakaji T, Takenaga S, Kuroha M, Izuta T (2002) Photosynthetic response of *Pinus densiflora* seedlings to high nitrogen load. Environ Sci 9:269–282

Nakaji T, Yonekura T, Kuroha M, Takenaga S, Izuta T (2005) Growth, annual ring structure and nutrient status of Japanese red pine and Japanese cedar seedlings after three years of excessive N load. Phyton (Ann Bot) 45:457–464

Nihlgård B (1985) The ammonium hypothesis – an additional explanation to the forest dieback in Europe. Ambio 14:2–8

Nilsson J (1986) Critical loads for nitrogen and sulfur, vol 11. The Nordic Council of Ministers, Rep, Copenhagen, p 232

Ohrui K, Mitchell M (1997) Nitrogen saturation in Japanese forested watersheds. Ecol Appl 7:391–401

Okita T, Murano K, Matsumoto M, Totsuka T (1993) Determination of dry deposition velocities to forest canopy from measurements of throughfall, stemflow and the vertical distribution of aerosol and gaseous species. Environ Sci 2:103–111

Schulze E-D (1989) Air pollution and forest decline in a spruce (Picea abies) forest. Science 244:776–783

Seith B, George E, Marschner H, Wallenda T, Schaeffer C, Einig W, Wingler A, Hampp R (1996) Effects of varied soil nitrogen supply on Norway spruce (*Picea abies* [L.] Karst.) I. Shoot and root growth and nutrient uptake. Plant Soil 184:291–298

Shen J, Liu X, Fangmeier A, Zhang F (2014) Enrichment of atmospheric ammonia and ammonium in the North China plain. In: Sutton MA, Mason, KE, Sheppard LJ, Sverdrup H, Haeuber R, Hicks WK (eds) Nitrogen deposition, critical loads and biodiversity. Springer, pp 57–65. Dordrecht Heidelberg London New York

Skeffington RA, Wilson EJ (1988) Excess nitrogen deposition: issues for consideration. Environ Pollut 54:159–184

Sogn TA, Abrahamsen G (1998) Effects of N and S deposition on leaching from an acid forest soil and growth of Scots pine (*Pinus sylvestris* L.) after 5 years of treatment. For Ecol Manag 103:177–190

Stoddard JL (1994) Long-term changes in watershed retention of nitrogen. In: Baker LA (ed) Environmental chemistry of lakes and reservoirs. American Chemical Society, Washington, DC, pp 223–284

Ulrich B, Sumner ME (eds) (1991) Soil acidity. Springer-Verlag, Berlin

Van Breemen N, Burrough PA, Velthorst EJ, Van Dobben HF, De Wit T, Ridder TB, Reijnders HFR (1982) Soil acidification from atmospheric ammonium sulphate in forest canopy throughfall. Nature 299:548–551

Wallenda T, Kottke I (1998) Nitrogen deposition and ectomycorrhizas. New Phytol 139:169–187

Wang G, Liu F (2014) Carbon allocation of Chinese pine seedlings along a nitrogen addition gradient. For Ecol Manag 334:114–121

WHO (2000) Effects of airborne nitrogen pollutants on vegetation: critical loads. In: WHO Regional Publications, European Series, No. 91, Copenhagen, Denmark

Wright RF, Roelofs JGM, Bredemeier M, Blanck K, Boxman AW, Emmett BA, Gundersen P, Hultberg H, Kjonaas OJ, Moldan F, Tietema A, Van Breemen N, Van Dijk HFG (1995) NITREX: responses of coniferous forest ecosystems to experimentally changed deposition of nitrogen. For Ecol Manage 71:163–169

Wilson EJ, Skeffington RA (1994a) The effects of excess nitrogen deposition on young Norway spruce trees. Part I the soil. Environ Pollut 86:141–151

Wilson EJ, Skeffington RA (1994b) The effects of excess nitrogen deposition on young Norway spruce trees. Part II the vegetation. Environ Pollut 86:153–160

Wöllecke J, Munzenberger B, Huttl RF (1999) Some effects of N on ectomycorrhizal diversity of Scots pine (*Pinus sylvestris* L.) in Northeastern Germany. Water Air Soil Pollut 116:135–140

Part VI
Effects of Aerosol on Plants

Chapter 19
Effects of Aerosol Particles on Plants

Masahiro Yamaguchi and Takeshi Izuta

Abstract Because aerosol particles in the atmosphere scatter and absorb incoming solar radiation, they can indirectly affect plant productivity by reducing incoming solar radiation. Conversely, aerosol particles affect plants after deposition from the atmosphere. In this chapter, we introduce both the possible indirect effects of aerosol particles on plant productivity, via reducing incoming solar radiation, and the direct effects of particulate matter deposited on the leaf surface. A wide range of experimental studies has shown the physical effects of particulate matter deposited on the leaf surface – leaf shading, increased leaf temperature, stomatal plugging, and interference with stomatal closure – that can occur in the field in Asia. Leaf shading by particulate matter reduces the net photosynthetic rate under low light conditions. The increase in leaf temperature caused by particulate matter increases or decreases the net photosynthetic rate under air temperatures below or above the optimum for photosynthesis, respectively. The light extinction coefficient of particulate matter is an important factor in the increase in leaf temperature. Stomatal plugging decreases gas diffusivity during the light period, and this can reduce the net photosynthetic rate. Interference with stomatal closure induces unexpected water loss at night, which can reduce drought tolerance. Although the chemical effects of dust depend on its chemical properties (e.g., pH), the effect of dust-induced injury cannot be explained by the effect of a single major dust component only, suggesting that dust components have an interactive effect. Currently, however, little information is available on the combined effects of the components of particulate matter.

Keywords Aerosol particle • Global dimming • Increasing leaf temperature • Interference with stomatal closure • Net photosynthetic rate • Plugging of stomata • Leaf shading • Stomatal diffusive conductance (resistance)

M. Yamaguchi (✉)
Graduate School of Fisheries and Environmental Sciences, Nagasaki University,
Nagasaki, Nagasaki 852-8521, Japan
e-mail: masah-ya@nagasaki-u.ac.jp

T. Izuta
Institute of Agriculture, Tokyo University of Agriculture and Technology,
Fuchu, Tokyo 183-8509, Japan
e-mail: izuta@cc.tuat.ac.jp

T. Izuta (ed.), *Air Pollution Impacts on Plants in East Asia*,
DOI 10.1007/978-4-431-56438-6_19

19.1 Introduction

Aerosol particles are defined as solid or liquid particles suspended in the atmosphere and they include a variety of chemical compounds with a size range between a few nanometers and tens of a micrometer (Horvath 2000). In the atmosphere, aerosol particles scatter and absorb incoming solar radiation, thus reducing the amount of solar radiation reaching the Earth's surface (Stanhill and Cohen 2001). Because the primary requirement for photosynthesis is the absorption of radiation in the leaves (Larcher 2003), aerosol particles can indirectly affect plant productivity by scattering and absorbing solar radiation (Stanhill and Cohen 2001). Conversely, after their deposition from the atmosphere, aerosol particles can directly affect plants. Aerosol particles are deposited not only on the leaf surface but also on the soil. Deposition of aerosol particles on the soil causes changes in its chemical properties, such as acidification and contamination by heavy metals included in the particles, thus indirectly affecting plant productivity (Grantz et al. 2003); this was partly described in the previous chapters (Chaps. 17 and 18). Deposition of aerosol particles on the leaf surface affects plants chemically and/or physically, depending on the properties of the deposited aerosol particles (Grantz et al. 2003).

The chemical and physical properties of aerosol particles differ considerably depending on the source of the particles (Horvath 2000). Aerosol particles are either emitted directly, such as from soil erosion, sea spray, fly ash, soot from diesel engines, or iron oxide fumes from the steel industry, or are produced in the atmosphere by the oxidation and reaction of chemical compounds such as sulfur dioxide and ammonia (Horvath 2000). Particles with a diameter greater than 1 μm are emitted from soil erosion, cement dust, and fly ash etc. (Horvath 2000). However, particles of less than 1 μm mainly consist of the end-products of condensation on existing particles and particles produced by gas-to-particle conversion (Horvath 2000). A wide range of studies on plant responses to the deposition of aerosol particles, especially coarse particulate matter with a diameter above 1 μm, has been reported. In this chapter, we introduce the possible indirect effects of aerosol particles on plant productivity, via the reduction of incoming solar radiation, and we outline the direct effects of particulate matter deposited on the leaf surface.

19.2 Possible Indirect Effects of Aerosol Particles on Plant Productivity

Aerosol particles scatter and absorb incoming solar radiation, and thus reduce the amount of solar radiation reaching the Earth's surface; so-called global dimming (Stanhill and Cohen 2001). From 1973 to 2007, surface solar radiation decreased substantially over South and East Asia, South America, Australia, and Africa (Wang et al. 2009), although the tendency differed regionally within Asia (Norris and Wild 2009; Wild 2012). Because aerosol particles can directly attenuate surface solar radiation, by scattering and absorbing solar radiation, or indirectly attenuate it by acting as cloud condensation nuclei, it has been suggested that these particles have a prominent role in the reduction of surface solar radiation (Wild 2012). Because the primary requirement for photosynthesis is the absorption of radiation in the leaves (Larcher 2003), aerosol particles can indirectly affect plant productivity via the scattering and absorption of solar radiation (Stanhill and Cohen 2001).

Photosynthesis is influenced by a number of external factors – the availability of radiation, CO_2, temperature, and the supply of water and mineral nutrients (Larcher 2003). Chameides et al. (1999) reported that when photosynthesis was not limited by environmental factors such as water supply and nutrients (e.g., in managed farms), atmospheric aerosols and regional haze in China induced up to a 30 % reduction in incoming solar radiation and yield reductions of rice and wheat estimated at up to 30 %. In many climates, however, theoretical studies have indicated that small decreases in direct radiation, if accompanied by an increase in the fraction of diffuse radiation, would increase photosynthesis (Stanhill and Cohen 2001). For example, Roderick et al. (2001) showed that vegetation was directly sensitive to changes in the diffuse fraction, and that photosynthesis within vegetation canopies was more efficient under diffuse light conditions. Aerosol particles scatter solar radiation and thus increase the diffuse fraction, and Mercado et al. (2009) estimated that variations in the diffuse fraction enhanced the land carbon sink by approximately 25 % between 1960 and 1999. Although this would not be the case for heavily polluted areas or areas under dark cloudy skies (UNEP 2011), future scenarios involving reductions in aerosol emissions suggest that the diffuse-radiation enhancement of the land carbon sink will decline (Ciais et al. 2013).

19.3 Direct Effects of Particulate Matter Deposited on the Leaf Surface

The deposition of aerosol particles on the leaf surface can affect plants physically and/or chemically, depending on the properties of the deposited aerosol particles (Grantz et al. 2003). In this section, we summarize the physical effects of particulate matter deposited on the leaf surface – leaf shading, increased leaf temperature, plugging of stomata, and interference with stomatal closure (Table 19.1) – followed by the chemical effects.

19.3.1 Leaf Shading

Several researchers have reported the shading effects of particulate matter on plants. In an experimental study, Thompson et al. (1984) exposed *Viburnum tinus* to car exhaust dust with a particle size range of 1–10 μm. When the adaxial side of the leaf was exposed to exhaust dust, the net photosynthetic rate (*A*) was reduced via shading. Although the amount of dust deposited by the exposure (up to 10 g m^{-2}) was much

Table 19.1 Summary of experimental studies on the physical effects of aerosol particles on plants

Physical effect	Reference	Particle type	Particle size	Deposition amount
Shading leaf	Thompson et al. (1984)	Car exhaust dust	1–10 μm	10 g m^{-2}
	Hirano et al. (1990, 1995)	Kanto loam powders (three types)	1.9, 8, and 30 μm (median)	3.82 g m^{-2}
		Black carbon	0.03–0.20 μm	1.5 g m^{-2}
Increasing leaf temperature	Chaston and Doley (2006)	Coal dust, overburden, and flyash	120, 70, and 120 μm, respectively (median)	10 g m^{-2}
	Hirano et al. (1991, 1995)	Kanto loam powders (three types)	1.9, 8, and 30 μm (median)	2.9 g m^{-2}
		Black carbon	0.03–0.20 μm	1.3 g m^{-2}
Plugging of stomata and interference with stomatal closure	Flückiger et al. (1979)	Silica gel	5, 15, 20–32, and 32–60 μm	10 g m^{-2}
	Thompson et al. (1984)	Car exhaust dust	1–10 μm	10 g m^{-2}
	Hirano et al. (1991, 1995)	Kanto loam powders (three types)	1.9, 8, and 30 μm (median)	1.2 g m^{-2}

higher than that observed in the field (about 2 g m^{-2}), these results indicated that dust loading on the leaf surface could shade the leaves. Hirano et al. (1990, 1995) exposed cucumber (*Cucumis sativus* L.) and kidney bean (*Phaseolus vulgaris* L.) to chemically inert dusts, including three kinds of Kanto loam powder (KL) with different median sizes (1.9, 8, and 30 μm) and black carbon (BC) with a size range of 0.03–0.20 μm. The deposition amounts of KL and BC were up to 3.82 g m^{-2} and 1.5 g m^{-2}, respectively. The dust exposure reduced *A* values in *C. sativus* and *P. vulgaris*, especially under relatively low photosynthetic photon flux density (PPFD). When *A* was plotted against the PPFD reaching the leaf surface by considering the illuminated PPFD and the shading ratio of dust deposited on the leaf surface, the light response curves of *A* were similar in dust-exposed and non-exposed leaves in both species. These results indicated that dust deposited on the leaf surface absorbs light, shades plant leaves, and reduces *A* under low light conditions (Hirano et al. 1990, 1995).

19.3.2 Increasing Leaf Temperature

The absorption of solar radiation by particulate matter deposited on the leaf surface can induce an increase in leaf temperature. In an experimental study done by Chaston and Doley (2006), *Harpullia pendula*, *Eucalyptus tereticornis*, and

Metrosideros tomentosa were exposed to coal dust, overburden, and fly ash at up to 10 g m^{-2}. Exposure to fly ash increased leaf temperature in *M. tomentosa*, but not in the other two species. Both coal and overburden increased leaf temperature in all three species, but the degree of increase was higher in leaves exposed to coal than in those exposed to overburden. This difference can be attributed to the differences in light extinction coefficients among coal dust, overburden, and fly ash. Hirano et al. (1991, 1995) exposed *C. sativus* to KL and BC at up to 2.9 and 1.3 g m^{-2}, respectively, and measured leaf temperature under various light conditions. The increase in leaf temperature from dust exposure became greater with increases in light intensity and the amount of dust deposited on the leaf surface. The increase in leaf temperature was greater in plants exposed to the dark-colored BC than the light-colored KL (Hirano et al. 1995). Therefore, it can be seen that the light extinction coefficient of particulate matter is an important factor for increasing leaf temperature.

The light-saturated net photosynthetic rate (A_{sat}) was increased by BC exposure when the air temperature was below 25 °C. Conversely, when the air temperature was above 30 °C, A_{sat} was decreased by BC exposure. These results indicate that BC deposited on the leaf surface absorbs light and increases the leaf temperature, resulting in an increase or decrease in A_{sat} under air temperatures below or above the optimum for photosynthesis, respectively (Hirano et al. 1991, 1995). Furthermore, because an increase in leaf temperature increases the vapor pressure deficit between the air and the leaves, the leaf temperature increase induced by BC exposure increased the transpiration rate in *C. sativus* leaves (Hirano et al. 1991, 1995).

19.3.3 Plugging of Stomata and Interference with Stomatal Closure

In an observational study conducted in a polluted valley site around a phurnacite plant, Ricks and Williams (1974) measured stomatal diffusive resistance and water relations, and observed the leaf surfaces of *Quercus petraea* plants subjected to a differential gradient of particulate pollution. They detected a reduction in the maximal diffusion resistance at night in leaves in the polluted area compared with values in leaves in an unpolluted rural site. Observations of leaf surfaces using scanning electron microscopy (SEM) revealed the plugging of stomata by particles in the polluted area. The magnitude of the reduction in maximal diffusion resistance at night was related to the proportion of plugged stomata, suggesting interference with stomatal closure due to the plugging of stomata with particles. Several researchers have conducted experimental studies on stomatal malfunction due to the deposition of particulate matter on the foliar surface. In an experimental study, Flückiger et al. (1979) exposed the abaxial sides of aspen leaves to silica gel with different particle sizes (5–60 μm at 10 g m^{-2}) during the morning (when the stomata opened). Exposure to silica gel slightly reduced the stomatal diffusive resistance of the leaves during light hours, and significantly reduced it at night. The reductions were greater in leaves exposed to small particles of 5 μm, which was less than the stomatal pore

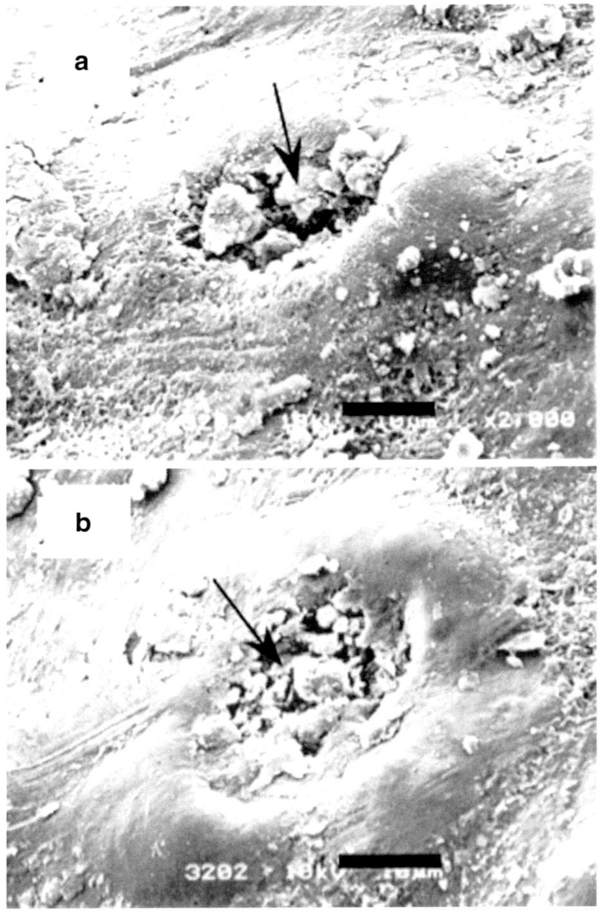

Fig. 19.1 Scanning electron micrographs of aerosols deposited in and around the stomata of 2-year-old needles of *C. japonica* from urban areas in Japan. *Scale bars* are 10 μm. *Arrows* in (**a, b**) indicate Fe-rich and Ti-rich particles, respectively, as determined by X-ray spectra (Takamatsu et al. 2001) (© 2008 Canadian Science Publishing or its licensors. Reproduced with permission)

size, suggesting the particles interfered with stomatal closure at night. Thompson et al. (1984) exposed *V. tinus* to car exhaust dust with a size range of 1–10 μm at up to 10 g m^{-2}. When the abaxial sides of the leaves were exposed, the net photosynthetic rate was reduced via an increase in leaf gas diffusion resistance. Hirano et al. (1991, 1995) exposed *C. sativus* to KL of different median sizes (1.9, 8, and 30 μm at 1.2 g m^{-2}) during the dark or light period (i.e., when stomata were closed or open, respectively). Exposure to KL during the light period decreased stomatal conductance (g_s) under light conditions and increased g_s under dark conditions, whereas exposure to KL during the dark period did not affect g_s under either condition. These results suggest that particulate matter deposited on the leaf surface plugs stomata during the daytime and interferes with stomatal closure at night.

Several researchers have reported stomatal plugging in field observation studies. In Japan, Sase et al. (1998) and Takamatsu et al. (2001) collected needles of *Cryptomeria japonica* from an urban area (Tokyo and Saitama Prefecture), where *C. japonica* is declining, and a mountainous area (Ibaraki Prefecture), where *C. japonica* is healthy. Observation of the needle surfaces using SEM revealed that the stomata in *C. japonica* needles grown in the urban area were plugged by particulate matter (Fig. 19.1).

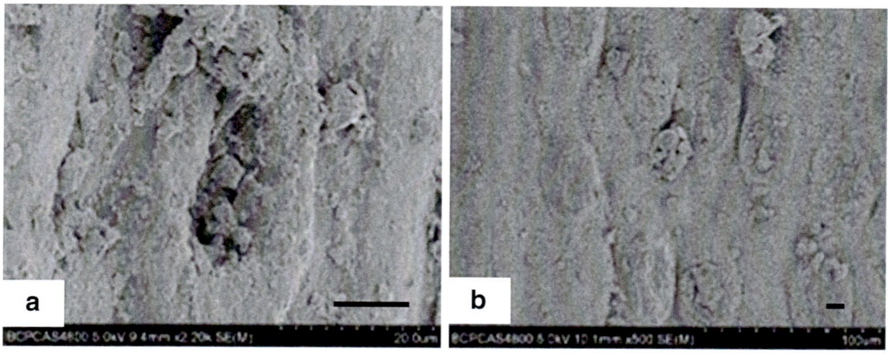

Fig. 19.2 Scanning electron micrographs of particulate matter on the leaf surfaces of evergreen tree species growing in Beijing, China. *Scale bars* are 10 μm. (**a**) Particulate matter in and around the stoma. (**b**) Particulate matter in the furrowed area (Reprinted from Song et al. (2015) with permission from Elsevier)

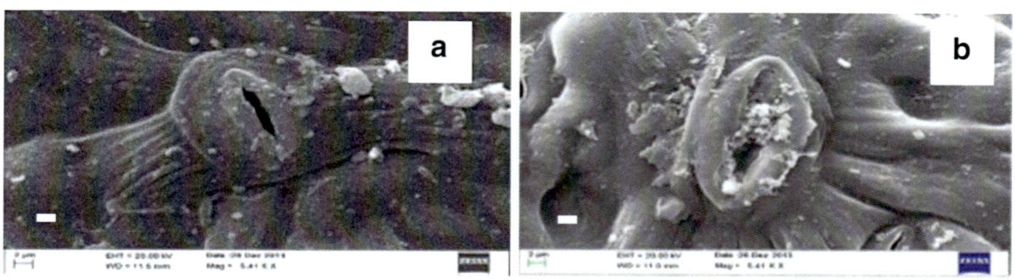

Fig. 19.3 Scanning electron micrographs of stomata of *Morus alba* from residential (**a**) and industrial (**b**) sites in Delhi, India. *Scale bars* are 2 μm. Stomatal plugging by dust is clearly depicted in (**b**) (Reprinted from Gupta et al. (2015b) with permission from Springer)

The minimum transpiration rate, which is the rate of transpiration through the cuticle and from incompletely closed stomatal pores, in needles collected from the urban area was higher than that in needles collected from the mountainous area. Based on these results, the authors suggested that stomatal malfunction may accelerate water loss and may be a significant factor in the decline of *C. japonica* in urban areas of Japan (Sase et al. 1998; Takamatsu et al. 2001). Song et al. (2015) observed the surfaces of the leaves or needles of five evergreen tree species – *Pinus tabulaeformis*, *P. bungeana*, *Platycladus orientalis*, and *Juniperus formosana* (conifers) and *Euonymus japonicus* (a broadleaved tree) – growing in Beijing, China. They found particulate matter with a size less than 2 μm in and around the stomatal pores (Fig. 19.2), suggesting that particulate matter up to 2 μm can get into the stomatal cavity. Gupta et al. (2015a, b) collected arjuna (*Terminalia arjuna*) and mulberry (*Morus alba*) leaves from residential and industrial sites in Delhi, India. In both species, plugging of stomata by dust was observed in leaves from the industrial sites (Fig. 19.3). Saha and Padhy (2012) reported that transpiration rates in the morning in the leaves of *Shorea robusta* were lower in plants growing in a forest polluted with stone dust than the rates in the leaves of plants in a non-polluted forest in the Birbhum district in West Bengal, India. In their study, observation of the leaf surface using SEM revealed stomatal plugging due to heavy deposition of dust, suggesting the malfunction of stomata in the leaves of *S. robusta* growing in polluted forests.

19.3.4 Chemical Effects

Darley (1966) exposed bean plants to cement-kiln dust, which is alkaline, with a pH above 12. Exposure to cement-kiln dust with water mist induced a reduction in leaf CO_2 exchange, and the bean plants showed leaf injury such as wilting and necrosis (Darley 1966; Lerman and Darley 1975). Kohno et al. (1977) exposed kidney bean (*P. vulgaris*) to dust derived from fossil fuel-fired large boilers in thermal power stations. Dust exposure induced leaf injury that appeared as reddish-brown or black-brown necrotic spots. The degree of injury was greater when the pH and electrical conductivity of the dust solution were lower and higher, respectively. To clarify the effects of dust components on the development of injury, various solutions including dust components were applied to kidney bean leaves (Kohno et al. 1979). Treatment with solutions of the major individual components of dust, such as ammonium sulfate, caused leaf injury, but the degree of injury was far less than that induced by the dust solution. The degree of leaf injury induced by ammonium sulfate solution became synergistically greater with the addition of vanadyl sulfate, a minor component of the dust. These results suggest that synergistic interaction between ammonium sulfate and a low concentration of vanadium in dust plays an important role in the development of leaf injury (Kohno et al. 1979). At the present time, however, little information is available on the combined effects of the components of particulate matter.

Martin et al. (1992) exposed bur oak (*Quercus macrocarpa*), soybean (*Glycine max*), and maize (*Zea mays*) to KNO_3, NH_4NO_3, or $(NH_4)_2SO_4$ particles (0.25 μm in diameter) for up to 5 h. The CO_2 uptake rates were significantly reduced in bur oak by exposure to KNO_3 and in maize by exposure to $(NH_4)_2SO_4$, and the rates were significantly increased in soybean by exposure to NH_4NO_3. Because the observed statistically significant changes in CO_2 uptake rates were small, the authors concluded that acute exposure of plants to KNO_3, NH_4NO_3, and $(NH_4)_2SO_4$ resulted in minor or no alterations of photosynthesis. In an experimental study by Chevone et al. (1986), exposure to a sulfuric acid aerosol at a concentration of 500 $\mu g\ m^{-3}$ for 4 h did not induce visible foliar injury or reduce chlorophyll content in soybean or pinto bean leaves. Gmur et al. (1983) exposed pinto bean (*P. vulgaris*) to $(NH_4)_2SO_4$ with a submicron size at concentrations of up to 26 $mg\ m^{-3}$ for 3 weeks. Although no significant effects of $(NH_4)_2SO_4$ on growth parameters were detected, the $(NH_4)_2SO_4$ exposure decreased leaf diffusive resistance and induced visible foliar injury. Because internal leaf injury occurred initially in spongy mesophyll cells in substomatal cavities, the authors suggested that submicrometer $(NH_4)_2SO_4$ entered the abaxial stomata and permeated the leaf preferentially through the spongy mesophyll. Godzik et al. (1979) measured metal elements in the leaves of *Quercus robur* and the needles of *Pinus sylvestris* growing in rural areas and growing near zinc and lead smelters. The concentrations of metal elements such as Fe, Zn, and Pb in the leaves or needles of both species from polluted areas were higher than those in rural areas. Washing the leaves and needles collected from polluted areas with chloroform, followed by ultrasonic treatment, removed the particulate matter deposited on the leaf surface. Although the removal of particulate matter from the leaf surface by washing reduced the Fe concentration to the concentration observed in the leaves collected from rural areas, the concentrations of Zn and Pb were still high. Because the stomata in *P. sylvestris* needles are covered with fine wax crystals, Godzik et al. (1979) suggested that the route of entry for dust particles into the leaf or needle was of little significance for tissue contamination with Zn and Pb, and that tissue contamination occurred indirectly (e.g., through a solution). Burkhardt (2010) suggested that hygroscopic particles such as $(NH_4)_2SO_4$ deposited on the leaf surface absorb water and deliquesce, and their solutes can be taken up by leaves via stomata; this is described in Chap. 20 in detail. However, it is currently unclear whether particulate matter with a diameter less than the stomatal pore size can directly enter leaves via the stomata.

19.4 Conclusion

Aerosol particles can indirectly affect plant productivity by scattering and absorbing solar radiation, although photosynthesis is influenced by a number of external factors. A wide range of studies have reported the physical effects of particulate matter deposited on the leaf surface – including leaf shading, increased leaf temperature, stomatal plugging, and interference with stomatal closure – that can occur in the field in Asia. Leaf shading by particulate matter reduces the net photosynthetic rate under low light conditions. The light extinction coefficient of particulate matter is an important factor in the extent of increase in leaf temperature. The increase in leaf temperature induced by particulate matter increases or decreases the net photosynthetic rate under air temperatures below or above the optimum for photosynthesis, respectively. Plugging of stomata decreases gas diffusivity during the light period, and this can reduce the net photosynthetic rate. Interference with stomatal closure induces unexpected water loss at night, which can reduce drought tolerance. Although the chemical effects of dust depend on its chemical properties (e.g., pH), the effect of dust-induced injury cannot be explained by the effect of a single major component, suggesting that dust components have an interactive effect. At present, however, little information is available on the combined effect of the components of particulate matter. In addition, it is unclear whether particulate matter with a diameter less than the stomatal pore size can directly enter leaves via the stomata.

References

Burkhardt J (2010) Hygroscopic particles on leaves: nutrients or desiccants? Ecol Monogr 80:369–399

Chameides WL, Yu H, Liu SC, Bergin M, Zhou X, Mearns L, Wang G, Kiang CS, Saylor RD, Luo C, Huang Y, Steiner A, Giorgi F (1999) Case study of the effects of atmospheric aerosols and

regional haze on agriculture: an opportunity to enhance crop yields in China through emission controls? Proc Natl Acad Sci U S A 96:13626–13633
Chaston K, Doley D (2006) Mineral particulates and vegetation: effects of coal dust, overburden and flyash on light interception and leaf temperature. Clean Air Environ Qual 40:40–44
Chevone BI, Herzfeld DE, Krupa SV, Chappelka AH (1986) Direct effects of atmospheric sulfate deposition on vegetation. J Air Pollut Control Assoc 35:813–815
Ciais P, Sabine C, Bala G, Bopp L, Brovkin V, Canadell J, Chhabra A, DeFries R, Galloway J, Heimann M, Jones C, Le Quéré C, Myneni RB, Piao S, Thornton P et al (2013) Carbon and other biogeochemical cycles. In: Stocker TF, Qin D, Plattner G-K, Tignor M, Allen SK, Boschung J, Nauels A, Xia Y, Bex V, Midgley PM (eds) Climate change 2013: the physical science basis, Contribution of Working Group I to the Fifth Assessment Report of the Intergovernmental Panel on Climate Change. Cambridge University Press, Cambridge
Darley EF (1966) Studies on the effect of cement-kiln dust on vegetation. J Air Pollut Control Assoc 16:145–150
Flückiger W, Oertli JJ, Flückiger H (1979) Relationship between stomatal diffusive resistance and various applied particle sizes on leaf surfaces. Z Pflanzenphysiol 91:173–175
Gmur NF, Evans LS, Cunningham EA (1983) Effects of ammonium sulfate aerosols on vegetation—II. Mode of entry and responses of vegetation. Atmos Environ 17:715–721
Godzik S, Florkowski T, Piorek S, Sassen MMA (1979) An attempt to determine the tissue contamination of *Quercus robur* L. and *Pinus silvestris* foliage by particulates from zinc and lead smelters. Environ Pollut 18:97–106
Grantz DA, Garner JHB, Johnson DW (2003) Ecological effects of particulate matter. Environ Int 29:213–239
Gupta GP, Kumar B, Singh S, Kulshrestha UC (2015a) Urban climate and its effect on biochemical and morphological characteristics of Arjun (*Terminalia arjuna*) plant in National Capital Region Delhi. Chem Ecol 31:524–538
Gupta GP, Singh S, Kumar B, Kulshrestha UC (2015b) Industrial dust sulphate and its effects on biochemical and morphological characteristics of Morus (*Morus alba*) plant in NCR Delhi. Environ Monit Assess 187:67
Hirano T, Kiyota M, Kitaya Y, Aiga I (1990) The physical effects of dust on photosynthetic rate of plant leaves. J Agric Meteorol 46:1–7 (in Japanese with English summary)
Hirano T, Kiyota M, Aiga I (1991) The effects of dust by covering and plugging stomata and by increasing leaf temperature on photosynthetic rate of plant leaves. J Agric Meteorol 46:215–222 (in Japanese with English summary)
Hirano T, Kiyota M, Aiga I (1995) Physical effects of dust on leaf physiology of cucumber and kidney bean plants. Environ Pollut 89:255–261
Horvath H (2000) Aerosols – an introduction. J Environ Radioact 51:5–25
Kohno Y, Takanashi S, Ishikawa H (1977) Studies on the effects of the dust on plants (1) effects of pH and electric conductivity of the dust solution on bean leaves. CRIEPI Rep No. 477001:2–20. (In Japanese with English summary)
Kohno Y, Takanashi S, Ishikawa H (1979) Studies on the effects of the dust on plants (3) effects of the water soluble components of the dust on bean leaves. CRIEPI Rep No. 479001:1–14. (In Japanese with English summary)
Larcher W (2003) In: Larcher W (ed) Physiological plant ecology, 4th edn. Springer, Berlin
Lerman SL, Darley EF (1975) Particulates. In: Mudd JB, Kozlowski TT (eds) Responses of plants to air pollution. Academic, New York, pp 141–158
Martin CE, Gravatt DA, Loeschen VS (1992) Photosynthetic responses of three species to acute exposures of nitrate- and sulphate-containing aerosols. Atmos Environ 26:381–391
Mercado LM, Bellouin N, Sitch S, Boucher O, Huntingford C, Wild M, Cox PM (2009) Impact of changes in diffuse radiation on the global land carbon sink. Nature 458:1014–1017
Norris JR, Wild M (2009) Trends in aerosol radiative effects over China and Japan inferred from observed cloud cover, solar "dimming" and solar "brightening". J Geophys Res 114:D00D15

Ricks GR, Williams RJH (1974) Effects of atmospheric pollution on deciduous woodland part 2: effects of particulate matter upon stomatal diffusion resistance in leaves of *Quercus petraea* (Mattuschka) Leibl. Environ Pollut 6:87–109
Roderick ML, Farquhar GD, Berry SL, Noble IR (2001) On the direct effect of clouds and atmospheric particles on the productivity and structure of vegetation. Oecologia 129:21–30
Saha DC, Padhy PK (2012) Effect of particulate pollution on rate of transpiration in *Shorea robusta* at Lalpahari forest. Trees 26:1215–1223
Sase H, Takamatsu T, Yoshida T, Inubushi K (1998) Changes in properties of epicuticular wax and the related water loss in Japanese cedar (*Cryptomeria japonica*) affected by anthropogenic environmental factors. Can J For Res 28:546–556
Song Y, Maher BA, Li F, Wang X, Sun X, Zhang H (2015) Particulate matter deposited on leaf of five evergreen species in Beijing, China: source identification and size distribution. Atmos Environ 105:53–60
Stanhill G, Cohen S (2001) Global diming: a review of the evidence for a widespread and signify reduction in global radiation with discussion of its probable causes and possible agricultural consequences. Agric For Meteorol 107:255–278
Takamatsu T, Sase H, Takeda J (2001) Some physiological properties of *Cryptomeria japonica* leaves from Kanto, Japan: potential factors causing tree decline. Can J For Res 31(4):663–672
Thompson JR, Mueller PW, Flückiger W, Rutter AJ (1984) The effects of dust on photosynthesis and its significance for roadside plants. Environ Pollut 34:171–190
UNEP (2011) Integrated assessment of black carbon and tropospheric ozone: summary for decision makers. United Nations Environment Programme and World Meterological Association, Nairobi, 38 pp
Wang K, Dickinson RE, Liang S (2009) Clear sky visibility has decreased over land globally from 1973 to 2007. Science 323:1468–1470
Wild M (2012) Enlightening global diming and brightening. Bull Am Meteorol Soc 93:27–37

Chapter 20
Effects of Black Carbon and Ammonium Sulfate Particles on Plants

Masahiro Yamaguchi and Takeshi Izuta

Abstract Asian countries face the problem of transboundary air pollution, including that of particulate matter with a diameter smaller than 2.5 μm ($PM_{2.5}$). The deposition velocities of $PM_{2.5}$ particles, such as black carbon (BC) and ammonium sulfate (AMS), onto forests are substantially high. In this chapter, we summarize the dynamics of BC and AMS deposited on leaf surfaces and we introduce experimental studies of their long-term effects on four representative Japanese forest tree species. Experimental studies of the acute effects of BC on plants have revealed its physical effects – stomatal plugging, leaf shading, and increased leaf temperature. These effects were not observed in an experimental study on the long-term effects of BC on four Japanese forest tree species, possibly because of the low amount of BC deposited on the leaf surfaces. The amounts of BC deposited on leaf surfaces differed considerably between experimental and field observation studies. Long-term AMS exposure did not significantly affect the growth of the four representative Japanese forest tree species, but significantly increased and decreased the net photosynthetic rate in current-year and previous-year needles, respectively, of *Cryptomeria japonica* seedlings. Hygroscopic particles such as AMS deposited on the leaf surface can establish a continuous liquid water connection between the leaf interior and surface, which enables the bidirectional transport of water and solutes between them. AMS exposure significantly increased the concentrations of NH_4^+, free amino acids, and total soluble protein in the current-year needles of *C. japonica*, suggesting that AMS deposited on the needle surfaces deliquesced, was absorbed into the needles, and was metabolized.

M. Yamaguchi (✉)
Graduate School of Fisheries and Environmental Sciences, Nagasaki University, Nagasaki, Nagasaki 852-8521, Japan
e-mail: masah-ya@nagasaki-u.ac.jp

T. Izuta
Institute of Agriculture, Tokyo University of Agriculture and Technology, Fuchu, Tokyo 183-8509, Japan
e-mail: izuta@cc.tuat.ac.jp

T. Izuta (ed.), *Air Pollution Impacts on Plants in East Asia*,
DOI 10.1007/978-4-431-56438-6_20

Keywords Ammonium sulfate • Black carbon • Fine particulate matter ($PM_{2.5}$) • Forest tree species • Growth • Hygroscopic particles • Physiological function • Plugging of stomata • Leaf shading

20.1 Introduction

In East and South Asia, trends in aerosol optical depth retrieved from satellite instruments have revealed a significant increase in the ground-level concentration of fine particulate matter with a diameter smaller than 2.5 μm ($PM_{2.5}$) from 1998 to 2012 (Boys et al. 2014). Lu et al. (2011) found that the increase in satellite-derived $PM_{2.5}$ could be attributed to the increase in emissions of sulfur dioxide (SO_2), the precursor of ammonium sulfate (AMS), and carbonaceous aerosols such as black carbon (BC; also known as elemental carbon, EC). China and India largely account for these emissions in Asia (Ohara et al. 2007; Lu et al. 2011). Although the wide application of flue-gas desulfurization equipment in power plants has decreased national SO_2 emissions in China (Lu et al. 2010), the levels of SO_2 emission and $PM_{2.5}$ are still high (Lu et al. 2011; Cao et al. 2012; Liu et al. 2015). In India, emissions of SO_2 and BC have been increasing, and high concentrations of $PM_{2.5}$ have been observed (Lu et al. 2011; Tiwari et al. 2015). $PM_{2.5}$ has a long lifetime in the atmosphere because of its low deposition velocity, and thus it is a transboundary air pollutant (Horvath 2000; Fowler 2002; Colvile 2002). Therefore, Asian countries face the problem of transboundary air pollution, including that with $PM_{2.5}$ (Takami et al. 2007; Begum et al. 2011; Begum et al. 2013; Kaneyasu et al. 2014). The deposition velocities of $PM_{2.5}$ onto forests are substantially higher than those onto other land surfaces, such as those with short vegetation (Fowler 2002; Petroff et al. 2008; Matsuda et al. 2010). Although the widespread threat of $PM_{2.5}$ to vegetation is not clear because of the variety of compounds it contains (Grantz et al. 2003), it is important to clarify the long-term effects of $PM_{2.5}$ such as BC and AMS on the growth and physiological functions of forest trees native to Asia. In this chapter, therefore, we summarize the dynamics of BC and AMS deposited on leaf surfaces, and introduce an experimental study on their long-term effects on four representative Japanese forest tree species.

20.2 Black Carbon

Black carbon (BC) is a component of $PM_{2.5}$ and its sources are the combustion of fossil fuels and biomass burning (WHO 2012). Because BC is chemically inert, it has been suggested that BC deposited on leaf surfaces has physical effects on plants. In this section, we summarize the physical effects of BC deposition on the leaf surface, and introduce an experimental study on the long-term effects of BC on the growth and gas exchange rates of four representative Japanese forest tree species, performed by Yamaguchi et al. (2012a).

20.2.1 Physical Effect of Black Carbon Deposited on Leaf Surface

Hirano et al. (1990, 1991, 1995) have reported the physical effects of black carbon (BC) on leaf gas exchange rates in cucumber (*Cucumis sativus* L.) and kidney bean (*Phaseolus vulgaris* L.) plants. Using a fluidized bed-type dust generator, *C. sativus* and *P. vulgaris* were exposed to BC with a size range of 0.03–0.20 μm within 3 min. The amount of BC deposition per unit leaf area was estimated to be up to 1.5 g m^{-2}. The exposure to BC reduced the net photosynthetic rate (A) in *C. sativus* and *P. vulgaris*, especially under relatively low light conditions (i.e., below the light intensity at which A is saturated with light). When A was plotted against the photosynthetic photon flux density (PPFD) reaching the leaf surface, which was calculated using the illuminated PPFD and the shading ratio of BC deposited on the leaf surface, the light response curves of A were similar in BC-exposed and non-exposed leaves in both species. These results indicate that BC deposited on the leaf surface absorbs light, shades the leaves, and reduces A under low light conditions (Hirano et al. 1990). Under light-saturated conditions, the leaf temperature of BC-exposed leaves in *C. sativus* was higher than that of non-exposed leaves (Hirano et al. 1995). When the air temperature was below 25 °C, the light-saturated net photosynthetic rate (A_{sat}) was increased by BC exposure. Conversely, when the air temperature was above 30 °C, A_{sat} was decreased by exposure to BC. These results indicate that BC deposited on the leaf surface absorbs light and increases the leaf temperature, resulting in an increase or decrease in A_{sat} under air temperatures below or above the optimum air temperature for photosynthesis, respectively (Hirano et al. 1995). Hirano et al. (1991, 1995) also reported that exposure to BC increased the transpiration rate in the leaves of *C. sativus*, a finding which could be explained by an increase in vapor pressure deficit between the air and leaves due to the increase in leaf temperature. In contrast, Hirano et al. (1991, 1995) reported that exposure to soil dusts (Kanto loam powder) of different median sizes (1.9, 8, and 30 μm in diameter) induced a reduction in stomatal conductance (g_s) under light conditions and an increase in g_s under dark conditions. Dust-induced changes in g_s were observed when the exposure was conducted under light conditions, but not under dark conditions, suggesting that stomata were plugged by the soil dusts. Because the plugging effect was greater when the size of the dust was smaller (Hirano et al. 1991, 1995), exposure to BC with a submicron size under light conditions could induce the plugging of stomata.

20.2.2 Experimental Study of the Effect of Black Carbon on Forest Trees

Yamaguchi et al. (2012a) conducted an experimental study of the long-term effects of BC on the growth and gas exchange rates of seedlings of *Fagus crenata* (a deciduous broad-leaved tree), *Castanopsis sieboldii* (an evergreen broad-leaved tree),

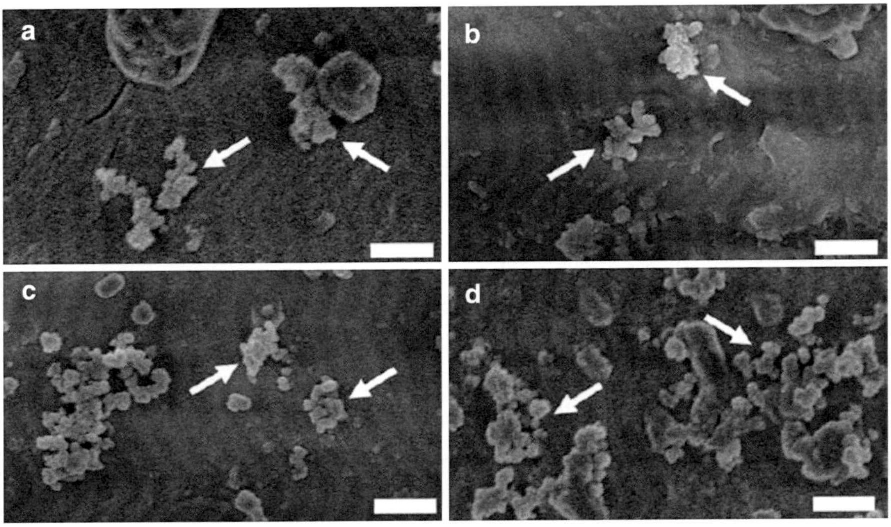

Fig. 20.1 Field-emission scanning electron micrographs showing black carbon (BC) particles deposited on the leaves or needles of (**a**) *Fagus crenata*, (**b**) *Castanopsis sieboldii*, (**c**) *Larix kaempferi*, and (**d**) *Cryptomeria japonica* seedlings grown under BC treatment. *Arrows* indicate BC particles originating from the BC exposure. *Bars* = 500 nm (Reprinted from Yamaguchi et al. 2012a with permission of Asian Journal of Atmospheric Environment)

Larix kaempferi (a deciduous conifer), and *Cryptomeria japonica* (an evergreen conifer). The seedlings were planted in 2 L pots and grown in phytotron chambers (2 × 2 × 2 m glasshouses) for two consecutive growing seasons. In the chambers, the seedlings were exposed to BC of submicron size, generated by an aerosol generator system based on an electrostatic spray (Lenggoro et al. 2002) and an ultrasonic nebulizer (Wang et al. 2008). The BC exposures were conducted within 1 h every 2 days during the first growing season and every day during the second one. Observations of leaf surfaces, using field-emission scanning electron microscopy (e.g., Yamane et al. 2012), revealed the deposition of BC of submicron size generated by the aerosol generator system (Fig. 20.1).

Exposure to BC in the four tree species did not significantly affect either the whole-plant dry mass measured at the end of the experiment or increases of plant height and stem base diameter. These results indicated that BC exposure for two growing seasons did not affect the growth of these species. The light-response curves of photosynthesis and the difference in temperature between the leaf and air under light-saturated conditions did not significantly differ between BC-exposed and non-exposed plants. These results indicated that BC deposited on the leaf surface did not absorb irradiation light and did not increase leaf temperature. BC had no significant effects on stomatal conductance (g_s) and on the response of g_s to an increase in vapor pressure deficit. These results indicated that BC deposition on the leaf surface did not induce stomatal plugging.

To discuss the different results for BC effects between those of Hirano et al. (1995) and those of Yamaguchi et al. (2012a), and the effects of BC on forest trees growing in the field, it is important to consider the amount of BC deposition on the leaf surface (M_{BC}). Yamaguchi et al. (2012b) reported an optical method for measuring M_{BC}. The leaves or needles of *F. crenata*, *C. sieboldii*, *L. kaempferi*, and *C. japonica* seedlings were harvested and washed with deionized water and then washed with chloroform to extract the BC deposited on the leaf or needle surface. The extracted BC was collected on a quartz-fiber filter. There was a significant positive correlation between the amount of BC collected on the filter, measured by a thermal optical method, and the absorbance of the filter at a wavelength of 580 nm (A_{580}). Therefore, the amount of BC on the filter could be predicted from the A_{580} value.

Using the optical method, Yamaguchi et al. (2012a) reported that M_{BC} values after BC exposure in their experiment were 0.13, 0.69, 0.32, and 0.58 mg m^{-2} total leaf area (TLA) in *F. crenata*, *C. sieboldii*, *L. kaempferi* and *C. japonica* seedlings, respectively. In the study by Hirano et al. (1995), the M_{BC} values were about 1,000 times higher than those in the study of Yamaguchi et al. (2012a). Therefore, the different results of these experimental studies on the effect of BC on plants reported by Hirano et al. (1995) and Yamaguchi et al. (2012a) can be attributed to the considerable difference in M_{BC}. In contrast, in a field observational study at a cool-temperate larch forest in Hokkaido, Japan, Fukazawa et al. (2012) reported M_{BC} values of 23 and 6 mg m^{-2} on a leaf area basis, in the upper and lower parts of hybrid larch (*Larix gmelinii* × *L. kaempferi*), respectively, and a value of 2 mg m^{-2} on a leaf area basis in Sasa bamboo, the understory vegetation. Sase et al. (2012) reported that the M_{BC} of *C. japonica* in central Japan was 26.8 mg m^{-2} TLA. In Tokyo, Japan, the M_{BC} of *Quercus serrata* at canopy height was 10–15 mg m^{-2} TLA (Hara et al. 2014). In northeastern Thailand, the M_{BC} values of a dry evergreen forest (averaged over three tree species, *Shorea henryana*, *Hopea ferrea*, and *Dipterocarpus turbinatus*), the M_{BC} values of *Pterocarpus macrocarpus* in a dry deciduous forest, and the M_{BC} values of *Acacia mangium* (a representative plantation species) were, respectively, 10.0, 9.3, and 10.8 mg m^{-2} TLA (Sase et al. 2012). Matsuda et al. (2012) reported that the deposition of elemental carbon from the atmosphere to tropical forest in Thailand was 0.34 mg C m^{-2} ground area day^{-1}, which corresponded to an M_{BC} of 8.9 mg C m^{-2} TLA, assuming that the leaf area index of the forest and the number of days during the leafy season were 3.5 and 183, respectively. These results indicate that M_{BC} values observed in the field are much lower than those in the acute exposure experiment reported by Hirano et al. (1995) and several tens of times higher than those produced by BC exposure in the long-term exposure experiment reported by Yamaguchi et al. (2012a). To clarify the effects of BC on the growth and gas exchange rates of forest tree species growing in the field, it is therefore necessary to clarify the effects of BC deposition on the leaf surface at a realistic level.

20.3 Ammonium Sulfate

A substantial proportion of primary gaseous pollutants, including SO_2 and NH_3, is transformed into submicron-size particles such as AMS (Fowler 2002). Unlike BC, AMS is a hygroscopic particle that can have not only physical but also chemical effects on plants. It has been suggested that hygroscopic particles such as AMS, ammonium hydrogen sulfate (NH_4HSO_4), and sodium chloride (NaCl) deposited on the foliar surface absorb water vapor, deliquesce, and can be taken up from the leaf surface as solutes. In this section, we summarize the proposed dynamics of hygroscopic particles such as AMS deposited on the leaf surface, and we introduce an experimental study of the long-term effect of AMS on the growth and physiological functions of four representative Japanese forest tree species (Yamaguchi et al. 2014).

20.3.1 Dynamics of Hygroscopic Particles Deposited on the Leaf Surface

Burkhardt and Eiden (1994) and Burkhardt (1995) reported a positive correlation between relative air humidity and electrical conductance (i.e., wettability) on the surfaces of the attached needles of a 40-year-old Norway spruce (*Picea abies* (L.) Karst.) in the field and on the surfaces of detached needles of Norway spruce and Scotch pine (*Pinus sylvestris* L.) inside a dew chamber in which the relative air humidity could be controlled. The correlation decreased or disappeared when the needles were washed thoroughly with deionized water, suggesting that hygroscopic salts on the needle surface, which were deposited in the field and removed by washing, absorb water vapor and form a thin water film on the needle surface (Burkhardt and Eiden 1994; Burkhardt 1995).

Burkhardt et al. (1999) exposed amphistomatous *Vicia faba* and hypostomatous *Aegopodium podagraria* to submicrometer $NaNO_3$ particles, a type of hygroscopic particle. They reported a strong correlation between stomatal conductance to water vapor (g_s) and the electrical conductance of the leaf surface at low relative air humidity (35 %) under various light and CO_2 levels. This correlation was found on the abaxial side, but not on the adaxial side (i.e., the astomatous side) of *A. podagraria* leaves. These results suggested that hygroscopic particles deposited on the leaf surface could absorb transpired water vapor from stomata, and then deliquesce (Burkhardt et al. 1999).

Burkhardt et al. (2012) observed the dynamics of hygroscopic particles (NaCl and $NaClO_3$) on hydrophobic leaf surfaces (isolated tomato cuticles) under changes in relative air humidity in a chamber in which relative air humidity could be controlled, using an environmental scanning electron microscope. They reported that several humidity cycles caused repetition of the deliquescence and efflorescence of the particles, spatial expansion of the particles, and an increase in the wettability of the leaf surface. Burkhardt and Pariyar (2014) reported similar observations on Scotch pine needles, and noted that hygroscopic particles ($(NH_4)_2SO_4$) expanded into the epistomatal chambers in the needles. In their report, the minimum epidermal conductances of Scotch pine needles sprayed with hygroscopic particles were higher than those of needles sprayed with H_2O, suggesting the establishment of a continuous liquid water connection from the leaf apoplast to the leaf surface, which was called "hydraulic activation of stomata" (HAS) (Burkhardt 2010). HAS enables the bidirectional transport of water and solutes between the leaf interior and leaf surface (Burkhardt 2010) (Fig. 20.2).

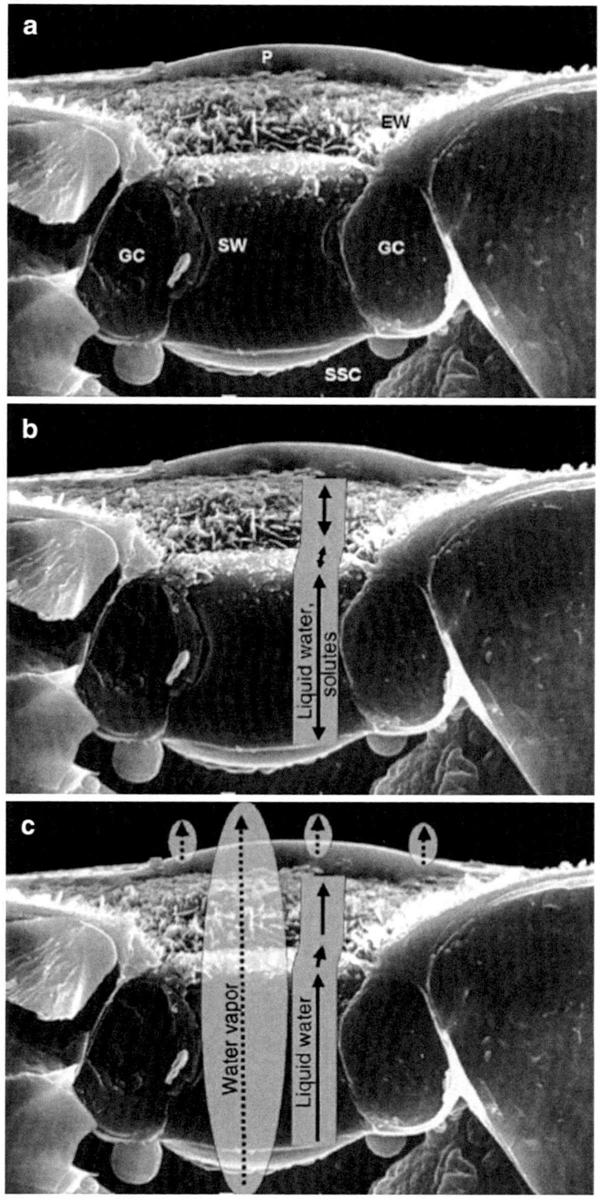

Fig. 20.2 Cryo-scanning electron microscopy image of a stomatal pore of *Allium cepa* with an amorphous salt particle on the surface. (**a**) Original picture, showing epicuticular waxes (EW), guard cells (GC), the stomatal wall (SW), the substomatal cavity (SSC), and the particle (P). The dimension of the pore is about 15 μm. (**b**, **c**) Hydraulic activation of stomata (HAS), with different scenarios of water and solute flow. (**b**) HAS is established, enabling stomatal transport of solutes and liquid water (*solid arrows*) in either direction. (**c**) Wicking of water, caused by HAS and salt on the surface. Liquid water flows along the stomatal walls to the salt particle on the leaf surface, where it evaporates. This is the "wick" part of a system with split transpiration, in addition to transpired water vapor (*dotted arrows*) (Reprinted from Burkhardt (2010) with permission of John Wiley & Sons, Inc.)

Pariyar et al. (2013) conducted an experimental study on the effect of ambient levels of particulate matter on water relations in sunflower (*Helianthus annuus*) and bean (*Vicia faba*) plants, using greenhouses equipped with or without a high-efficiency particulate air filter. In their study, ambient levels of particulate matter increased g_s and the transpiration rate, resulting in lower leaf water potential and osmotic potential. Burkhardt (2010) suggested that excessive HAS is caused by the over-accumulation of particles, which may work as "desiccants", reducing the drought tolerance of plants. Furthermore, Burkhardt (2010) hypothesized that deliquescent hygroscopic particles were amorphous and thus may have been misinterpreted as "degraded waxes", which have often been observed in forest declines caused by air pollution.

20.3.2 Experimental Study of the Effect of Ammonium Sulfate on Forest Trees

Yamaguchi et al. (2014) conducted an experimental study of the long-term effect of AMS particles on the growth and physiological functions of *F. crenata*, *C. sieboldii*, *L. kaempferi*, and *C. japonica* seedlings. The seedlings were planted in 2-L pots and grown in phytotron chambers ($2 \times 2 \times 2$ m glasshouses) for two consecutive growing seasons. In the chambers, the seedlings were exposed to AMS of submicron size generated from 0.1 % (w/v) AMS solution; this solution was atomized with an ultrasonic nebulizer, and then dried in the gas-phase with a homemade tubular-type aerosol drying system equipped with granular silica gels (e.g., Hori et al. 2003). AMS exposure for about 1 h was conducted every day during the first growing season and twice per day during the second growing season. Observation of the leaf surface, using field-emission scanning electron microscopy and elemental analysis of deposited particles by energy dispersive X-ray spectrometry, revealed the deposition of AMS (Fig. 20.3).

The exposure to AMS did not significantly affect whole-plant dry mass measured at the end of the experiment. The increases in daily mean sulfate concentration in $PM_{2.5}$ inside the chamber during the first and second growing seasons were 2.73 and 4.32 μg SO_4^{2-} m^{-3}, respectively. In Japan, the annual mean concentration of sulfate in particulate matter was 1.02–4.76 μg SO_4^{2-} m^{-3} in 2011 (Network Center for EANET 2013). These results suggested that exposure to AMS at ambient levels for two growing seasons did not significantly affect the whole-plant growth of *F. crenata*, *C. sieboldii*, *L. kaempferi*, or *C. japonica* seedlings. Because AMS exposures of about 1 h were conducted once or twice per day in the study byYamaguchi et al. (2014), the AMS concentrations inside the chamber were high only during the exposure time (i.e., for about 1 h), resulting in an unrealistic AMS exposure regime. Therefore, further experimental study with a more realistic exposure regime is required to clarify the effects of AMS on forest trees growing in the field.

The AMS exposure did not significantly affect the net photosynthetic rate under light-saturated conditions (A_{sat}) in the leaves or needles of *F. crenata*, *C. sieboldii*, and *L. kaempferi* seedlings. Conversely, in *C. japonica* seedlings, AMS exposure significantly reduced A_{sat} in previous-year needles and significantly increased A_{sat} in current-year needles (Fig. 20.4). These results indicated that the AMS sensitivity of net photosynthesis in *C. japonica* seedlings was higher than that in *F. crenata*, *C. sieboldii*, and *L. kaempferi* seedlings. Several researchers have suggested that the deposition velocity of particulate matter onto the foliar surface is higher in coniferous

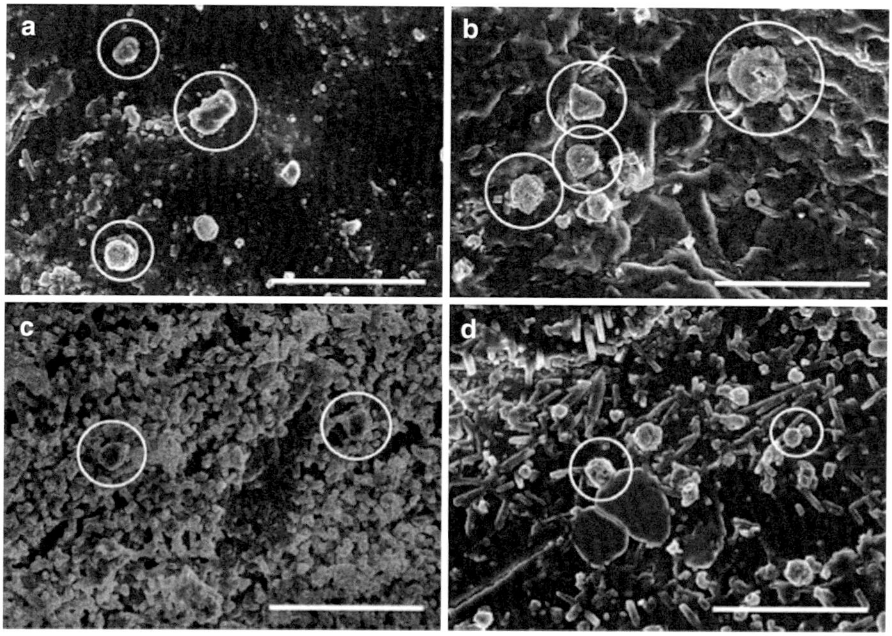

Fig. 20.3 Field-emission scanning electron micrographs showing ammonium sulfate (AMS) particles (*circled*) on the foliar surfaces of (**a**) *Fagus crenata*, (**b**) *Castanopsis sieboldii*, (**c**) *Larix kaempferi*, and (**d**) *Cryptomeria japonica* seedlings grown under AMS treatment. *Bars* = 5 μm (Reprinted from Yamaguchi et al. (2014) with permission from Elsevier)

tree species because of the complex structure of the shoots and the smaller leaves in conifers compared with broad-leaved tree species (Beckett et al. 2000; Freer-Smith et al. 2004; Hwang et al. 2011). Therefore, the higher AMS sensitivity of net photosynthesis in *C. japonica* needles could have been caused by a larger amount of AMS on the needle surface resulting from a higher deposition velocity (Yamaguchi et al. 2014).

AMS-induced increases in stomatal conductance and ribulose 1,5-bisphosphate carboxylase/oxygenase (Rubisco) concentrations were observed in current-year needles of *C. japonica* seedlings, and this could have induced the increase in A_{sat}. As described above, hygroscopic particles such as AMS deposited on the leaf surface can cause HAS, which enables the bidirectional transport of water and solutes, such as SO_4^{2-} and NH_4^+, between the leaf surface and leaf interior via the stomata (Burkhardt 2010). It has been shown that solutes on the leaf surface can be taken up by the leaf through the waxy cuticle (Bielenberg and BassiriRad 2005). In the study by Yamaguchi et al. (2014), AMS exposure significantly increased the concentrations of not only NH_4^+ but also of N compounds such as free amino acids and total soluble protein in current-year needles of *C. japonica* (Fig. 20.5). These results suggested that AMS deposited on the surface of the current-year needles of *C. japonica* deliquesced, was absorbed into the needles via the stomata and/or via the waxy

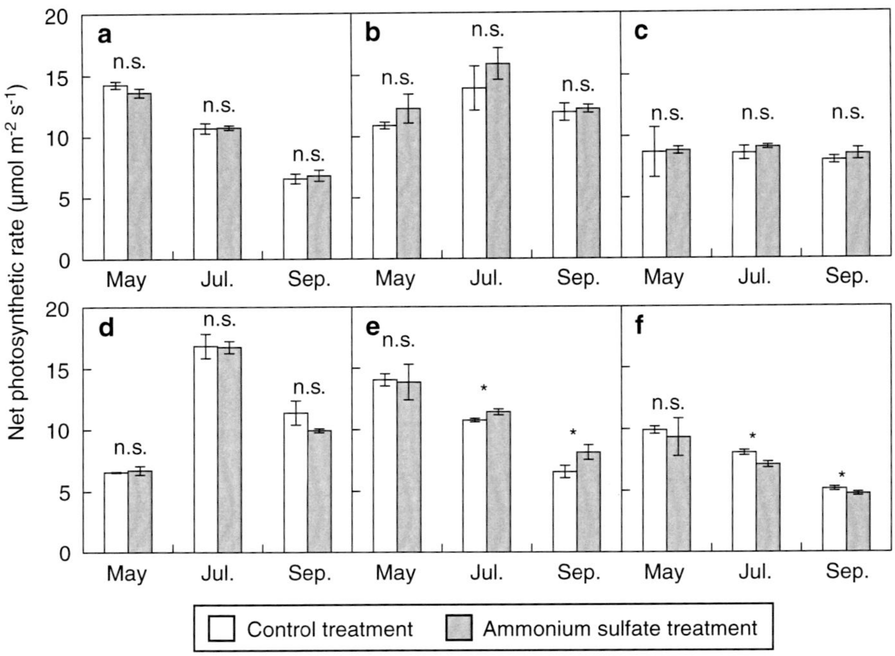

Fig. 20.4 Effects of ammonium sulfate (AMS) exposure on the net photosynthetic rate in (**a**) *Fagus crenata* leaves, (**b**) current-year and (**c**) previous-year leaves of *Castanopsis sieboldii*, (**d**) *Larix kaempferi* needles, and (**e**) current-year and (**f**) previous-year needles of *Cryptomeria japonica* in May, July, and September in the second growing season. Student's *t*-test: * $p<0.05$, *n.s.* not significant (Reprinted from Yamaguchi et al. (2014) with permission from Elsevier)

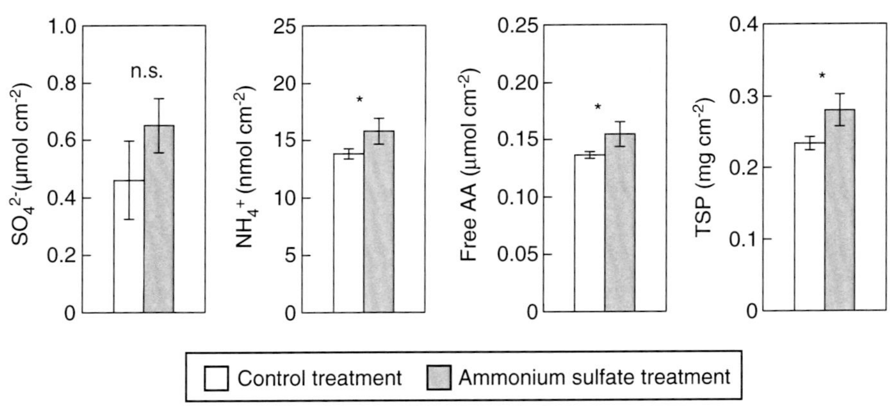

Fig. 20.5 Effects of ammonium sulfate (AMS) exposure on the concentrations of SO_4^{2-}, NH_4^+, free amino acids (AA), and total soluble protein (TSP) in current-year needles of *Cryptomeria japonica* seedlings in July in the second growing season. Student's *t*-test: * $p<0.05$, *n.s.* not significant (Reprinted from Yamaguchi et al. (2014) with permission from Elsevier)

cuticle, and was metabolized into amino acids and protein in the needle, resulting in higher net photosynthesis capacity.

In the previous-year needles of *C. japonica* seedlings, AMS exposure induced a significant reduction in the Rubisco concentration, which could account for the reduction in A_{sat} (Yamaguchi et al. 2014). Several researchers have reported that fertilization shortens the leaf lifespan (Reich et al. 1992; Balster and Marshall 2000; Cordell et al. 2001). Because AMS exposure was suggested to have a fertilization effect in the current-year needles of *C. japonica*, the reduction in Rubisco concentration in the previous-year needles may have been caused by a shortened needle lifespan, due to the fertilization effect. In contrast, Martin et al. (1992) reported a reduction in the net photosynthetic rate in *Zea mays* leaves after acute exposure to AMS. Therefore, because of the longer exposure period in previous-year needles than in current-year needles of *C. japonica* seedlings, the reduction in Rubisco concentration could also be interpreted as an adverse effect of excessive AMS deposition, which might cause excessive HAS and reduce drought tolerance (Burkhardt 2010). Further study is required to clarify the mechanisms underlying the reduction in net photosynthesis caused by AMS exposure.

20.4 Conclusion and Future Perspectives

Experimental studies of the acute effects of BC on plants revealed that BC deposition on the leaf surface induced physical effects – stomatal plugging, leaf shading, and increased leaf temperature. However, in an experimental study of the long-term effects of BC on four representative Japanese forest trees, no physical effects of BC were observed. Because there were considerable differences in the amount of BC deposition on the leaf surface not only between the short-term and long-term experiments, but also between the experimental and field observational studies, it is unclear whether BC has physical effects on forest trees growing in the field. To evaluate the effects of BC on the growth and gas exchange rates of forest tree species in the field, it is therefore necessary to clarify the effects of BC deposition on the leaf surface at a realistic level.

An experimental study of the long-term effects of AMS revealed that AMS exposure at ambient levels for two growing seasons did not significantly affect the whole-plant growth of *F. crenata*, *C. sieboldii*, *L. kaempferi*, or *C. japonica* seedlings. Conversely, the AMS sensitivity of net photosynthesis in *C. japonica* needles was higher than that in the leaves or needles of *F. crenata*, *C. sieboldii*, and *L. kaempferi* seedlings, and this could be a result of higher AMS deposition velocity onto the needle surface of *C. japonica* seedlings. In the current-year needles of *C. japonica*, AMS deposited on the surface could deliquesce, be absorbed into the needles via the stomata and/or the waxy cuticle, and be metabolized into amino acids and protein in the needles, resulting in higher net photosynthesis capacity. Conversely, in the previous-year *C. japonica* needles, it is unclear why net photosynthesis was reduced by AMS exposure. Further study is required to clarify whether the AMS-induced reduction in net photosynthetic rate in the previous-year needles was caused by the excessive deposition of AMS, which may reduce drought tolerance by HAS, and/or by a shortening of the leaf lifespan through the fertilization effect of AMS on the current-year needles.

References

Balster NJ, Marshall JD (2000) Decreased needle longevity of fertilized Douglas-fir and grand fir in the northern Rockies. Tree Physiol 20:1191–7

Beckett KP, Freer-Smith PH, Taylor G (2000) Particulate pollution capture by urban trees: effect of species and windspeed. Glob Change Biol 6:995–1003

Begum BA, Biswas SK, Pandit GG, Saradhi IV, Waheed S, Siddique N, Shirani Seneviratne MC, Cohen DD, Markwitz A, Hopke PK (2011) Long-range transport of soil dust and smoke pollution in the South Asian region. Atmos Pollut Res 2:151–7

Begum BA, Hopke PK, Markwitz A (2013) Air pollution by fine particulate matter in Bangladesh. Atmos Pollut Res 4:75–86

Bielenberg DG, BassiriRad H (2005) Nutrient acquisition of terrestrial plants in a changing climate. In: BassiriRad H (ed) Nutrient acquisition by plants – an ecological perspective. Springer, Berlin, pp 311–30

Boys BL, Martin RV, van Donkelaar A, MacDonell RJ, Hsu NC, Cooper MJ, Yantosca RM, Lu Z, Streets DG, Zhang Q, Wang SW (2014) Fifteen-year global time series of satellite-derived fine particulate matter. Environ Sci Tech 48:11109–18

Burkhardt J (1995) Hygroscopic salts on the leaf surface as a possible cause of forest decline symptoms. Water Air Soil Pollut 85:1245–50

Burkhardt J (2010) Hygroscopic particles on leaves: nutrients or desiccants? Ecol Monogr 80:369–99

Burkhardt J, Eiden R (1994) Thin water films on coniferous needles. Atmos Environ 28:2001–17

Burkhardt J, Pariyar S (2014) Particulate pollutants are capable to 'degrade' epicuticular waxes and to decrease the drought tolerance of Scots pine (*Pinus sylvestris* L.). Environ Pollut 184:659–67

Burkhardt J, Kaiser H, Goldbach H, Kappen L (1999) Measurements of electrical leaf surface conductance reveal recondensation of transpired water vapour on leaf surfaces. Plant Cell Environ 22:189–96

Burkhardt J, Basi S, Pariyar S, Hunsche M (2012) Stomatal penetration by aqueous solutions – an update involving leaf surface particles. New Phytol 196:774–87

Cao J-J, Shen Z-X, Chow JC, Watson JG, Lee S-C, Tie X-X, Ho K-F, Wang G-H, Han Y-M (2012) Winter and summer $PM_{2.5}$ chemical compositions in fourteen Chinese cities. J Air Waste Manage 62:1214–26

Colvile RN (2002) Emissions, dispersion and atmospheric transformation. In: Bell JNB, Treshow M (eds) Air pollution and plant life, 2nd edn. Wiley, England, pp 23–42

Cordell S, Goldstein G, Meinzer FC, Vitousek PM (2001) Regulation of leaf life-span and nutrient-use efficiency of *Metrosideros polymorpha* trees at two extremes of a long chronosequence in Hawaii. Oecologia 127:198–206

Fowler D (2002) Pollutant deposition and uptake by vegetation. In: Bell JNB, Treshow M (eds) Air pollution and plant life, 2nd edn. Wiley, England, pp 43–68

Freer-Smith PH, El-Khatib AA, Taylor G (2004) Capture of particulate pollution by trees: a comparison of species typical of semi-arid areas (*Ficus nitida* and *Eucalyptus globulus*) with European and North American species. Water Air Soil Pollut 155:173–87

Fukazawa T, Murao N, Sato H, Takahashi M, Akiyama M, Yamaguchi T, Noguchi I, Takahashi H, Kozuka C, Sakai R, Takagi K, Fujinuma Y, Saigusa N, Matsuda K (2012) Deposition of

aerosols on leaves in a cool-temperate larch forest in Northern Hokkaido, Japan. Asian J Atmos Environ 6:281–7
Grantz DA, Garner JHB, Johnson DW (2003) Ecological effects of particulate matter. Environ Int 29:213–39
Hara H, Kashiwakura T, Kitayama K, Bellingrath-Kimura SD, Yoshida T, Takayanagi M, Yamagata S, Murao N, Okouchi H, Ogata H (2014) Foliar rinse study of atmospheric black carbon deposition to leaves of konara oak (*Quercus serrata*) stands. Atmos Environ 97:511–8
Hirano T, Kiyota M, Kitaya Y, Aiga I (1990) The physical effects of dust on photosynthetic rate of plant leaves. J Agric Meteorol 46:1–7 (in Japanese with English summary)
Hirano T, Kiyota M, Aiga I (1991) The effects of dust by covering and plugging stomata and by increasing leaf temperature on photosynthetic rate of plant leaves. J Agric Meteorol 46:215–22 (in Japanese with English summary)
Hirano T, Kiyota M, Aiga I (1995) Physical effects of dust on leaf physiology of cucumber and kidney bean plants. Environ Pollut 89:255–61
Hori M, Ohta S, Murao N, Yamagata S (2003) Activation capability of water soluble organic substances as CCN. J Aerosol Sci 34:419–48
Horvath H (2000) Aerosols – an introduction. J Environ Radioact 51:5–25
Hwang H-J, Yook S-J, Ahn K-H (2011) Experimental investigation of submicron and ultrafine soot particle removal by tree leaves. Atmos Environ 45:6987–94
Kaneyasu N, Yamamoto S, Sato K, Takami A, Hayashi M, Hara K, Kawamoto K, Okuda T, Hatakeyama S (2014) Impact of long-range transport of aerosols on the $PM_{2.5}$ composition at a major metropolitan area in the northern Kyusyu area of Japan. Atmos Environ 97:416–25
Lenggoro IW, Xia B, Okuyama K (2002) Sizing of colloidal nanoparticles by electrospray and differential mobility analyzer methods. Langmuir 18:4584–91
Liu Z, Hu B, Wang L, Wu F, Gao W, Wang Y (2015) Seasonal and diurnal variation in particulate matter (PM_{10} and $PM_{2.5}$) at an urban site of Beijing: analyses from a 9-year study. Environ Sci Pollut Res 22:627–42
Lu Z, Streets DG, Zhang Q, Wang S, Carmichael GR, Cheng YF, Wei C, Chin M, Diehl T, Tan Q (2010) Sulfur dioxide emissions in China and sulfur trends in East Asia since 2000. Atmos Chem Phys 10:6311–31
Lu Z, Zhang Q, Streets DG (2011) Sulfur dioxide and primary carbonaceous aerosol emissions in China and India, 1996–2010. Atmos Chem Phys 11:9839–64
Martin CE, Gravatt DA, Loeschen VS (1992) Photosynthetic responses of three species to acute exposures of nitrate- and sulphate-containing aerosols. Atmos Environ 26:381–91
Matsuda K, Fujimura Y, Hayashi K, Takahashi A, Nakaya K (2010) Deposition velocity of $PM_{2.5}$ sulfate in the summer above a deciduous forest in central Japan. Atmos Environ 44:4582–7
Matsuda K, Sase H, Murao N, Fukazawa T, Khoomsub K, Chanonmuang P, Visaratana T, Khummongkol P (2012) Dry and wet deposition of elemental carbon on a tropical forest in Thailand. Atmos Environ 54:282–7
Network Center for EANET (2013) Acid Deposition Monitoring Network in East Asia (EANET) Data report 2011, Network Center for EANET, Niigata, pp 180, 190
Ohara T, Akimoto H, Kurokawa J, Horii N, Yamaji K, Yan X, Hayasaka T (2007) An Asian emission inventory of anthropogenic emission sources for the period 1980–2020. Atmos Chem Phys 7:4419–44
Pariyar S, Eichert T, Goldbach HE, Hunsche M, Burkhardt J (2013) The exclusion of ambient aerosols changes the water relations of sunflower (*Helianthus annuus*) and bean (*Vicia faba*) plants. Environ Exp Bot 88:43–53
Petroff A, Mailliat A, Amielh M, Anselmet F (2008) Aerosol dry deposition on vegetative canopies. Part II: a new modelling approach and applications. Atmos Environ 42:3654–83
Reich PB, Walters MB, Ellsworth DS (1992) Leaf life-span in relation to leaf, plant, and stand characteristics among diverse ecosystems. Ecol Monogr 62:365–92
Sase H, Matsuda K, Visaratana T, Garivait H, Yamashita N, Kietvuttinon B, Hongthong B, Luangjame J, Khummongkol P, Shindo J, Endo T, Sato K, Uchiyama S, Miyazawa M, Nakata

M, Lenggoro IW (2012) Deposition process of sulfate and elemental carbon in Japanese and Thai forest. Asian J Atmos Environ 6:246–58

Takami A, Miyoshi T, Shimono A, Kaneyasu N, Kato S, Kajii Y, Hatakeyama S (2007) Transport of anthropogenic aerosols from Asia and subsequent chemical transformation. J Geophys Res 112:D22S31

Tiwari S, Hopke PK, Pipal AS, Srivastava AK, Bisht SS, Tiwari S, Singh AK, Soni VK, Attri SD (2015) Intra-urban variability of particulate matter ($PM_{2.5}$ and PM_{10}) and its relationship with optical properties of aerosols over Delhi, India. Atmos Res 166:223–32

Wang W-N, Purwanto A, Lenggoro IW, Okuyama K, Chang H, Jang HD (2008) Investigation on the correlations between droplet and particle size distribution in ultrasonic spray pyrolysis. Ind Eng Chem Res 47:1650–9

WHO (World Health Organization) (2012) Health effects of black carbon. WHO Regional Office for Europe, Bonn

Yamaguchi M, Otani Y, Takeda K, Lenggoro IW, Ishida A, Yazaki K, Noguchi K, Sase H, Murao N, Nakaba S, Yamane K, Kuroda K, Sano Y, Funada R, Izuta T (2012a) Effects of long-term exposure to black carbon particles on growth and gas exchange rates of *Fagus crenata*, *Castanopsis sieboldii*, *Larix kaempferi* and *Cryptomeria japonica* seedlings. Asian J Atmos Environ 6:259–67

Yamaguchi M, Takeda K, Otani Y, Murao N, Sase H, Lenggoro IW, Yazaki K, Noguchi K, Ishida A, Izuta T (2012b) Optical method for measuring deposition amount of black carbon particles on foliar surface. Asian J Atmos Environ 6:268–74

Yamaguchi M, Otani Y, Li P, Nagao H, Lenggoro IW, Ishida A, Yazaki K, Noguchi K, Nakaba S, Yamane K, Kuroda K, Sano Y, Funada R, Izuta T (2014) Effects of long-term exposure to ammonium sulfate particles on growth and gas exchange rates of *Fagus crenata*, *Castanopsis sieboldii*, *Larix kaempferi* and *Cryptomeria japonica* seedlings. Atmos Environ 97:493–500

Yamane K, Nakaba S, Yamaguchi M, Kuroda K, Sano Y, Lenggoro IW, Izuta T, Funada R (2012) Visualization of artificially deposited submicron-sized aerosol particles on the surfaces of leaves and needles in trees. Asian J Atmos Environ 6:275–80

Chapter 21
Dry Deposition of Aerosols onto Forest

Kazuhide Matsuda

Abstract Recent knowledge of the dry deposition of aerosols onto forests, focusing especially on field measurements in East Asia, is summarized with respect to aerosol impact on the vegetation in the region. Direct measurements of aerosol deposition in Japan by means of the gradient method and the relaxed eddy accumulation method were introduced together with their principles. They indicate that the levels of deposition velocity of $PM_{2.5}$ sulfate were in agreement with experimental results in North America and Europe and were higher than those calculated by theoretical models. The deposition velocity was probably influenced by humidity due to the growth in size of hygroscopic aerosols. The calculations taking into account the effect of growth of hygroscopic aerosols were in reasonable agreement with the measurements. A complex terrain contributed to increasing the range of measured deposition velocities. However, the measurements agreed with calculated deposition velocities on average for about 2 weeks. Measurements of vertical profiles of $PM_{2.5}$ components in a forest indicate that the dry deposition of NH_4NO_3 particles was more efficient than that of $(NH_4)_2SO_4$ particles, probably due to the volatilization processes of NH_4NO_3. Finally, a case study on the dry deposition of elemental carbon estimated by the inference method in a tropical forest in Thailand was introduced.

Keywords Deposition velocity • Flux • $PM_{2.5}$ • Sulfate • Nitrate • Elemental carbon • Hygroscopic aerosol • Complex terrain • Resistance model

21.1 Introduction

Aerosols are an important component of air pollution in East Asia. Transboundary air pollution of $PM_{2.5}$ and the impact of deposition on the ecosystem are of particular concern. However, dry deposition of air pollutants has rarely been known due to the lack of field measurements on dry deposition flux and deposition velocity in the region. In the first decade of this century, field studies on the dry deposition of gases such as O_3

K. Matsuda
Faculty of Agriculture Field Science Center, Tokyo University of Agriculture and Technology, Fuchu, Tokyo 183-8509, Japan
e-mail: kmatsuda@cc.tuat.ac.jp

T. Izuta (ed.), *Air Pollution Impacts on Plants in East Asia*,
DOI 10.1007/978-4-431-56438-6_21

and SO_2 by direct measurement methods were implemented in some countries in East Asia, for example, in Japan, China, Taiwan and Thailand (Matsuda et al. 2010). Following these studies, a research project titled "The Impacts of Aerosols in East Asia on Plants and Human Health Project (ASEPH)" was implemented from 2008 to 2012, resulting in several field studies on the direct measurement of aerosol deposition onto forests. Petroff et al. (2008) reviewed more than 50 papers representing the current knowledge of aerosol dry deposition on vegetative canopies. Almost all of the papers reviewed were based on field studies in Europe and North America, and there were no papers derived from Asia. Currently, we do have, however, several papers that were produced as part of the ASEPH project. Based on these field studies, recent knowledge is summarized to elaborate upon the assessment methodologies presently being used to assess the impact of aerosols on the vegetation in the region.

21.2 Directly Measured Deposition Velocity

In order to study processes of aerosol deposition or to validate models estimating deposition velocity, direct measurements of aerosol flux are required. A variety of direct measurement methods have been summarized in previous texts (e.g., Erisman and Draaijers 1995; Seinfeld and Pandis 2006; Foken 2008). Among the methods, micrometeorological methods (e.g., eddy correlation, relaxed eddy accumulation, and gradient) are suitable for measuring aerosol flux onto forest. In East Asia, the gradient method has often been used, and more recently, the relaxed eddy accumulation (REA) method is also being used. The REA method is an alternative to the eddy correlation method, which provides the most direct measure of flux. Since the REA method avoids the need for a fast-response analyzer, it is appropriate for measuring the flux of aerosol components that are difficult to measure quickly. Flux measurements by these methods are carried out at a sampling point above the forest canopy and within the constant flux layer by means of an experimental tower constructed in the forest. In flat, homogeneous terrain, the flux represents the average vertical flux over the upwind fetch (Erisman and Draaijers 1995). An illustration of the instrumentation used for micrometeorological measurements at the Field Museum Tamakyuryo (FM Tama) in suburban Tokyo is shown in Fig. 21.1.

21.2.1 *Gradient Method*

In the gradient method (Erisman and Draaijers 1995), flux, F, is computed using the following equation:

$$F=-ku^{*}\Delta c/\left[ln\left((z_2-d)/(z_1-d)\right)-\Psi_h\left((z_2-d)/L\right)+\Psi_h\left((z_1-d)/L\right)\right], \quad (21.1)$$

where k is the Von Karman constant, u^* is the friction velocity, Δc represents the difference in concentration between heights z_1 and z_2 above the canopy, L is the

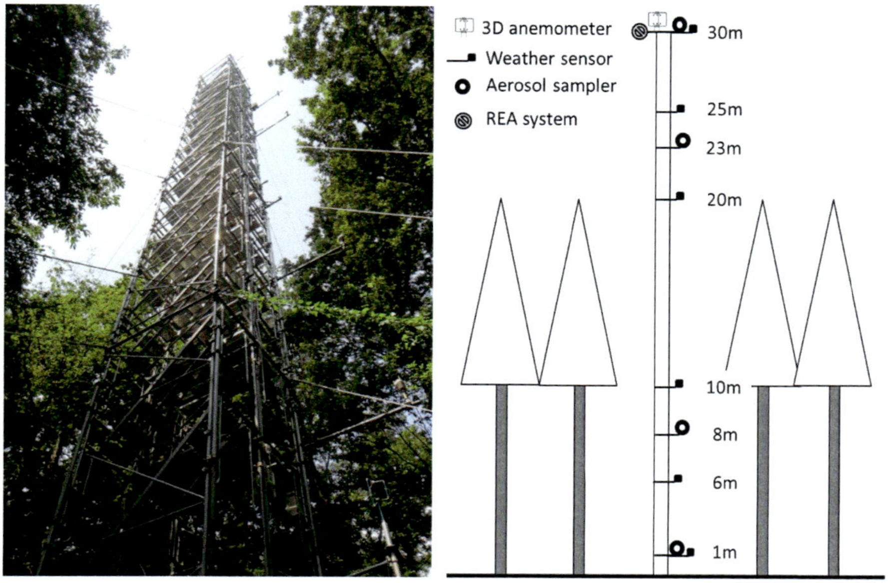

Fig. 21.1 Illustration of instrumentation used for micrometeorological measurements at the Field Museum Tamakyuryo (FM Tama)

Monin-Obukhov length, Ψ_h is the integrated stability correction function for heat, and d is the zero-plane displacement height. In the case of the illustration (Fig. 21.1), z_1 and z_2 represent 23 m and 30 m, respectively. The deposition velocity, V_d, is determined by the following equation:

$$V_d = -F / C, \tag{21.2}$$

where C is the concentration at height z_2.

The gradient method based on the Monin-Obukhov theory is valid in the inertial sub-layer ($z >> z_0$) (Businger 1986). When measurements are performed closer to the canopy top, in the roughness sub-layer, a height-dependent correction factor can be used (Wyers and Duyzer 1997). The precision of flux determined by the gradient method strongly depends on the precision of Δc. An illustration of the Δc measurements for $PM_{2.5}$ sulfate at a deciduous forest in Nagano, Japan is shown in Fig. 21.2 based on the study of Matsuda et al. (2010). Matsuda et al. (2010) discusses significant detection of Δc and the height-dependent correction factor in the roughness sub-layer.

21.2.2 Relaxed Eddy Accumulation Method

The REA method (Foken 2008) is currently used for determining the air-surface exchange of gas- and aerosol-phase sulfur and nitrogen compounds (e.g., Meyers et al. 2006). Matsuda et al. (2015) applies the method to determine $PM_{2.5}$ sulfate flux

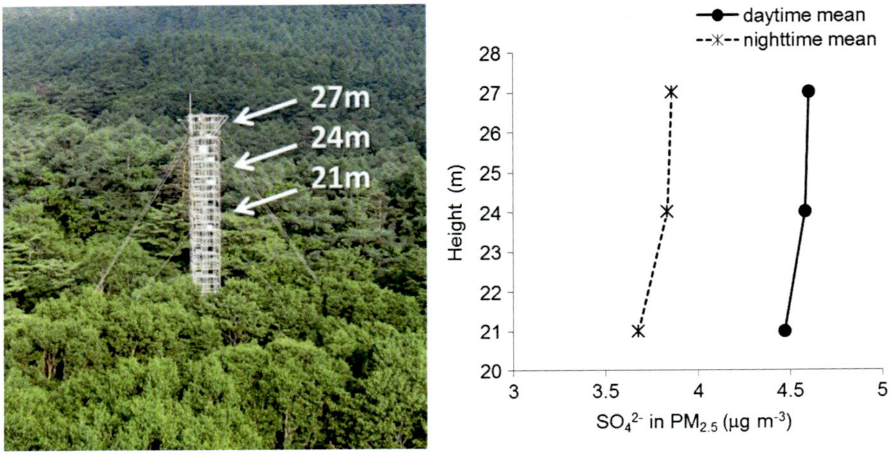

Fig. 21.2 Illustration of concentration gradient measurement of $PM_{2.5}$ sulfate at a deciduous forest at the Kitasaku site

at a mixed forest in FM Tama (Fig. 21.1). The REA sampling system for $PM_{2.5}$ components is shown in Fig. 21.3. Aerosol samples of updraft and downdraft were collected onto separate filters, and a constant flow rate was controlled by a mass flow controller following flow through a $PM_{2.5}$ cyclone. The updraft or downdraft was determined by the 3D sonic anemometer.

The fluxes of $PM_{2.5}$ sulfate are determined as

$$F = \beta\sigma_w\left(C_u - C_d\right) \tag{21.3}$$

where β is an empirical coefficient, σ_w is the standard deviation of the vertical wind speed, and C_u and C_d are the average concentrations of sulfate in the updraft and downdraft sample filters, respectively. The switching frequency of the REA sampling system needs to be fast depending on the eddy above the canopy (e.g., 10 Hz). "Deadband" sampling to maximize the difference between C_u and C_d can be adopted in some cases (Meyers et al. 2006). Matsuda et al. (2015) did not employ deadband sampling because of the reduction of the sampling loss caused by frequent switching. The empirical coefficient β is calculated by using the following equation with data collected by the 3D sonic anemometer:

$$\beta = F_h\left(\sigma_w\left(T_u - T_d\right)\right)^{-1} \tag{21.4}$$

where F_h is the sensible heat flux ($w'T'$) determined using the eddy covariance method; w' and T' are the variations of the vertical wind speed and temperature, respectively; and T_u and T_d are the average temperatures of the updrafts and downdrafts, respectively.

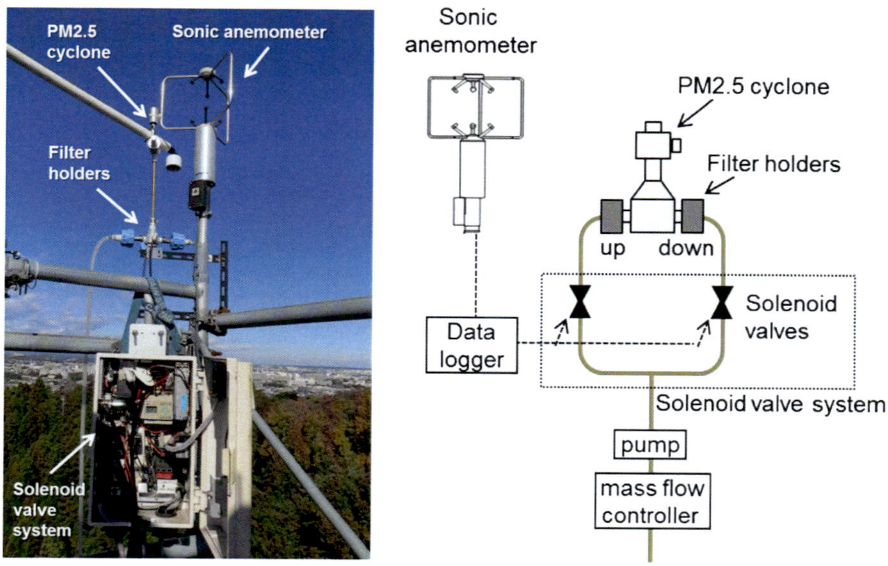

Fig. 21.3 Illustration of REA sampling system for $PM_{2.5}$ components

21.2.3 Field Measurements in Japan

Several direct measurements of submicron aerosol deposition onto forests were carried out in Japan (Table 21.1). Directly measured deposition velocities above forest in Japan corroborate the experimental results in Europe and North America (Petroff et al. 2008; Pryor et al. 2008). On the other hand, the measured deposition velocities (Table 21.1) were higher than those calculated by theoretical models, which was less than 0.1 cm s^{-1} for submicron aerosols (e.g., Slinn 1982). Typically, sulfate and nitrate in $PM_{2.5}$ are known to have a maximum peak in the range of the accumulation mode, i.e., 0.1–1 μm in diameter. In the range of the nucleation and Aitken mode (<0.1 μm), the deposition velocity is high because the particles are strongly affected by Brownian motion (Fig. 21.4). In the range of the coarse mode (>1 μm), deposition velocity is also high because the particles are strongly affected by gravitational settling (Fig. 21.4). In contrast, particles in the range of the accumulation mode, where neither Brownian motion nor gravitational settling is efficient for dry deposition, theoretically do not have strong removal mechanisms from the atmosphere. Nevertheless, it is widely known that theoretical calculations tend to underestimate the deposition velocity of submicron particles measured in natural environments, especially above forest canopies (Fig. 21.4) (e.g., Erisman et al. 1997; Petroff et al. 2008; Pryor et al. 2008). In addition, the deposition velocity of submicron nitrate was higher than that of submicron sulfate in some measurements in Japan (Takahashi and Wakamatsu 2004; Yamazaki et al. 2015), which was possibly caused by the differences in their chemical properties (see Sect. 21.4).

Table 21.1 Directly measured deposition velocities onto forests in Japan

Surface	Season	Method	Deposition velocity [cm s^{-1}] SO_4^{2-} ($PM_{2.5}$)	NO_3^- ($PM_{2.5}$)	Reference
Coniferous forest, Gunma (Akagi)	Feb.–Oct.	Gradient	0.71	1.24	Takahashi and Wakamatsu (2004)
Deciduous forest, Nagano (Kitasaku)	July	Gradient	0.9 (day)	–	Matsuda et al. (2010)
			0.3 (night)		
Mixed hilly forest, Tokyo (FM Tama)	Jul.–Aug., Nov.	REA	0.50	–	Matsuda et al. (2015)

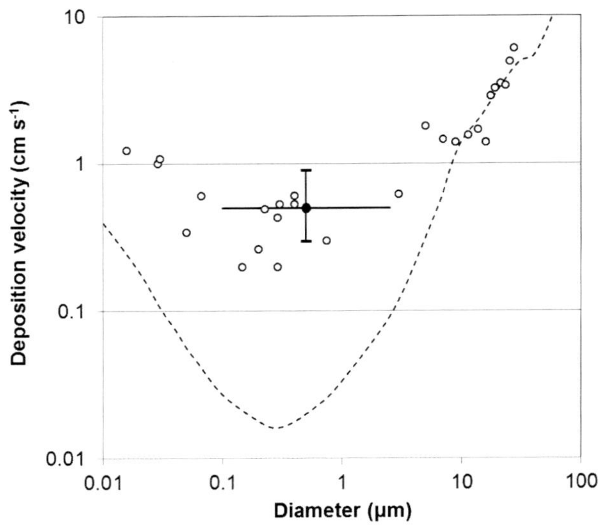

Fig. 21.4 Relationship between deposition velocity of particles above forest and their diameter. *Dashed line* indicates a theoretical prediction based on Slinn (1982). *Open* and *closed* symbols refer to measurements summarized in Petroff et al. (2008) and in Table 21.1 (SO_4^{2-}), respectively

21.3 Parameterizations of Deposition Velocity

21.3.1 Effect of Humidity

To evaluate regional atmospheric deposition, the inferential method is adopted in regional monitoring networks such as CASTNet (Clean Air Status and Trends Network, US EPA) and EANET (Acid Deposition Monitoring Network in East Asia). The inferential method estimates the dry deposition based on the following equation:

$$F = -CV_d \tag{21.5}$$

where F, C, and V_d are the same as the terms of Eq. 21.2. The deposition velocity for aerosols is parameterized by the following equation (Ruijgrok et al. 1997):

$$V_d = \left(R_a + V_{ds}^{-1}\right)^{-1} + V_g \tag{21.6}$$

where R_a is the aerodynamic resistance, V_{ds} is the surface deposition velocity, and V_g is the gravitational settling velocity. The surface deposition velocity depends on the collection efficiency with which the surface collects particles. In theoretical models, the collection efficiency is modeled, in addition to several driving forces (e.g., Brownian motion, interception, inertial impaction, etc.), depending on the particle size. On the other hand, some parameterizations were developed by being fit empirically to the field measurement data or by simplifying theoretical models.

Matsuda et al. (2010) consider the applicability of parameterizations of the surface deposition velocity (V_{ds}). Two parameterizations (Wesely et al. 1985; Ruijgrok et al. 1997) were compared with the measurements by the gradient method at a deciduous forest at the Kitasaku site in Nagano, Japan (Fig. 21.2). Aerodynamic resistance (R_a) was determined using micrometeorological data based on that of Erisman and Draaijers (1995). For the V_d estimations of $PM_{2.5}$ sulfate, values of gravitational settling velocity (V_g) less than 0.01 cm s^{-1} (Seinfeld and Pandis 2006) were ignored due to their relatively small contribution to the deposition velocity. The V_{ds} parameterization of Wesely et al. (1985) is an empirical fit to the field experiment data above grassland in North America. This is parameterized as:

$$\begin{aligned} V_{ds} &= u^*/500, \text{neutral or stable conditions,} \\ V_{ds} &= (u^*/500)\left[1+\left(300/(-L)\right)^{2/3}\right], \text{unstable conditions.} \end{aligned} \tag{21.7}$$

The Ruijgrok et al. (1997) parameterization is a simplified parameterization base on the Slinn's model, additionally taking into account the effect of growth in size of hygroscopic particles due to high relative humidities. This is parameterized as:

$$V_{ds} = \left(u^{*2}/u_h\right)E \tag{21.8}$$

where u_h is the wind speed at the top of the canopy, and E is the total collection efficiency with which the canopy captures particles, as defined by Ruijgrok et al. (1997). E depends on u^*, relative humidity, and surface conditions (wet or dry). Katata et al. (2014) carried out numerical simulations of $PM_{2.5}$ sulfate deposition using a multi-layer atmosphere-SOiL-VEGetation model (SOLVEG) in accordance with the Kitasaku measurements mentioned above. To investigate the impact of hygroscopic growth on dry deposition onto forest canopies, Katata et al. (2014) took into account the hygroscopic growth factor, *Gf*, representing the degree of water uptake by aerosols in the simulations of SOLVEG.

Based on Matsuda et al. (2010) and Katata et al. (2014), measured and calculated deposition velocities of $PM_{2.5}$ sulfate at the Kitasaku site are compiled in Fig. 21.5. Between the two parameterizations, the Ruijgrok et al. (1997) parameterization (Ruijgrok97) reproduced the measurements relatively well, while the Wesely et al. (1985) parameterization (Wesely85) underestimated the measurements (Fig. 21.5). Regarding the simulations of SOLVEG, the simulations with

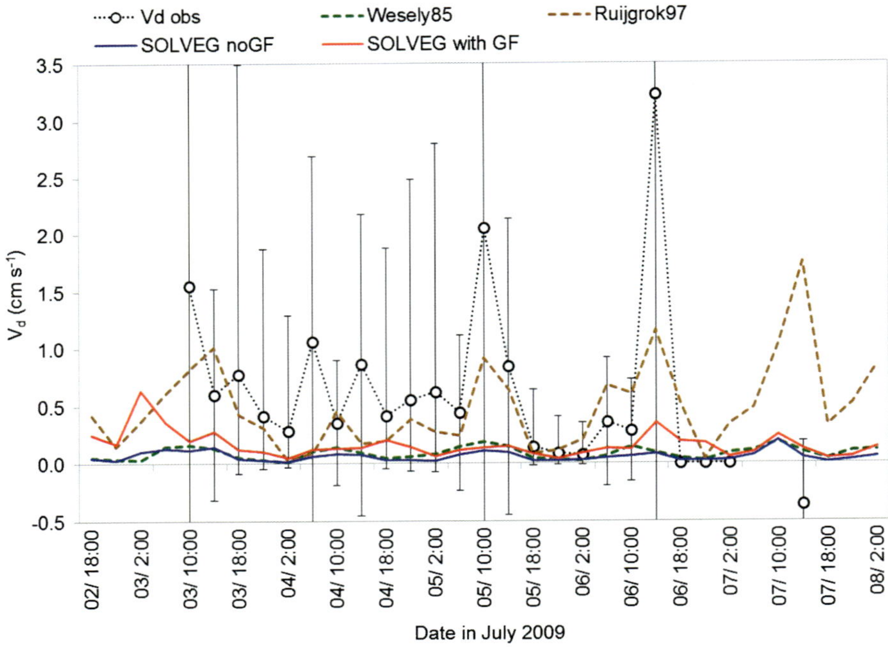

Fig. 21.5 Deposition velocities of $PM_{2.5}$ sulfate measured by the gradient method (Vd obs) and calculated by two parameterizations (Ruijgrok97, Wesely85) (Matsuda et al. 2010) and by SOLVEG with ("with GF") and without ("no GF") particle growth (Katata et al. 2014)

particle growth ("SOLVEG with GF") were closer to the measurements than those without particle growth ("SOLVEG no GF"), although both simulations underestimated the measurements (Fig. 21.5). Deposition velocities of "SOLVEG no GF" were in good agreement with those of "Wesley85". Both simulations, which do not take into account particle growth, probably represent the above-mentioned theoretical values. In the case of "SOLVEG no GF", the calculated deposition velocity averaged for daytime periods was 0.07 cm s^{-1}, which was one-tenth of the observed velocity (0.56 cm s^{-1}). In the case of "SOLVEG with GF", temporal variations of $PM_{2.5}$ sulfate flux increased under very humid conditions (RH>90 %). This tendency also appeared in calculations by the parameterization of Ruijgrok et al. (1997). As a result, mean deposition velocity during the daytime doubled (0.13 cm s^{-1}) in calculations with particle growth. This indicates that particle growth can significantly affect the deposition velocity of $PM_{2.5}$ sulfate under humid conditions. Underestimations still remain in calculations for the range because the model did not include the additional particle deposition processes. Katata et al. (2014) suggest additional processes due to thermophoresis, diffusiophoresis, the Stefan flow effect, turbophoresis, electrophoresis, and the effects of surface characteristics of leaves (micro-roughness, wetness, rebound, and resuspension). Further improvements of the model and comparisons with detailed datasets are required.

21.3.2 Effect of Complex Terrain

Complex terrain probably influences the aerodynamic resistance of dry deposition processes. Hicks and Meyers (1988) suggest that the influence of complex terrain on the deposition of SO_2, O_3, and aerosols, which are largely controlled by surface factors (surface deposition velocity or surface resistance), is small compared with those for which aerodynamic factors are more important, such as HNO_3. Recent field measurements show large values of aerosol deposition velocity (Fig. 21.4). It possibly means the aerosol depositions are controlled by not only the surface but also aerodynamic factors. In order to measure fluxes above sloping terrain, the mean streamline coordinate system of the eddy correlation method, which is the most useful coordinate system for analysis above the terrain, is available (Wilczak et al. 2001). The basic idea of the system could be applied to the REA method.

Matsuda et al. (2015) evaluated the dry deposition and deposition velocity of $PM_{2.5}$ sulfate above a forest on complex terrain at the FM Tama site (Fig. 21.1) by means of the REA method (Fig. 21.3). Over sloping terrain, the average vertical wind speed is not always approximately zero. Therefore, we determined a base level for the variables using the mean value of the vertical wind speed, and then we let the mean value change with time. In particular, in the REA sampling, a 10-min mean vertical wind speed was adopted as the base level to be used for the next 10-min interval for the difference between the updraft and downdraft. Measured deposition velocities of $PM_{2.5}$ sulfate were compared with those calculated by a resistance model (Eq. 21.6) using the Ruijgrok et al. (1997) parameterization (Eq. 21.8). The measurements show that the complex terrain contributed to an expansion of the range of deposition velocities measured by the REA method. Therefore, the increased range of deposition velocities was associated with flux footprints from complex terrain. Calculated deposition velocities, however, agreed with measurements on average for about 2 weeks. The good agreement between the measured and calculated fluxes (or deposition velocity) at the FM Tama site is possibly because the deposition of sulfate is largely controlled by the surface resistance. EANET's dry deposition monitoring uses a filter pack with weekly or biweekly samplings to measure the gas and aerosol concentrations. Therefore, weekly or biweekly averages of deposition velocity are used to estimate dry depositions. This study suggests that there might be no significant influence of complex terrain on the deposition estimations for $PM_{2.5}$ sulfate using weekly or biweekly averages of deposition velocity.

21.4 Dry Deposition of Sulfate and Nitrate Aerosols onto Forest

Takahashi and Wakamatsu (2004) report a significant difference in deposition velocity between sulfate and nitrate in submicron aerosols onto forest (Table 21.1). Yamazaki et al. (2015) discusses the difference based on the measurements of

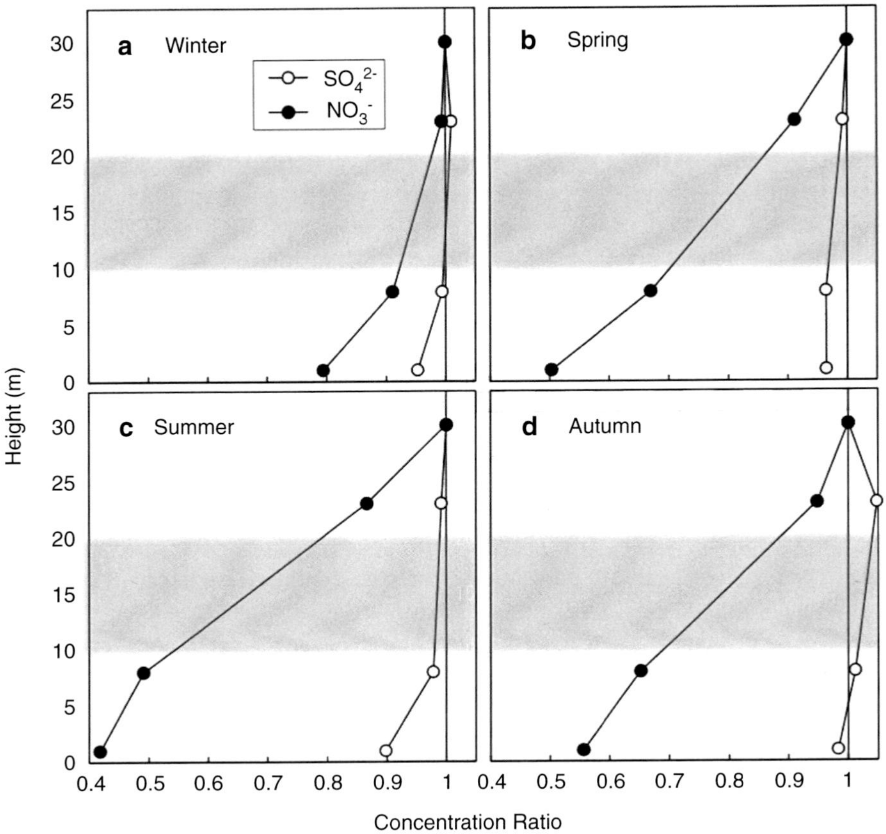

Fig. 21.6 Normalized vertical profiles of NO_3^- (*closed circles*) and SO_4^{2-} (*open circles*) concentrations in $PM_{2.5}$ from December 2012 to November 2013 at the FM Tama site. The concentrations of NO_3^- and SO_4^{2-} at 30 m are regarded as 1. (**a–d**) Show averages in winter (December–February), spring (March–May), summer (June–August), and autumn (September–November), respectively. *Gray* layers indicate canopies (Yamazaki et al. 2015)

vertical profiles of SO_4^{2-} and NO_3^- in $PM_{2.5}$ at the FM Tama site (Fig. 21.1) during 1 year from December 2012 to November 2013. Weekly samplings of $PM_{2.5}$ were carried out at four heights (two heights above canopy, two heights below canopy). Major inorganic aerosols through the year consisted of NH_4NO_3 and $(NH_4)_2SO_4$ at the four heights. Significant differences between SO_4^{2-} and NO_3^- profiles were observed (Fig. 21.6). NO_3^- through the year decreased toward the forest floor compared with SO_4^{2-}. In winter and spring, the air temperature of the forest floor was higher than that of other heights. On the other hand, in summer and autumn, the air temperature of the forest floor was lower than that of other heights. The concentration profiles and air temperatures indicated that in winter and spring volatilization of NH_4NO_3 caused by the higher temperature on the floor led to the rapid decrease of NO_3^-, whereas in summer and autumn the volatilization of NH_4NO_3 was caused

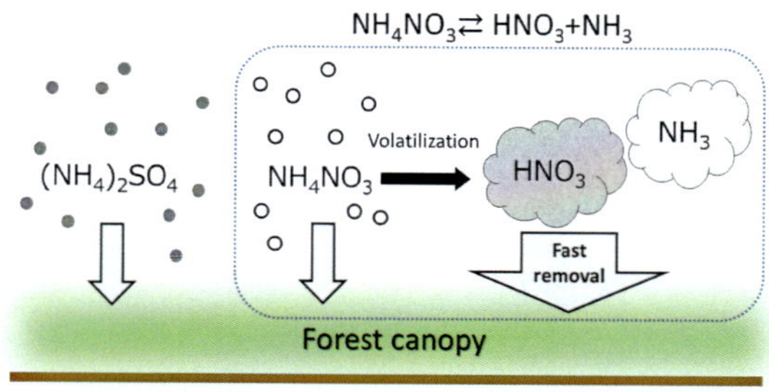

Fig. 21.7 Illustration of dry deposition processes of ammonium sulfate and ammonium nitrate

by the low concentration of HNO_3 brought about by its significant removal to leaves. The volatilization process indicates that the dry deposition of ammonium nitrate to forests is more efficient than that of ammonium sulfate (Fig. 21.7). Matsuda et al. (2015) report that the dry deposition amount of $PM_{2.5}$ sulfate estimated by the inferential method was in good agreement with that observed by a direct measurement method (relaxed eddy accumulation) at the same site. Since the volatilization process is not taken into account in the inferential method, the current dry deposition amount of ammonium nitrogen is possibly underestimated.

21.5 A Case Study: Elemental Carbon Deposition onto a Tropical Forest

A significant amount of aerosol elemental carbon (EC) is emitted from the Asian region, mainly due to fuel combustion and biomass burning. Biomass burning and the associated production of aerosols containing EC occur primarily in tropical regions (Matsuda et al. 2010). In order to understand the influence of the emissions of EC on the regional or global budget of carbon with respect to its climate impact, it is important to quantify the atmospheric deposition of EC in the region. In this section, a case study of the estimation of EC deposition onto a tropical forest is introduced.

Dry and wet deposition of EC was investigated in a tropical deciduous forest in Sakaerat, Thailand (Matsuda et al. 2012). Micro-meteorological measurements and monthly sampling of $PM_{2.5}$ aerosols were carried out continuously from 2010 to 2011 at the top of an experimental tower 38 m above the ground (Fig. 21.8). The dry deposition was estimated by the inferential method using a resistance model (Eq. 21.6) with the Ruijgrok et al. (1997) parameterization (Eq. 21.8). For the measurement of wet deposition, biweekly sampling of EC directly filtrated from rainwater was carried out continuously over the 2 years (Fig. 21.8). Dry deposition peaked during the period from February to March, when concentration and deposition velocity were both high, and decreased during the wet season around June–October, when the concentration or deposition velocity was low (Fig. 21.9). Wet deposition peaked around February–April because of the washout effect of high EC in the atmosphere of the late dry season (Fig. 21.9). Wet deposition was somewhat high

Fig. 21.8 Experimental tower for dry deposition measurements (*left*) and precipitation sampling site (*right*) in Sakaerat, Thailand

Table 21.2 Estimations of dry and wet deposition of elemental carbon (black carbon)

Location	Dry deposition ($mg\ m^{-2}\ d^{-1}$)	Wet deposition ($mg\ m^{-2}\ d^{-1}$)	Reference
Tropical forest, Sakaerat, Thailand	0.58	0.05	Matsuda et al. (2012)
Background sites, Europe	0.00–0.02	0.01–0.10	Cerqueira et al. (2010)
Global oceans (30°S –30°N)	0.01–0.02	0.07–0.18	Jurado et al. (2008)

from August to October with increased rainfall. Both dry and wet deposition increased in the leafless season and decreased in the leafy season, respectively.

Cerqueira et al. (2010) estimated EC dry and wet deposition based on the measurement of aerosols in the atmosphere and precipitation at background sites in Europe (Table 21.2). Jurado et al. (2008) simulated dry and wet deposition of black carbon using a global model over the oceans (Table 21.2). Wet deposition in this study was in agreement with these reports (Cerqueira et al. 2010; Jurado et al. 2008); however, dry deposition in this study was significantly higher. The major reason for the difference in dry deposition was the difference in deposition velocity. Both reports aimed to estimate regional or global depositions and therefore estimated deposition velocity for large surfaces such as the sea based on theoretical global models. The deposition velocities were less than 0.1 cm s^{-1}, e.g., 0.025 cm s^{-1} in Cerqueira et al. (2010), which is a typical value from the theoretical simulation for submicron aerosols. On the other hand, this study aimed to estimate deposition at a specific tropical forest and therefore adopted the Ruijgrok et al. (1997) parameterization, which reproduced the measurements relatively well (Fig. 21.5). Moreover, the high concentration in the site also caused the high dry deposition. In this study, the dry deposition of EC was about

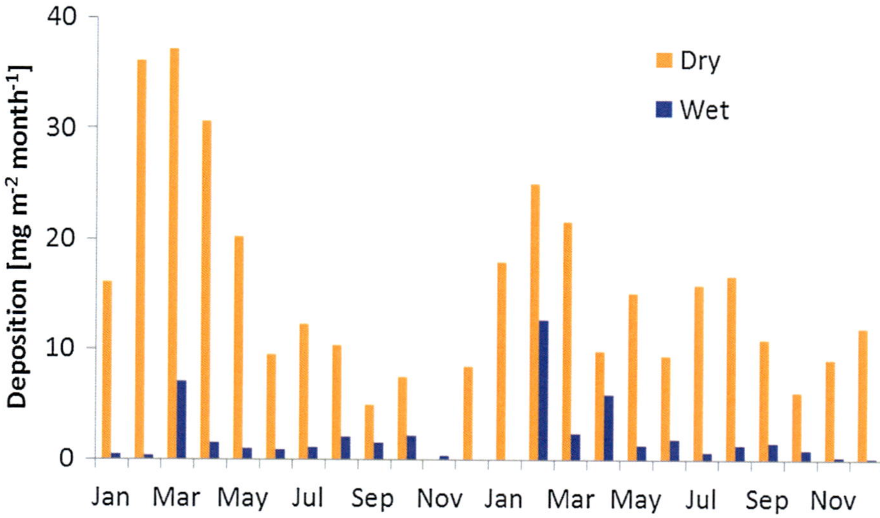

Fig. 21.9 Variations of dry and wet deposition of EC at Sakaerat site, Thailand, from 2010 to 2011. Dry deposition in November 2010 could not be quantified due to a problem with the measurements

ten times higher than the wet deposition. This indicates that the majority of EC deposition onto tropical forests in Southeast Asia is possibly due to the dry disposition process.

21.6 Concluding Summary

Recent knowledge on the dry deposition of aerosols onto forests, focusing especially on field measurements in East Asia, was introduced. Directly measured deposition velocities above forest in Japan corroborate experimental results in Europe and North America and were higher than those calculated by theoretical models. The effect of growth of hygroscopic aerosols in humid conditions is one of the reasons of the discrepancy between the measurements and the models. That is because the calculations taking into account the effect were in reasonable agreement with the measurements, compared with those not taking into account. A complex terrain contributed to increasing the range of measured deposition velocities. The increased range was associated with flux footprints from complex terrain. The calculated deposition velocities, however, agreed with the measurements on average for about 2 weeks. The removal of NH_4NO_3 particles from the atmosphere was more efficient than that of $(NH_4)_2SO_4$ particles, probably because of the volatilization processes of NH_4NO_3 particles. Therefore, the chemical properties of particles probably have a large effect upon deposition velocities, as well as the physical properties. Further measurements are required to improve our understanding of dry deposition processes and to improve the models.

References

Businger JA (1986) Evaluation of the accuracy with which dry deposition can be measured with current micrometeorological techniques. J Clim Appl Meteorol 25:1100–1124
Cerqueira M, Pio C, Legrand M, Puxbaum H, Kasper-Gieblc A, Afonso J, Preunkert S, Gelencsér A, Fialho P (2010) Particulate carbon in precipitation at European back ground sites. Aerosol Sci 41:51–61
Erisman JW, Draaijers GPJ (1995) Atmospheric deposition in relation to acidification and eutrophication, vol 63, Studies in Environmental Science. ELSEVIER, Amsterdam, pp 55–75
Erisman JW, Draaijers G, Duyzer J, Hofshreuder P, van Leeuwen NFM, Römer F, Ruijgrok W, Wyers P, Gallagher M (1997) Particle deposition to forests-summary of results and application. Atmos Environ 31:321–332
Foken T (2008) Micrometeorology. Springer, Berlin, pp 105–151
Hicks BB, Meyers TP (1988) Measuring and modelling dry deposition in mountainous areas. Acid deposition at high elevation sites. Kluwer Academic Publishers, Boston, pp 541–552
Jurado E, Dachs J, Duarte CM, Simo R (2008) Atmospheric deposition of organic and black carbon to the global oceans. Atmos Environ 42:7931–7939
Katata G, Kajino M, Matsuda K, Takahashi A, Nakaya K (2014) A numerical study of the effects of aerosol hygroscopic properties to dry deposition on a broad-leaved forest. Atmos Environ 97:501–510
Matsuda K, Fujimura Y, Hayashi K, Takahashi A, Nakaya K (2010) Deposition velocity of $PM_{2.5}$ sulfate in the summer above a deciduous forest in central Japan. Atmos Environ 44:4582–4587
Matsuda K, Sase H, Murao N, Fukazawa T, Khoomsub K, Chanonmuang P, Visaratana T, Khummongkol P (2012) Dry and wet deposition of elemental carbon on a tropical forest in Thailand. Atmos Environ 54:282–287
Matsuda K, Watanabe I, Mizukami K, Ban S, Takahashi A (2015) Dry deposition of $PM_{2.5}$ sulfate above a hilly forest using relaxed eddy accumulation. Atmos Environ 107:255–261
Meyers TP, Luke WT, Meisinger JJ (2006) Fluxes of ammonia and sulfate over maize using relaxed eddy accumulation. Agric For Meteorol 136:203–213
Petroff A, Mailliat A, Amielh M, Anselmet F (2008) Aerosol dry deposition on vegetative canopies. Part I: review of present knowledge. Atmos Environ 42:3625–3653
Pryor SC, Gallagher M, Sievering H, Larsen SE, Barthelmie RJ, Birsan F, Nemitz E, Rinne J, Kulmala M, Grönholm T, Taipale R, Vesala T (2008) A review of measurement and modelling results of particle atmosphere–surface exchange. Tellus 60B:42–75
Ruijgrok W, Tieben H, Eisinga P (1997) The dry deposition of particles to a forest canopy: a comparison of model and experimental results. Atmos Environ 31:399–415
Seinfeld JH, Pandis SN (2006) Atmospheric chemistry and physics, from air pollution to climate change (second edition). WILEY-INTERSCIENCE, Hoboken, p 407–411
Slinn WGN (1982) Predictions for particle deposition to vegetative canopies. Atmos Environ 16:1785–1794
Takahashi A, Wakamatsu T (2004) Estimation of deposition velocity of particles to a forest using the concentration gradient method. J Jpn Soc Atmos Environ 39:53–61 (in Japanese)
Wesely ML, Cook DR, Hart RL (1985) Measurement and parameterization of particulate sulfur dry deposition over grass. J Geophys Res 90:2131–2143
Wilczak JM, Oncley SP, Stage SA (2001) Sonic anemometer tilt correction algorithms. Bound-Layer Meteorol 99:127–150
Wyers GP, Duyzer JH (1997) Micrometeorological measurement of the dry deposition flux of sulphate and nitrate aerosols to coniferous forest. Atmos Environ 31:333–343
Yamazaki T, Takahashi A, Matsuda K (2015) Differences of dry deposition between sulfate and nitrate in $PM_{2.5}$ to a forest in suburban Tokyo by vertical profile observations. J Jpn Soc Atmos Environ 50:167–175 (in Japanese)

FSC
www.fsc.org
MIX
Papier | Fördert gute Waldnutzung
FSC® C083411

Kolibri 360 GmbH
Ferdinand-Jühlke-Straße 7
99095 Erfurt, Deutschland
produktsicherheit@kolibri360.de